Insights and Innovations in Gallbladder Carcinogenesis

This book comprehensively explores the latest advancements in multiomics strategies for diagnosing gallbladder cancer. The book delves into the genetic mutations and molecular pathways critical for the onset and progression of gallbladder cancer. Through targeted gene sequencing and genomic profiling, it identifies both somatic and rare germline mutations, pinpointing driver genes and predisposing mutations. The subsequent section examines the proteins and molecular pathways altered in gallbladder cancer. It uncovers dysregulated proteins, identifies tumor biomarkers, and evaluates circulating markers for early detection. The book further investigates the impact of infections like *Salmonella enterica* of the serotype Paratyphi A on gallbladder tissue, revealing genotoxic effects and potential biomarkers. It also explores the correlation between the gallbladder microbiome, physiological state, and biliary metabolic profile. Toward the end, the book introduces cutting-edge techniques like machine learning-based CT imaging for survival prediction and examines the role of epigenetic pathways and noncoding RNAs in gallbladder cancer. This section also discusses the potential of targeting mTOR pathways and the significance of inflammatory and immune markers, offering a glimpse into the future of gallbladder cancer diagnosis and treatment. This book is a valuable resource for oncologists and researchers working in the fields of oncology, genomics, proteomics, metabolomics, bioinformatics, molecular biology, and translational medicine.

Insights and Innovations in Gallbladder Carcinogenesis

A Multi-Omics Perspective

Edited By
Vijay Kumar Shukla
Manoj Pandey
Ruhi Dixit

CRC Press
Taylor & Francis Group
Boca Raton London New York

CRC Press is an imprint of the
Taylor & Francis Group, an **informa** business

Designed Cover Image: ShutterStock

First edition published 2026
by CRC Press
2385 NW Executive Center Drive, Suite 320, Boca Raton FL 33431

and by CRC Press
4 Park Square, Milton Park, Abingdon, Oxon, OX14 4RN

CRC Press is an imprint of Taylor & Francis Group, LLC

ISBN: 978-1-032-82782-7 (hbk)
ISBN: 978-1-032-82783-4 (pbk)
ISBN: 978-1-003-50620-1 (ebk)

DOI: 10.1201/9781003506201

Typeset in Times
by Apex CoVantage, LLC

Contents

Preface

Gallbladder cancer (GBC) is the most common malignancy of the biliary tract and ranks fifth among gastrointestinal cancers. Due to its asymptomatic nature, it is frequently diagnosed at an advanced stage, resulting in poor prognosis and a 5-year survival rate below 5%. While surgery remains the most effective treatment, early detection is essential for improving outcomes. Histologically, adenocarcinoma is the predominant type, accounting for 70–90% of cases. Understanding the biological and prognostic variables that influence GBC progression is vital. In this context, multiomics approaches—including genomics, transcriptomics, proteomics, and metabolomics—offer powerful tools to explore disease origin and molecular pathways, enabling more accurate diagnosis and prognostication.

The integration of multiomics technologies is transforming gallbladder cancer research by linking molecular profiles with targeted therapeutic strategies. Combining data across various biological layers helps uncover key mutations, altered signaling pathways, and potential biomarkers. This holistic view enhances our ability to classify tumors, predict outcomes, and design personalized treatments. Despite the growing global burden of GBC, multiomics research remains underutilized in education and clinical translation. This volume addresses that gap by offering an in-depth overview of how omics-driven discoveries can contribute to early detection, disease monitoring, and future cancer therapies, emphasizing the promise of personalized medicine in gallbladder cancer management.

Vijay Kumar Shukla
Manoj Pandey
Ruhi Dixit
Editors

About the Editors

Dr. Vijay Kumar Shukla is the ex-rector of Banaras Hindu University and a former professor of general surgery at the Institute of Medical Sciences, Banaras Hindu University (BHU), Varanasi. Internationally recognized for his work in gallbladder cancer and wound healing, he has published over 380 research papers and served as a visiting professor at reputed international institutions. He has received numerous awards, including the Lifetime Achievement Award from the World Union of Wound Healing Societies (Toronto, 2008), and fellowships from NAMS and FCWCS, Malaysia. Dr. Shukla has held several key positions, including President of the Indian Society of Wound Management, Member of the ICMR Task Force on Gallbladder Cancer, and Governing Council Member of ASI, contributing extensively to academic leadership and healthcare policy.

Dr. Manoj Pandey is a professor in the Department of Surgical Oncology at the Institute of Medical Sciences, BHU, Varanasi. A renowned expert in gallbladder cancer research, Dr. Pandey's dedication to advancing cancer care is evident through his extensive body of work, comprising over 357 scientific publications in reputed journals. He has been honored with several prestigious awards including the Dr. U. C. Gupta Memorial Award for Excellence in Surgery. He has also served as a member of the ICMR Task Force for the review of cancer management in India. Dr. Pandey currently serves as the Principal Investigator for both the Health Technology Assessment in India (HTAIn) Resource Centre and the DIAMOnDS project, highlighting his pivotal role in evidence-based healthcare and translational research at the national level.

Dr. Ruhi Dixit is currently serving as Scientist-C at the Health Technology Assessment India (HTAIn) Resource Centre, Institute of Medical Sciences, BHU, Varanasi, India. She has a strong research interest in cancer biology, particularly gallbladder cancer, and has made notable contributions to the field through her work on genomic and molecular biomarkers. Alongside her cancer research, she is also actively engaged in Health Technology Assessment projects, contributing to evidence-based healthcare policy and decision-making. Dr. Dixit has authored several publications in high-impact, peer-reviewed international journals and serves as both editor and reviewer for reputed double-blind international journals, reflecting her academic integrity and active involvement in the global scientific community.

1 Gallbladder Carcinoma
Estimation of Risk Factors and Identification of Prognostic Features

Consolato M. Sergi[1,2]*
[1]Children's Hospital of Eastern Ontario, Anatomic Pathology, Canada
[2]University of Ottawa and Alberta, Ottawa, ON, and Edmonton, AB, Canada
*Corresponding author: CSergi@cheo.on.ca

1.1 INTRODUCTION

Among malignancies of the biliary tract, gallbladder cancer (GBC) is the most prevalent malignancy and ranks fifth overall [1]. Because GBC does not manifest with any particular symptoms, it is often undiagnosed until it has progressed to an advanced stage. A bad prognosis characterizes this stage, and the 5-year survival rate is below 5% when GBC is diagnosed late [2]. Thus, early identification of this neoplasm is crucial for the survival of these patients. Surgery is still the gold standard for treating GBC, and it can guarantee a good prognosis, even though chemotherapy has advanced [3]. On the histopathological level, adenocarcinoma accounts for around 70–90% of cases, 5% of cases have additional subtypes, such as papillary tumors, and 2–10% have squamous or adenosquamous histopathology [4]. Variables that affect the development of GBC and the prognosis of individuals with it are outlined in this chapter, along with the most frequently encountered prognostic features.

This narrative review was written following the principles of PRISMA 2020, but it does not mirror a systematic or scoping review. Web of Science, PubMed, Elsevier-Scopus, Google Scholar, Cochrane Library, and Embase were among the databases scoured for articles about gallbladder cancer. The research was conducted in January 2025 and then again in March 2025 before the report was completed [5, 6]. The included papers were published between 2000 and 2024 and were written in either English or French. GBC, surgery, resection, prognostic variables, and invasion were utilized. Once we found these studies, we analyzed the following variables: incidence, age, race, gender, anatomical abnormalities, cholelithiasis association, and gallbladder polyp association. Moreover, we emphasize our interest in chronic infectious disorders and exposure to environmental toxins, including alcohol, nicotine, and aflatoxins. Prognostic markers were also discussed and evaluated. They included TNM staging, resection status, lymph node status, distant metastases, perineural invasion, and lymphovascular invasion.

The results of this narrative review will be subclassified into sections, including the epidemiology of GBC, the association of GBC with cholelithiasis, the eventual disruptions in the biliopancreatic duct, gallbladder precancerous conditions, long-term infections, overweight and obesity, chronic alcohol habit and tabagism, chemical substance exposure, and GBC staging and subsections.

The global incidence rate of GBC is 1.2 per 100,000 people, with 122,491 newly diagnosed cases in 2022; nevertheless, the number of deaths reached 89,055, confirming the aggressiveness of the disease, according to GLOBOCAN 2022 [7]. The incidence of GBC is 1.4 per 100,000 females and

DOI: 10.1201/9781003506201-1

0.89 per 100,000 males, indicating that women are more likely to be affected by this malignancy. Asia accounts for 71.9% of GBC instances globally, with Europe following at 10.3%, Latin America at 8.5%, and Africa at 4.5%. Compared to North America, where both the incidence and death rates are low, Latin America has the highest rates: 7.6 per 100,000 in Bolivia, 5.7 per 100,000 in Chile, and 2.6 per 100,000 in Peru [8–10]. India is responsible for 10% of the world's GBC cases; the frequency is lower in southern and western India but higher in northern regions (10–22/100,000 population). Incidence rates in these areas have been rising consistently for both sexes [11]. Specifically, the sixth and seventh decades of life are associated with an increased incidence of GBC [12, 13]. According to multiple studies, women are 2–4 times more likely to be exposed to GBC than males [14–17]. Several factors may contribute to this. They may include hormonal imbalances (estrogen increases cholesterol synthesis and saturates bile content, and progesterone slows down the gallbladder's motility, leading to the formation of stones and bile sludge) and the frequency of pregnancies and multiple births.

1.2 CHOLELITHIASIS AND GBC

As early as 1924, Leitch postulated in an experimental investigation that GBC developed due to the persistent mechanical irritation of the gallbladder mucosa caused by the presence of stones [18]. It is now commonly understood that stones, as a pathological state, can cause chronic inflammation, which in turn can initiate carcinogenesis through the metaplasia-dysplasia-carcinoma process, at least for some patients. Nevertheless, there is debate that chronic inflammation is not enough to start the carcinogenic process, and a first input should be present, making chronic inflammation a critical promotion factor.

However, leukocyte populations, including lymphocytes and macrophages, are crucial to expedite oncogenesis. Leukocytes are recruited by inflammatory mediators' release, which in turn increases reactive oxygen species and COX-2 overexpression. Deregulation of tumor suppressor genes and proteins follows this occurrence, allowing oncogenic mutations to surface and cancer to proceed from simple polyps to carcinomas [19]. Cholelithiasis precedes chronic cholecystitis, and stone formation is associated with certain risk factors. Age, gender, ethnicity, and genetic predisposition are examples of nonmodifiable risk factors; overweight, diabetes, dyslipidemia, food, and specific drugs are examples of modifiable risk factors [20].

Stones are present in 75–90% of GBC patients [20]; however, only 3% of stone patients will develop GBC [13, 21, 22]. Chronic inflammation and alterations in the mucosa increase the risk of GBC in patients with a cholelithiasis history of more than 20 years with stones larger than 3 cm [12, 21, 23]. Abdominal ultrasonography has a 90% specificity for detecting gallbladder stones. Ultrasonography is a routine procedure that may be carried out at the patient's bedside; it is noninvasive, inexpensive, and gives information in real time. Ultrasonography can notice stones as little as 2 mm, with the precision of the data collected depending on the operator. Since most stones are isoattenuated compared to the surrounding bile, they are invisible to computed tomography (CT), making this technique inferior to ultrasound. Although more expensive than CT or ultrasonography, magnetic resonance imaging is another useful approach for stone detection. Other symptoms and consequences of gallstones, such as cholecystitis, pancreatitis, stones in the common bile duct, or related tumor processes, can be detected using imaging methods that do not exclude one another [24].

The last stage of chronic cholecystitis is known as porcelain gallbladder (PG), which is characterized by gallbladder calcification and thickness of the walls. It is associated with cholelithiasis in 95% of cases, and most patients do not experience symptoms [25]. The estimated incidence of PG seems to be oriented in less than 1% [8–10]. Chronic inflammation of the mucosa is the cause of GBC in patients harboring PG [26, 27]. Some earlier articles indicate that the incidence of GBC related to PG varies between 12.5% and 61% in the evaluated literature. The variation of this data is most often due to the decrease in the autopsy rate. Thus, missing PG is due to the decreased rate of autopsies as quality control. With 1,050 instances and reports, Calomino et al. conducted a study on the subject and found that 2.6% of cases had GBC. There is some disagreement between investigators over the best surgical method for treating PG; while some advocate open cholecystectomy for more complex cases, others suggest laparoscopic cholecystectomy for less complex cases. In our opinion, to prevent tumor dissemination, an open approach is recommended whenever GBC is suspected in a patient with PG [25]. However, some writers advocate against routine cholecystectomy screenings in the absence of obvious gallbladder problems.

1.3 DISRUPTIONS IN THE BILIOPANCREATIC DUCT

A lengthy common duct not directly affected by the Oddi sphincter is a consequence of biliopancreatic ductal junction abnormalities or bilio-pancreatic maljunctions (BPMs). This congenital disorder also increases the risk of GBC [28, 29]. Arnolds initially documented congenital bile duct dilatation and biliopancreatic ductal junction abnormalities in an autopsy case in 1906 [30]. Patients with congenital bile duct dilatation are more likely to experience symptoms, such as abdominal pain, jaundice, vomiting, or fever, according to the literature on BPMs [28, 31]. The hyperplasia-dysplasia-carcinoma sequence is activated when pancreatic juice reflux, increased pancreatic duct pressure, and biliary stasis cause a chronic inflammatory process similar to that caused by stones [28]. The gallbladder and bile duct typically contain elevated amounts of amylase [29, 32, 33]. Individuals without accompanying common bile duct dilatation are more likely to acquire GBC, while patients with BPMs are more likely to get biliary tract malignancies overall. The diagnosis of a BPM can be achieved with endoscopic ultrasonography, magnetic resonance cholangiopancreatography (MRCP), percutaneous transhepatic cholangiography, or endoscopic retrograde cholangiopancreatography (3D-DIC-CT) [34].

In order to determine if BPMs are associated with GBC, Deng et al. [35] conducted a meta-analysis. There was a profile of younger women who did not have gallstones, and patients who had BPMs but no common bile duct dilatation were more likely to acquire GBC. In contrast to the 3.1% and 6.9% of cases with extrahepatic cholangiocellular carcinoma, respectively, 37.4% of patients without dilatation and 13.4% of patients with dilatation in a statewide Japanese survey of 2651 adult patients with BPMs were found to have GBC [32]. In their study, Kamisawa et al. [36] looked at 145 BPM patients; 73 had a dilatation, whereas 72 did not. According to their findings, 68% of patients without dilatation and 21% of cases with dilatation had GBC.

The Japan Biliary Association and the Japanese Study Group on Pancreaticobiliary Maljunction developed clinical practice recommendations for the approach and management of BPMs in 2012. Prophylactic cholecystectomy is recommended to prevent gallstones in individuals with BPM who do not have congenital common bile duct dilatation, regardless of whether they are experiencing symptoms or not [31]. Gallbladder mucosa damage from persistent inflammation from reflux and stasis may help explain why GBC is more common than biliary tract cancer in people with BPMs. Patients with BPMs and cancer had significantly greater levels of γ-H2AX expression in their biliary tracts, which is a sign of DNA degradation, a failure of the repair response, and the development of DNA double-strand breaks. DNA damage markers, including double-strand breaks, are identified during carcinogenesis. According to Kuiraishi et al. [37], the median positive proportions for γ-H2AX were 17.3% and 9.7%, respectively, in patients with BPMs without dilatation and with dilatation, suggesting that γ-H2AX expression is significantly stronger in the former group. It appears that biopsies could be used to predict GBC and bile duct cancer, and γ-H2AX could play a role in early cancer detection as a prognostic indicator in patients with BPMs. Kamakura et al. [38] linked mucosal lesions with high levels of γ-H2AX expression, especially in the gallbladder. Finally, a workable plan for early identification is required for BPMs because they are high-risk congenital defects for GBC.

Another cause of GBC is covert pancreatic juice reflux. High amylase levels in the gallbladder and the bile duct and hyperplastic and dysplastic mucosal lesions produce GBC in these instances, even while the pancreaticobiliary junction is physically normal [39–41]. Possible causes of occult reflux include problems with the Oddi sphincter, such as dyskinesia or stenosis. However, the exact process is yet undiscovered. Among patients with benign biliary illness, 36.6% had occult reflux incidence, 27.7% had related cholangiocellular carcinoma, and 62.5% had GBC [41].

1.4 GALLBLADDER PRECANCEROUS CONDITIONS

Although this topic will be part of another chapter in the book, it is crucial to give some information. Most gallbladder polypoid lesions are asymptomatic and discovered by chance during regular abdominal ultrasonography [20]. Gallbladder polyps (GPs) are defined as benign lesions that originate from the mucosa. As the scope and application of imaging studies grow, polyp detection becomes more commonplace. To prevent malignant transformation, individuals with polyps measuring 10 mm or larger are typically sent to the surgical service for prophylactic cholecystectomy. Polyp prevalence rates on cholecystectomy specimens range from 2 to 12% [42], while the global prevalence rate of polyps in adults is 3 to 7% [43]. In 1970, Christensen and Ishak distinguished between benign and malignant polyps in GPs [43, 44]. Neoplastic polyps account for a small percentage of benign polyps (between 1% and 5% of cases), whereas non-neoplastic polyps such as cholesterol polyps account for 60–70% of cases [42]. Cholecystectomy is recommended for sessile polyps, polyps measuring 10 mm or more, polyp growth during follow-up measuring up to 10 mm, or accompanying symptoms, according to the European Joint Guidelines. The guidelines were initially created in 2017 and were revised beginning in May 2020. Any patient with polyps and a risk factor for primary sclerosing cholangitis, being over the age of 60, or being of Asian race should undergo a cholecystectomy. Polyps measuring 6–9 mm should have ultrasonography follow-up at 6, 12, and 24 months; polyps measuring less than 5 mm and free of risk factors do not require follow-up [45].

Abdominal ultrasonography can help with risk classification according to the Society of Radiologists in Ultrasound's guidelines. This information includes morphological characteristics such as size, polyp type (sessile versus pedunculated), and gallbladder wall thickness. These separate general practitioners into three groups according to their potential danger: very low risk, low risk, and uncertain risk [46]. Similarly, Tang et al. developed a different method of categorizing incidentally detected polyps into four groups: extremely low risk, low risk, intermediate risk, and high risk. This method included sonographic features such as size, shape, number of GPs, the gallbladder wall's appearance, vascularity, and the presence of stones or sludge. The goal was to aid in clinical management. Neoplastic polyps were thought to be more likely to form in gallbladders with a single oblate or round GP bigger than 12 mm, polyp vascularity, and gallbladder stones or sludge present. The following features were seen in malignant polyps: a size of 14 mm or larger, an oblate shape, a solitary tumor, a damaged gallbladder wall, and gallbladder stones or sludge. Malignant polyps were detected in 70.37% of cases in the high-risk group [47].

The identification of stones, solitary polyps, and sessile polyps, in addition to being 65 years old or older and 13 mm in size, was listed by Onda et al. [48] as a strong indicator of potential carcinogenesis. For example, a polyp could develop into a malignant tumor. In addition, for polyps less than 10 mm in size, the probability of malignant transformation was much lower than previously thought, according to a thorough meta-analysis of almost 67,000 individuals with GPs seen on abdominal ultrasounds [49]. Among 622,227 adults who were followed retrospectively for 20 years, 35,856 (5.8%) had polyps found by ultrasonography; however, only 19 of these adults went on to develop GBC. Rarely were polyps 10 mm or more related to GBC formation, while GBC development was nearly never associated with polyps less than 10 mm. The study's authors argue that the 10 mm limit has to be changed based on the results [50]. Additionally, according to the clinical guidelines of the American College of Gastroenterology, patients with primary sclerosing cholangitis, a condition linked to hepatobiliary cancers, should undergo prophylactic cholecystectomy if the polyp size reaches 8 mm; however, the guidelines of the European and American Associations for the Study of Liver Diseases advise cholecystectomy irrespective of the polyp size [51]. Figure 1.1 shows a schema of gallbladder carcinogenesis, including cholelithiasis as a major risk factor.

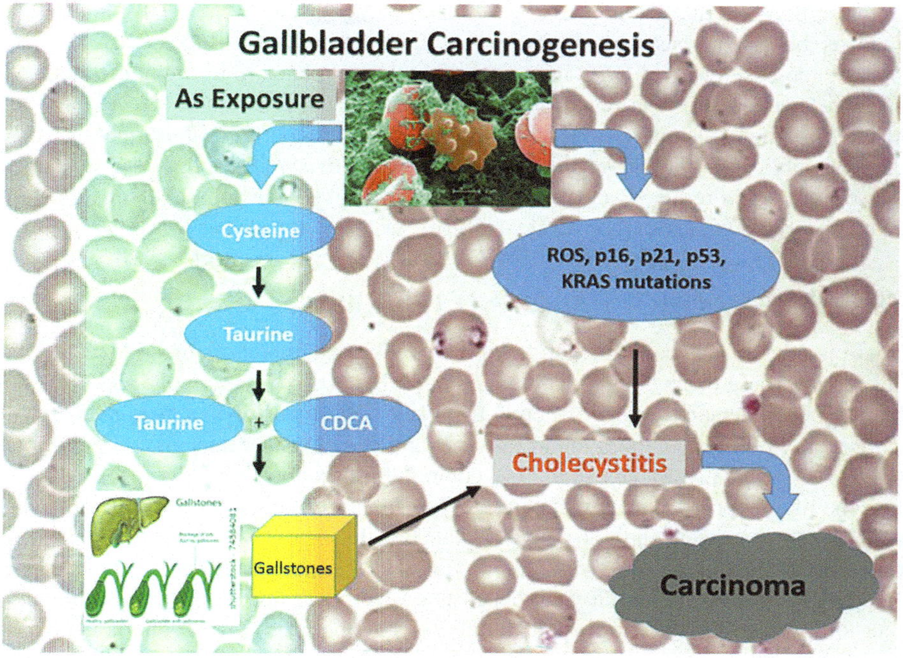

FIGURE 1.1 Gallbladder carcinogenesis.

Note: As, arsenic; ROS. reactive oxygen species; KRAS, the *KRAS* gene makes the K-Ras protein, which acts like a molecular switch. It is called KRAS because it was first identified as causing cancer in Kirsten rat sarcoma virus. CDCA (chenodeoxycholic acid) is a primary bile acid that is conjugated with the amino acid taurine to form taurochenodeoxycholic acid (TCDCA) before being secreted into bile.

1.5 LONG-TERM INFECTIONS

Pathogen infections are another potential risk factor; these infections might trigger the hyperplasia-dysplasia-carcinoma cascade by inducing chronic or acute inflammatory responses that harm the mucosa. Gastric cancer, peptic ulcers, cholelithiasis, cholecystitis, and biliary tract malignancies are all linked to *Helicobacter pylori*, a Gram-negative bacterium. In fact, *H. pylori* DNA has been extracted from GBC patient tissues [52]. We think that *H. pylori* may enter the biliary tree through portal circulation, bile reflux via the Oddi sphincter, or through a potential translocation. Another stressor on the gallbladder mucosa that contributes to the development of GBC is stone formation, which results from chronic inflammation [53]. Various *Helicobacter* species have been associated with GBC, and this microorganism seems to be related to a threefold greater risk of developing biliary tract cancers when isolated from bile samples [54]. Typhoid- and paratyphoid-fever-causing *Salmonella typhi* and *Salmonella paratyphi* are prevalent in GBC-endemic regions like India. *Salmonella* and GBC are known to be related mainly because *Salmonella* causes chronic inflammation, which damages the mucosa of the gallbladder. Antibodies to *Salmonella* have been detected in patients with GBC, particularly in males and those with gallstones (GS) [55]. *Salmonella typhi* makes specific carcinogenic chemicals that are involved in the formation of GBC through prolonged exposure. This risk factor has a double impact due to the *Salmonella* it harbors; the surface of GS makes colonization easier, which in turn makes treatment more challenging [11].

1.6 OVERWEIGHT AND OBESITY

As a matter of public health concern, obesity is known to increase the likelihood of developing cardiovascular disease, diabetes, gallstones, and several malignancies. According to some research, being overweight raises the risk of GBC by 10% and being obese by 58%. This is particularly true for women, as the risk increases by 4% for every 1 kg/m^2 of body mass index (BMI) [56]. An explanation for the increased risk of GBC in overweight women relates to hormonal status, specifically estrogen, which encourages the retention of cholesterol in the bile. As one's weight increases, so does the likelihood of developing stones and, thus, the risk of GBC [57]. Consequently, it is essential to educate the public and implement preventative measures to encourage regular exercise, a diet low in carbohydrates, and a reduced sugar intake (sugars are simple carbohydrates, while starches and fiber are considered complex carbohydrates) [58].

1.7 CHRONIC ALCOHOL HABIT AND TABAGISM

While there is some evidence that cigarette and alcohol usage increases the risk of cancer, there is less evidence that these behaviors increase the risk of GBC in those who smoke and drink. No matter how much alcohol a person drinks or whether they have GS, according to a meta-analysis carried out by Wenbin et al. [59] these authors found that smokers had a 45% greater risk of getting GBC than non-smokers. Alternatively, McGee et al. [60] compiled information from 26 retrospective studies to determine the likelihood of biliary tract cancer. They found that smoking is linked to the ampulla of Vater malignancies as well as intra- and extrahepatic bile duct neoplasms. Tobacco use was not shown to be associated with an increased risk of GBC. There was no conclusive evidence that alcohol intake increased the risk of GB or ampullary neoplasms; however, the same article did examine its role as a risk factor for intrahepatic and extrahepatic biliary tract neoplasms. The amount of alcohol consumed was measured in drinks per day, with 14 g of ethanol being contained in one drink. Hence, additional research is required to definitively link alcoholic beverages, cigarette use, and GBC before any conclusions can be drawn.

1.8 CONTACT WITH CHEMICAL SUBSTANCES

Aspergillus parasiticus, *Aspergillus flavus*, and, on a rarer occasion, *Aspergillus nomius* are fungi that create aflatoxins (AFs), a class of secondary metabolites [61–67]. These fungi are abundant in grains, dried fruits, nuts, spices, oilseeds, and other agricultural goods. Several domestic birds died after consuming Brazilian peanut flour in the 1960s, leading to the discovery of mycotoxins. About 20 different kinds of AFs are known to date, with AFB1, AFB2, AFG1, and AFG2 ranking highest. They have been identified as blue B1 and -2 and green G1 and -2 using chromatographic techniques [61, 62, 68]. Because AFs are classified as a Group 1 carcinogenic and mutagenic risk by the International Agency for Research on Cancer (IARC), it is important to examine the health effects of prolonged exposure to low levels of AFs. The fact that AFs have been linked to liver cancer suggests that they may also induce other types of cancer, including GBC, when they are excreted into bile and build up in the biliary tract [62, 64]. In a study of 84 individuals with GBC, Koshiol et al. examined the association between AFs and GBC. They discovered AFB1 in 50% of GBC cases [62]. Chile, Bolivia, and Peru all have high rates of GBC [65–67], and AFs could be involved in the progression of the disease. Red chili peppers from these countries also contain significant quantities of AF. Ochratoxin A (OA) is another mycotoxin that is particularly fond of grains, spices, and seeds; it is produced by the fungi *Penicillium* and *Aspergillus*. Peru, Bolivia, and Chile are South American countries where it is particularly abundant, and the menu on tables in these countries may contain this mycotoxin. When comparing dried red chili peppers to fresh ones, Ikoma et al. discovered much higher quantities of OA in the former. According to the authors, this mycotoxin could be involved in the progression of GBC. Postharvest conditions are crucial since this mycotoxin is not present in significant quantities in fresh chili peppers [65].

Arsenic (As) is one of the top ten substances that the World Health Organization considers a public health issue, and the IARC has categorized it as a human carcinogen [69, 70]. Groundwater in Chile, Argentina, Bangladesh, Mexico, India, China, Vietnam, and the United States has high amounts of arsenate, which is correlated with the high rates of groundwater contamination in these nations. Water, namely water used for drinking, water for irrigation of crops, and water used in food preparation, is a significant source of arsenic exposure. According to the World Health Organization, 10 µg/L of arsenate should be the maximum level in potable water (Arsenic. December 7, 2022. Available online: https://www.who.int/news-room/fact-sheets/detail/arsenic; last time accessed on May 26, 2025). More and more research is linking arsenic to an increased chance of harboring GBC. Groundwater contamination and chronic exposure have caused arsenic poisoning in a large portion of the population in Bihar, India, near the Ganges River, where GBC is common. Patients with GBC have been found to have high levels of arsenic in their hair, blood, stones, and gallbladder tissues [71]. Five female patients (14%) with GBC had elevated arsenic levels in a small cohort study in Chile. The study included 35 patients with GBC and 42 without GBC. In cases when GBC was not present, elevated arsenic concentrations were not detected [72]. In large-scale population research, cadmium, vanadium, chromium, copper, and molybdenum are among the additional heavy metals linked to GBC risk. It is now known that metals, including arsenic, boron, lithium, and molybdenum, may be present in gallstones. Their presence can increase the likelihood that the gallbladder will develop GBC, probably through several steps of carcinogenesis, which need to be investigated in detail [73].

1.9 GBC STAGING

An essential prognostic factor in GBC is that surgical treatment is now the sole option to attain long-term results. Surgery for GBC can vary from a straightforward cholecystectomy to more involved procedures such as prolonged cholecystectomy involving the liver, significant liver resection, resection of the main bile duct, or resection of adjacent organs. These variations are based on the preoperative staging of the cancer. Surgery type, tumor grading, perineural invasion, lymphovascular invasion, TNM (Tumor, Node, Metastasis) score, and R0 (no residual tumor) status are essential prognostic GBC variables affecting long-term survival. R0 means no residual tumor was found, indicating a complete resection with clear margins, while R1 means microscopic residual tumor was identified at the surgical margin following resection. Most of the data originate from retrospective research that analyzed the clinicopathological features of patients with GBC, and few studies precisely examine each component (Please see the College of American Pathologists [CAP]—Protocol for GBC).

The American Joint Committee on Cancer (AJCC) system is the gold standard for cancer staging worldwide. It incorporates the work of the International Union for Cancer Control and uses the TNM score, which considers the primary tumor's size and extent (T), the involvement of regional nodes (N), and the presence or absence of distant metastases (M). New adjustments and subclassifications of T2 into T2a—tumor invades peritoneal side peri-muscular connective tissue—and T2b—tumor invades hepatic side peri-muscular connective tissue—are introduced in the 8th edition of the AJCC compared to the 7th edition. There is less evidence of survival for T2b when compared to T2a. The N category also underwent modifications, with the staging of N now based on the number of positive nodes rather than their location. Hence, N1 indicates nodes ranging from 1 to 3, N2 signifies nodes ≥ 4 positive regional nodes, and for the pathologist to establish N, they must analyze a minimum of six nodes. In addition, the celiac, peripancreatic, superior mesenteric nodes, and other distant metastatic sites are framed as M1 in the 8th edition, replacing N2 in the 7th edition [74]. Regarding GBC, TNM is the most important prognostic indicator that can determine survival rates. The 5-year survival rate for Tis/T1a GBC is nearly 100%, and numerous authors have shown that a straightforward cholecystectomy, whether laparoscopic or open, is enough. However, regarding T1b, there is a lack of clarity and discussion on whether a basic or extended cholecystectomy is better. According to the NCCN guidelines, T1b radical cholecystectomy and regional lymphadenectomy are the procedures to be performed (Network NCC. Inc. [US]. Hepatobiliary Cancers. Available online: www.nccn.org/professionals/physician_gls/default.aspx# site; last time accessed May 26, 2025). Alternatively, the Japanese guidelines suggest a straightforward cholecystectomy, with T1b managed as a local illness extending into the muscle layer [75]. Despite the lack of evidence supporting the use of extended cholecystectomy for T1b over simple cholecystectomy, Lee et al. [76] still advise lymphadenectomy due to the 11% of cases reported with lymph node involvement in one study. Reports of 5-year survival rates of 93.7% and 95.5%, respectively, after a simple and extended cholecystectomy were shown in an international multicenter study by Kim et al. [77]. While 247 patients underwent a straightforward cholecystectomy, 217 underwent a radical cholecystectomy en bloc with hepatic resection and regional lymphadenectomy following the NCCN criteria, as shown in a retrospective study by Vo et al. [78]. Positive margins were found in 6.1% of patients who had a straightforward cholecystectomy and 2.3% of patients who had a combined radical cholecystectomy and lymphadenectomy. A total of 84.4% of patients who had radical cholecystectomy and regional lymphadenectomy had negative nodal status, 14.8% had positive nodes, and 1.8% had no way of knowing their nodal status. Among the cases that had adjuvant therapy, 53.1% had positive nodes, 9.4% had negative nodes, and 10.9% had previously undergone simple cholecystectomy. Adjuvant treatment resulted in positive margins and poorly or nonexistent tumor differentiation in the second patient. Radical cholecystectomy with regional lymphadenectomy patients had a higher 5-year survival rate (57.7% versus 48.3% for simple cholecystectomy patients). The results of this study suggest that T1b patients should undergo radical cholecystectomy and regional lymphadenectomy and that current guidelines should be followed more strictly.

Concerned about the lack of consensus on the best surgical approach for T1b GBC (simple cholecystectomy versus extended cholecystectomy), Wang et al. [79] sought to shed light on the matter by investigating the potential correlation between tumor size and lymph node metastasis with the hope that this information could guide surgical decision-making. A lymphadenectomy was performed on 127 of the 277 patients who were a part of this research. Of these patients, 23 had a tumor size of 1 cm or less without lymph node metastases, while 104 had a tumor size of 1 cm or more, with 15 patients showing lymph node metastases. The authors suggest a simple cholecystectomy for T1b tumors ≤ 1 cm and an extended cholecystectomy for those ≥ 1 cm because there were no survival differences between the groups with tumors ≤ 1 cm and those with tumors ≥ 1 cm when comparing the groups with and without tumors.

The OMEGA cohort research, which included 3676 patients from 133 sites who had GBC surgically removed, found that resection margin, N status, and overall survival were associated with a lower risk of recurrence. This study suggests regional lymphadenectomy as an alternative to liver resection for T1b, albeit it does not increase survival [80]. A recent study by Goel et al. [2] found that 21% of T1b patients tested positive for lymph nodes, highlighting the need to incorporate lymphadenectomy into standard therapy for improved staging. Ren et al., who also found that patients who underwent lymphadenectomy and cholecystectomy had superior survival rates, observed a similar result. Hepatectomy did not enhance survival rates for patients [81]. Finally, a thorough preoperative staging is essential to avoid underdiagnosis and incomplete treatment for patients with T1b, even if it is challenging to determine the optimal surgical approach.

1.9.1 Stage 2 GBC

The standard treatment for type 2 of GBC is an en bloc extended cholecystectomy with probably 2–3 cm of wedge hepatic resection or a partial hepatectomy of segments IVb-V of the liver. In addition, lymph nodes in the following areas should be removed: cystic duct, common hepatic duct, and hepatoduodenal ligament (i.e., hepatic artery and portal vein nodes, upper posterior pancreaticoduodenal nodes) [82]. In their study, Shindoh et al. [83] examined 437 patients from four high-volume facilities across Europe, Asia, and the Americas to see if the tumor site correlates with long-term survival. Tumors were categorized according to whether they were in the liver or the peritoneum. A lower 5-year survival rate (76% versus less than 50%) was described in cases with liver-sided tumors compared to peritoneal-sided tumors, likely due to the increased incidence of vascular invasion, lymph node invasion, and liver metastases in the former. Even after curative surgery, individuals with T2 hepatic-sided tumors are recommended to have adjuvant chemotherapy. Accordingly, cholecystectomy, liver resection, and lymphadenectomy are deemed to be beneficial for all T2s in terms of survival results. In the years after the AJCC introduced the T2 subclassification, many researchers have investigated the possibility of differences between T2a (peritoneal side) and T2b (liver side) tumors in terms of surgical approach, long-term results (survival and recurrence), and overall validity of the new classification. A meta-analysis was published by Kang et al. [84] to evaluate the best surgical strategy for each location of GBC and to find out the prognostic importance in patients with GBC located according to the AJCC subclassifications T2a and T2b. Seven studies included 1,789 patients in this meta-analysis. The overall 5-year survival rate for the T2b group was much lower than that of the T2a group (74–40%) compared to the other groups (96–60%). This range represents the range of survival rates found in the included studies. When comparing T2a and T2b, the lymph node metastatic rate was 26.6% and 36.6%, respectively. The T2b group had a higher incidence of perineural invasion and vascular and lymphatic invasions when contrasted with the T2a group. Although some studies did find an improvement in survival rates after extended cholecystectomy compared to simple cholecystectomy, the difference was not statistically significant. This could be due, in part, to the fact that these studies did not always utilize the same terminology or adhere to conventional surgical procedures when describing the procedures they conducted. For example, lymph node dissection was really considered part of a "simple cholecystectomy" in certain research. According to this meta-analysis, T2b indicates a poor prognosis in type 2 GBC. During the same time, Alrawashdeh et al. [85] compared T2a and T2b survival rates using a meta-analysis and systematic review of 15 retrospective studies involving 2,531 patients. From nine studies, we may infer that 81% of T2a patients and 79% of T2b patients underwent liver resection, which can take many forms (e.g., wedge resection, segmental hepatectomy, or hemihepatectomy). Lymph node dissection was performed on 90% of T2a patients and 88% of T2b patients in seven trials that used comparable data and methodology. Six studies found that patients with T2b tumors were more likely to experience relapse than those with T2a.

Additionally, seven studies looked at data on adjuvant chemotherapy administration and found that patients with T2b tumors were more likely to receive adjuvant therapy than those with T2a. According to nine studies that provided overall survival rates, T2b patients behave worse than T2a patients did. Additionally, six studies that looked at individual patient data revealed that T2a had 74.2% and 71.8% survival rates after 5 and 10 years, respectively, whereas T2b had 59.9% and 53.6%. The results of this meta-analysis corroborate those of other researchers, who found that T2a patients fared better in terms of overall survival compared to T2b patients. There was an improvement in the 5-year survival rate (82.9% versus 59.3%, respectively) for all T2 patients who had liver resection in this meta-analysis compared to those who did not. Postoperative problems were less common in patients who had wedge resections for T2 GBC tumors than those who received IVb-V liver segment resection, according to a recent meta-analysis by Matsui et al. [86]. Wedge resection is an effective method if the resection margins are negative; however, patients who underwent IVb-V liver segment resection fared better oncologically than those who underwent wedge resection. The cystic vein controls blood flow through the veins of segments IVb and V. This may explain why IVb-V liver segment excision is more effective: it eliminates the possibility of hidden liver metastases. Finally, T2 tumors should be treated with lymphadenectomy and liver resection for better long-term survival. It would be helpful if researchers could standardize the definitions of liver resection for easier work and more concrete results. And just like with T1b, rigorous preoperative staging is necessary.

1.9.2 T3-T4 GBC

As long as R0 resection is still possible, a patient with an advanced T stage can still have a curative-targeted operation, which sometimes involves removing more than one organ. Lymph node dissection, comprising cystic lymph nodes, main bile duct nodes, hepatoduodenal ligament, and higher posterior pancreaticoduodenal nodes, is suggested for T3 cases after an en bloc cholecystectomy or hepatectomy segments IVb-V. Postoperative complications and increased mortality are associated with extensive resections that are necessary to obtain cancer-free margins for T4 tumors. These procedures include gastrectomies, colon resections, main bile duct resections, pancreatoduodenectomy, and major liver resections with more than three segments. A palliative procedure may be necessary if the patient's condition or the severity of their illness make radical surgery impossible [82]. In research conducted by Birnbaum et al., 79% of patients showed T3 lesions, while 21% had T4 lesions. The rate of successful R0 resection was 86%, while the 3-year survival rate was 32%. Removing the pancreatoduodenal region of the gastrointestinal tract, major bile duct, right hemicolectomy, or distal gastrectomy were necessary for 69% of patients to attain R0 resection. No evidence of celiac disease or involvement of the retropancreatic lymph nodes was found to rule out curative surgery. There was a complete lack of survival among patients who needed a pancreatoduodenectomy at 2 years following surgery. Similarly, at 3 years post-op, there was zero survival rate for individuals who had stomach or colon resections. A poor prognosis for survival was observed when additional organs were removed in addition to the main bile duct and the liver during surgery. After large resections were examined in multicenter Dutch research [87, 88], patient outcomes included 33 patients with advanced GBC. Every patient underwent lymph node dissection and main bile duct resection; 16 patients successfully underwent R0 resection. Ten individuals survived for 2 years, 5 for 5, and the median survival was 12. A total of 8 months. A median of 11 months passed before 24 patients experienced a recurrence despite the substantial morbidity and mortality. Therefore, the authors advise that patients with advanced GBC undergo major surgery after carefully considering the risks and advantages. In another retrospective investigation, the effects of major surgery and poor prognostic variables on mortality were examined in 338 patients with T3-T4 GBC [89]. Each of the 338 patients was assigned to either a palliative or curative surgical group. In the curative group, there were 134 patients, whereas 204 individuals were in the palliative group. After 3 years and 5 years, the authors found that the curative surgery group had a survival percentage of 47.3% and 44.3%, respectively. At three and 5 years, the palliative group's survival rates were 8.3% and 7.7%, respectively, which is a very low figure. Assuming R0 was attainable, extended surgery also improved prognosis for T4 patients. Ascites, positive margin, poor histological differentiation, lymph node metastases, and other prognostic markers seem to have a detrimental impact on survival. Finally, surgical indication should be determined after thoroughly evaluating individuals with advanced GBC. Extensive surgery, despite the risk of postoperative complications and mortality, improves survival rates as long as R0 resection can be done in patients with good health.

1.10 UPDATE ON LYMPH NODES

The latest AJCC edition reinstates staging according to the number of inspected nodes, which is an important step in determining lymph node status before GBC surgery. Stage N1 contains 1 to 3 positive nodes, stage N2 > 4, and stage N3 requires at least six nodes. Studies demonstrate that the likelihood of lymph node metastasis increases with tumor size and the number of tissues invaded [79]. Lymph nodes are the most prevalent location of metastasis in GBC. The lymph node positivity rate for T1a ranges from 0% to 2.5%, for T1b from 5% to 16%, for T2 from 9% to 30%, and for T3 and T4, the lymph node metastasis rate is more than 62.7% [79]. Recommendations from the Japanese Society of Hepato-Biliary-Pancreatic Surgery and the Chinese Society of Biliary Surgery are utilized in the management of GBC, alongside the most popular American recommendations. The Chinese Medical Association has issued guidelines that suggest checking the para-aortic and pancreatic posterior nodes. Extensive lymph node dissection is necessary if the pancreatic posterior nodes test positive or ≥T3, and resection with curative aim is not warranted if the latter test is positive, as it indicates the presence of distant metastases [90]. Using data and matching criteria from 18 studies, Widmann et al. (2020) [91] conducted a meta-analysis to examine the effects of lymphadenectomy on patients with resected GBC. The data that were retrieved and analyzed demonstrated that lymphadenectomy influenced survival rates for T1b-T3 but had no effect on T1a GBC. There was a lack of data for the T4 group. The studies that discussed lymphadenectomy used the number of lymph nodes removed or the lymph node stations (D1 and D2) to identify the procedure. There are two types of lymph node removal procedures: D1 and D2. D1 involves a more limited dissection of lymph nodes around the hepatoduodenal ligament, hepatic artery, portal vein, cystic duct, and bile duct. In contrast, D2 involves a more thorough dissection of lymph nodes in the celiac, para-aortic, and pancreaticoduodenal areas. There was a survival benefit in seven of the included studies, and D2 lymph node dissection is suggested in advanced GBC; the number of nodes ranged from three to eight, depending on the extent of the dissection. Researchers also employ the lymph node ratio (LNR), which is the ratio of positive nodes to excised nodes, to improve prognostic information for GBC patients. Research has demonstrated that LNR can be used as a standalone predictor of prognosis, with greater values indicating a less favorable outcome, particularly in cases of advanced GBC [92]. When it comes to staging, the total lymph node count (TLNC) is crucial. For patients with GBC, Tsimiligras et al. (2019) [93] determined the minimal and optimal range of lymph node counts using a machine-based approach that depends on artificial intelligence and data acquired from the National Cancer Database. The patients with a median survival of 40 months with four to seven nodes removed (regardless of the T) fared better than those with fewer than four nodes removed (27% probability of death for the latter, which may indicate sub-staging), and this difference was statistically significant. Additionally, patients in the 4–7-node group with three or more lymph node metastases had a 16.5-month survival rate, compared to 90.6-month survival for patients without metastases. The fact that 67.5% of patients have fewer than four nodes—despite a minimum of four nodes being necessary to attain favorable results—and that a vast majority did not fulfill the AJCC suggested criteria suggests that achieving a total of six nodes can be hard.

In a retrospective review of ≥T1b cases, Leigh et al. [94] found it difficult to acquire a TLNC > 6 following portal lymphadenectomy. Eighty percent of patients in this study had fewer than six nodes found, with the group with six nodes or more showing an average of three positive nodes and the group with fewer than six nodes showing one positive node. The fact that there are fewer nodes in the area around the hepatoduodenal ligament could be the reason behind this; therefore, a more thorough lymphadenectomy would be necessary to reach many nodes greater than six. According to the investigators, regular peripancreatic lymph node dissection should be included for all patients with a T1b level of 1 or higher. A proper evaluation of the justification for the degree of lymph node dissection is necessary to avoid underdiagnosis in patients whose lymph node involvement is a significant prognostic factor. Surgeons at high-volume hospitals with experience in hepatobilio-pancreatic surgery and pathologists with experience in evaluating final diagnoses and staging are the best places to have GBC surgery. Surgery for GBC at a high-volume center is associated with better overall survival, fewer hospitalization days, and a lower risk of postoperative complications and perioperative morbidity, although no studies have evaluated the impact of surgeon experience or hospital volume on outcomes of GBC resections.

1.11 METASTASIS AT A DISTANCE

Adjuvant therapy has recently replaced surgical treatment as the recommended course of action for GBC with distant metastases; patients undergoing this treatment had an average survival rate of 6.7% to 12.5% after 5 years and a survival rate of 2 to 9 months [95, 96].

Patients with distant lymph node metastases have a median survival of 6 months, though this drops to 3 months for those with bone metastases, 4 months for those with liver metastases, 9 months for those with lung metastases, and a median survival of 4 months for the entire cohort. The reported survival for metastatic patients with GBC varies according to the site of metastasis. In the setting of complete removal, that is, when patients with liver and lymph node metastases have their main tumors surgically removed and chemotherapy is administrated, their survival rates improve. Those who have metastases to the bones or lungs should not undergo surgery. Patients with bone metastases who undergo concurrent chemotherapy have a chance of a 5-month survival increase. Patients without negative prognostic markers like jaundice or peritoneal spread who have modest liver metastases may be able to have a mild hepatectomy and adjuvant therapy. Patients' survival rates were higher after T2 minor hepatectomy, which required less blood loss and took less time to perform. Patients with locally advanced or metastatic cancer who were initially considered inoperable saw their disease stage decrease following neoadjuvant chemotherapy, which improved their prognosis and made R0 surgery more likely.

1.12 NEUROPATHY OR DAMAGE TO THE NERVE ENDS

Malignant cells that invade in, around, or through nerves are considered perineural invasion (PNI). PNI is an important component that affects the aggressiveness and prognosis of malignant tumors linked to poor survival rates. For PNI to take place, a number of things must come together. Initiation involves the tumor microenvironment, Schwann cells, and the creation of growth and chemostatic molecules that are particular to the tumor. There is a correlation between neurotrophin factors, such as TGF-β, and the development of new nerve cells and the maintenance of existing ones. Through receptors, additional cell adhesion molecules strengthen extracellular matrix and cell-cell bonds. All of these things help regulate and govern the cell cycle, and microRNA and long noncoding RNA are also involved. To confirm a diagnostic diagnosis of PNI, a microscopic examination must reveal tumor cells around the nerve on at least one-third of its circumference or the presence of malignant cells within any layer of the nerve sheath. Because of the dense network of nerve tissue surrounding the gallbladder and common bile duct, PNI is a common complication of GBC, particularly in the advanced stages of the disease, and it is a factor that affects the prognosis for individuals with GBC. Due to a lack of research examining PNI in GBC and its omission from the AJCC's 8th edition list of variables or prospective prognostic factors, the impact of PNI on patients' prognoses remains underexplored. Overall, PNI seems to be more common in T2 peritoneal tumors than T2 hepatic tumors, and it seems to be more common for stages T3–T4. Patients without PNI had a 34.7-month survival rate, while those with PNI had a 12-month survival rate and a 5-year survival rate of 7% for patients with PNI and 72% for patients without PNI. PNI can indeed be an independent prognostic factor [97]. They also found no association between PNI and lymphovascular invasion, suggesting that PNI is caused by primary tumor extension. Compared to the tumor's margins, PNI was greater in the tumor's center. There was also a correlation between PNI and invasion of the common bile duct; in cases when invasion is suspected, it is advised to conduct a frozen examination of the stump of the distal common bile duct. There was no correlation between lymphatic invasion and PNI, although Maruyama et al. [98] found that PNI was an independent prognostic predictor. Birnbaum et al. [99, 100] found that PNI is an independent prognostic predictor in a study of patients with advanced GBC. Patients with an initial PNI were more likely to have local recurrence and had a 6% survival rate at 5 years. Based on the institutions' accumulated 10-year data, a Chinese study analyzed 324 GBC surgery patients to determine the prognostic significance of PNI [97]. The results showed that 19.8% of patients tested positive for PNI, with most of those patients having advanced tumor stages and worse overall and disease-free survival rates compared to those without PNI. In terms of median survival time and median time without recurrence, the results were 19 months and 36 months, respectively. There was an increase in the incidence of lymph node metastases, lymphovascular invasion, liver invasion, and moderate-to-poor differentiation grade in PNI-positive individuals, along with preoperative jaundice and a lower R0 rate. In comparing patients with PNI who got adjuvant treatment and those who did not, the median survival for the former group was 24 months, while the disease-free period was 13 months compared to 6.5 months for the latter. Adjuvant treatment is recommended for PNI-positive patients, even though PNI did not greatly affect overall survival. However, PNI did have a big effect on early relapse. Therefore, PNI should be considered and included in the list of factors affecting GBC patients' survival rates. Additional research is required to confirm PNI as a highly relevant and substantial prognostic factor in GBC.

1.13 LYMPHOVASCULAR INVASION

Another risk factor linked to a worse prognosis for GBC patients is lymphovascular invasion (LVI). In retrospective studies on GBC, clinicopathological descriptions produced by authors provide the majority of the material. Still, the impact in terms of survival may not be completely investigated, and further studies are necessary. Based on the presence or absence of LVI, radical surgery for incidental GBC was recommended by Shibata et al. [101]. Dominguez et al. [102] identified 46.4% of 1649 patients with GBC as having LVI, and it was found to be the strongest predictor for lymph node metastasis, with a sensitivity of 57.1% and a specificity of 77.7%. In stage T1b, the median survival was 26 months compared to 76.2 months in those without LVI; in stage T2, it was 25.8 months compared to 42.6 months; and in stage T3, it was 11.9 months compared to 18.6 months. Additional research is required to determine the highly relevant and substantial function of LVI in GBC definitively.

1.13.1 R0 Condition

Patients with surgically fully operable GBC still only have R0 surgical treatment as a curative option. The difficulty of obtaining R0 and the necessity of multi-organ resections both increase with the progression of the GBC. There is an increased risk of postoperative complications and morbidity with multi-organ resections. R0 resections increase survival rates compared to R1 resections, and hospitals with knowledge in this area are ideal for this. Extensive oncological reoperation is necessary to obtain R0 for incidental GBC. If the tumor is ≥T1b, more testing is necessary to determine the right staging, and re-intervention may be necessary after histological diagnosis, all in an effort to achieve R0 resection. The less advanced T stage in incidental GBC patients likely explains why they have a greater overall survival rate compared to GBC patients. It is crucial to wait 4 to 8 weeks after a cholecystectomy before undergoing further surgery to cure the oncological condition. Inflammation from the prior operation makes surgery more challenging and prone to problems if performed before 4 weeks, and the condition advances and spreads beyond 8 weeks. Resection of the port site is not generally advised unless gallbladder perforation and bile leaking happen during the initial surgery; 10% of patients experience disease recurrence at the port location. The cystic duct histopathology study is also significant. A study conducted by Vega et al. [103] revealed that 17% of patients with Incidental Gallbladder Cancer (IGCC) had a positive margin. The researchers also found that main bile duct resection resulted in a 5-year survival rate of 75%, which was comparable to the 78% survival rate observed in patients with a negative cystic duct margin. Among patients who did not undergo primary bile duct resection and had a favorable margin, 26% survived. Despite the increased risk of complications like pancreatic fistula, liver failure, and bile fistula, major liver resection and pancreatoduodenectomy can be performed on selected patients with locally advanced GBC in order to achieve R0. This type of surgery is only performed by centers with extensive surgical experience and has a better survival rate than inoperable cases, ranging from 25% to 5 years [104].

The extent of liver resection, specifically wedge resection versus segment IVb-V hepatectomy, is controversial. Assuming negative margins and oncological outcomes similar to those seen in cases undergoing segment IVb-V hepatectomy, we may able to support that a wedge resection of 1–2 cm is enough. For GBCs in stages 2 and 3, as well as IGBCs that necessitate reoperation, wedge resection is still an option; however, it is not recommended for T1 instances, where a simple cholecystectomy would suffice, and T4 cases, where extensive hepatectomies are necessary. Finally, for a better long-term prognosis, it is required to do a thorough preoperative evaluation in order to reach R0.

1.14 FINAL REMARKS

Long-term survival rates for GBC are still low, and early detection is crucial for better outcomes because the disease is more treatable at an earlier stage. When thinking about prevention, it is important to focus on modifiable factors, including nutrition, lifestyle, and obesity. Additionally, targeted regional programs and processes to lessen exposure to heavy metals should be established. Particularly for those at risk, routine ultrasonographic investigations should be done to detect gallstones—a precursor to GBC. To avoid understaging or incorrect staging, a patient with GBC should undergo a comprehensive evaluation before treatment choices are decided. Additionally, it is advised to refer patients diagnosed with GBC to a reputable surgical facility. Patients have a better probability of survival if the surgeons are skilled and experienced and if a pathology report gives the final stage using synoptic formats. An operable case can be turned into an inoperable one with the help of neoadjuvant chemotherapy. Patients with GBC had the best likelihood of long-term survival when TNM staging, negative resection margin, and R0 status are considered as prognostic variables. To make research easier and more accessible, we also need to introduce new technologies, such as nanotechnology and RNA sequencing.

ACKNOWLEDGMENTS

The author would like to recognize the patients who suffered from gallbladder carcinoma and helped to support the research in this relatively rare condition.

FUNDING

This research has been funded by the generosity of the Children's Hospital of Eastern Ontario, Ottawa, Ontario, and the Stollery Children's Hospital Foundation and supporters of the Lois Hole Hospital for Women through the Women and Children's Health Research Institute (Grant ID #: 2096), Natural Science Foundation of Hubei Province for Hubei University of Technology (100-Talent Grant for Recruitment Program of Foreign Experts Total Funding: Digital PCR and NGS-based diagnosis for infection and oncology, 2017–2022), Österreichische Krebshilfe Tyrol (Krebsgesellschaft Tirol, Austrian Tyrolean Cancer Research Institute, 2007 and 2009—"DMBTI and cholangiocellular carcinomas" and "Hsp70 and HSPBP1 in carcinomas of the pancreas"), Austrian Research Fund (Fonds zur Förderung der wissenschaftlichen Forschung, FWF, Grant ID L313-B13), Canadian Foundation for Women's Health ("Early Fetal Heart-RES0000928"), Cancer Research Society (von Willebrand factor gene expression in cancer cells), Canadian Institutes of Health Research (Omega-3 Fatty Acids for Treatment of Intestinal Failure Associated Liver Disease: A Translational Research Study, 2011–2014, CIHR 232514), and the Saudi Cultural Bureau, Ottawa, Canada. The funders had no role in study design, data collection and analysis, decision to publish, or preparation of the manuscript.

REFERENCES

1. Rakić M, Patrlj L, Kopljar M, Kliček R, Kolovrat M, Loncar B, Busic Z. Gallbladder cancer. *Hepatobiliary Surg Nutr.* 2014;3(5):221–6. DOI: 10.3978/j.issn.2304–3881.2014.09.03
2. Goel M, Pandrowala S, Patel P, Patkar S. Node positivity in T1b gallbladder cancer: A high volume centre experience. *Eur J Surg Oncol.* 2022;48(7):1585–9. DOI: 10.1016/j.ejso.2022.03.013
3. Wang J, Liu F, Ma W, Hu H, Li F. Metastatic lymph node ratio as an important prognostic factor in advanced gallbladder carcinoma with at least 6 lymph nodes retrieved. *Langenbecks Arch Surg.* 2023;408(1):382. DOI: 10.1007/s00423-023-03119-5
4. Samuel S, Mukherjee S, Ammannagari N, Pokuri VK, Kuvshinoff B, Groman A, LeVea CM, Iyer R. Clinicopathological characteristics and outcomes of rare histologic subtypes of gallbladder cancer over two decades: A population-based study. *PLoS One.* 2018;13(6):e0198809. DOI: 10.1371/journal.pone.0198809

5. Tricco AC, Lillie E, Zarin W, O'Brien KK, Colquhoun H, Levac D, Moher D, Peters MDJ, Horsley T, Weeks L, Hempel S, Akl EA, Chang C, McGowan J, Stewart L, Hartling L, Aldcroft A, Wilson MG, Garritty C, Lewin S, Godfrey CM, Macdonald MT, Langlois EV, Soares-Weiser K, Moriarty J, Clifford T, Tunçalp Ö, Straus SE. PRISMA extension for scoping reviews (PRISMA-ScR): Checklist and explanation. *Ann Intern Med.* 2018;169(7):467–73. DOI: 10.7326/m18–0850
6. Odetola S, Feulefack J, Sergi CM. Eosinophilic esophagitis: Absolute eosinophilic count, peak eosinophilic count, and potential biomarkers of eosinophilic degranulation products-an in-depth systematic review. *Transl Pediatr.* 2024;13(3):474–83. DOI: 10.21037/tp-23–478
7. Filho AM, Laversanne M, Ferlay J, Colombet M, Piñeros M, Znaor A, Parkin DM, Soerjomataram I, Bray F. The GLOBOCAN 2022 cancer estimates: Data sources, methods, and a snapshot of the cancer burden worldwide. *Int J Cancer.* 2025;156(7):1336–46. DOI: 10.1002/ijc.35278
8. Kratz JD, Klein AB, Gray CB, Marten A, Vilu HL, Knight JF, Kumichel A, Ueno M. The epidemiology of biliary tract cancer and associated prevalence of MDM2 amplification: A targeted literature review. *Target Oncol.* 2024;19(6):833–44. DOI: 10.1007/s11523-024-01086-5
9. Rosato V, Negri E, Bosetti C, Malats N, Gomez-Rubio P, Consortium P, Maisonneuve P, Miller AB, Bueno-de-Mesquita HB, Baghurst PA, Zatonski W, Petersen GM, Scelo G, Holcatova I, Fabianova E, Serraino D, Olson SH, Vioque J, Lagiou P, Duell EJ, Boffetta P, La Vecchia C. Gallbladder disease, cholecystectomy, and pancreatic cancer risk in the International Pancreatic Cancer Case-Control Consortium (PanC4). *Eur J Cancer Prev.* 2020;29(5):408–15. DOI: 10.1097/CEJ.0000000000000572
10. Miranda-Filho A, Pineros M, Ferreccio C, Adsay V, Soerjomataram I, Bray F, Koshiol J. Gallbladder and extrahepatic bile duct cancers in the Americas: Incidence and mortality patterns and trends. *Int J Cancer.* 2020;147(4):978–89. DOI: 10.1002/ijc.32863
11. Dutta U, Bush N, Kalsi D, Popli P, Kapoor VK. Epidemiology of gallbladder cancer in India. *Chin Clin Oncol.* 2019;8(4):33. DOI: 10.21037/cco.2019.08.03
12. Halaseh SA, Halaseh S, Shakman R. A review of the etiology and epidemiology of gallbladder cancer: What you need to know. *Cureus.* 2022;14(8):e28260. DOI: 10.7759/cureus.28260
13. Hundal R, Shaffer EA. Gallbladder cancer: Epidemiology and outcome. *Clin Epidemiol.* 2014;6:99–109. DOI: 10.2147/clep.S37357
14. Singh J, Shukla D, Gupta S, Shrivastav BR, Tiwari PK. Clinical epidemiology of gallbladder cancer in North-Central India and association of immunological markers, NLR, MLR and PLR in the diagnostic/prognostic prediction of GBC. *Cancer Treat Res Commun.* 2021;28:100431. DOI: 10.1016/j.ctarc.2021.100431
15. Ouyang G, Liu Q, Wu Y, Liu Z, Lu W, Li S, Pan G, Chen X. The global, regional, and national burden of gallbladder and biliary tract cancer and its attributable risk factors in 195 countries and territories, 1990 to 2017: A systematic analysis for the Global Burden of Disease Study 2017. *Cancer.* 2021;127(13):2238–50. DOI: 10.1002/cncr.33476
16. Feroz Z, Gautam P, Tiwari S, Shukla GC, Kumar M. Survival analysis and prognostic factors of the carcinoma of gallbladder. *World J Surg Oncol.* 2022;20(1):403. DOI: 10.1186/s12957–022–02857-y
17. Jaruvongvanich V, Yang JD, Peeraphatdit T, Roberts LR. The incidence rates and survival of gallbladder cancer in the USA. *Eur J Cancer Prev.* 2019;28(1):1–9. DOI: 10.1097/cej.0000000000000402
18. Espinoza JA, Bizama C, García P, Ferreccio C, Javle M, Miquel JF, Koshiol J, Roa JC. The inflammatory inception of gallbladder cancer. *Biochim Biophys Acta.* 2016;1865(2):245–54. DOI: 10.1016/j.bbcan.2016.03.004
19. Sun Y, Yang Z, Lan X, Tan H. Neoplastic polyps in gallbladder: A retrospective study to determine risk factors and treatment strategy for gallbladder polyps. *Hepatobiliary Surg Nutr.* 2019;8(3):219–27. DOI: 10.21037/hbsn.2018.12.15
20. Stinton LM, Shaffer EA. Epidemiology of gallbladder disease: Cholelithiasis and cancer. *Gut Liver.* 2012;6(2):172–87. DOI: 10.5009/gnl.2012.6.2.172
21. Schmidt MA, Marcano-Bonilla L, Roberts LR. Gallbladder cancer: Epidemiology and genetic risk associations. *Chin Clin Oncol.* 2019;8(4):31. DOI: 10.21037/cco.2019.08.13
22. Misra S, Chaturvedi A, Misra NC, Sharma ID. Carcinoma of the gallbladder. *Lancet Oncol.* 2003;4(3):167–76. DOI: 10.1016/s1470–2045(03)01021–0
23. Ganeshan D, Kambadakone A, Nikolaidis P, Subbiah V, Subbiah IM, Devine C. Current update on gallbladder carcinoma. *Abdom Radiol (NY).* 2021;46(6):2474–89. DOI: 10.1007/s00261-020-02871-2
24. Murphy MC, Gibney B, Gillespie C, Hynes J, Bolster F. Gallstones top to toe: What the radiologist needs to know. *Insights Imaging.* 2020;11(1):13. DOI: 10.1186/s13244-019-0825-4
25. Iqbal S, Ahmad S, Saeed U, Al-Dabbagh M. Porcelain gallbladder: Often an overlooked entity. *Surg J (NY).* 2017;3(4):e145–e7. DOI: 10.1055/s-0037–1606546

26. Khan SM, Emile SH, Choudhry MS, Sumbal R. Tumor location and concurrent liver resection, impact survival in T2 gallbladder cancer: A meta-analysis of the literature. *Updates Surg*. 2021;73(5):1717–26. DOI: 10.1007/s13304–021–01150-z
27. Khan ZS, Livingston EH, Huerta S. Reassessing the need for prophylactic surgery in patients with porcelain gallbladder: Case series and systematic review of the literature. *Arch Surg*. 2011;146(10):1143–7. DOI: 10.1001/archsurg.2011.257
28. Kamisawa T, Kuruma S, Chiba K, Tabata T, Koizumi S, Kikuyama M. Biliary carcinogenesis in pancreaticobiliary maljunction. *J Gastroenterol*. 2017;52(2):158–63. DOI: 10.1007/s00535–016–1268-z
29. Kamisawa T, Kuruma S, Tabata T, Chiba K, Iwasaki S, Koizumi S, Kurata M, Honda G, Itoi T. Pancreaticobiliary maljunction and biliary cancer. *J Gastroenterol*. 2015;50(3):273–9. DOI: 10.1007/s00535-014-1015-2
30. Arnolds A. Mannskopfgrosse retentionszyste des choledochus. *Dthch Med Wochenschr*. 1906;32:1804.
31. Kamisawa T, Ando H, Suyama M, Shimada M, Morine Y, Shimada H, Working Committee of Clinical Practice Guidelines for Pancreaticobiliary M, Japanese Study Group on Pancreaticobiliary M. Japanese clinical practice guidelines for pancreaticobiliary maljunction. *J Gastroenterol*. 2012;47(7):731–59. DOI: 10.1007/s00535-012-0611-2
32. Morine Y, Shimada M, Takamatsu H, Araida T, Endo I, Kubota M, Toki A, Noda T, Matsumura T, Miyakawa S, Ishibashi H, Kamisawa T, Shimada H. Clinical features of pancreaticobiliary maljunction: Update analysis of 2nd Japan-nationwide survey. *J Hepatobiliary Pancreat Sci*. 2013;20(5):472–80. DOI: 10.1007/s00534-013-0606-2
33. Kimura K, Ohto M, Saisho H, Unozawa T, Tsuchiya Y, Morita M, Ebara M, Matsutani S, Okuda K. Association of gallbladder carcinoma and anomalous pancreaticobiliary ductal union. *Gastroenterology*. 1985;89(6):1258–65. DOI: 10.1016/0016–5085(85)90641–9
34. Nishiwaki M, Toyoda M, Oishi Y, Ishida S, Horiuchi SI, Makino-Itou H, Kimura T, Ohno SI, Ohkura T, Enosawa S, Akutsu H, Nakazawa A, Kasahara M, Kiyono T, Umezawa A. Immortalization of human hepatocytes from biliary atresia with CDK4(R24C), cyclin D1, and TERT for cytochrome P450 induction testing. *Sci Rep*. 2020;10(1):17503. DOI: 10.1038/s41598-020-73992-3
35. Deng YL, Cheng NS, Lin YX, Zhou RX, Yang C, Jin YW, Xiong XZ. Relationship between pancreaticobiliary maljunction and gallbladder carcinoma: Meta-analysis. *Hepatobiliary Pancreat Dis Int*. 2011;10(6):570–80. DOI: 10.1016/s1499–3872(11)60098–2
36. Kamisawa T, Honda G. Pancreaticobiliary maljunction: Markedly high risk for biliary cancer. *Digestion*. 2019;99(2):123–5. DOI: 10.1159/000490816
37. Kuraishi Y, Uehara T, Muraki T, Iwaya M, Kinugawa Y, Nakajima T, Watanabe T, Miyagawa Y, Umemura T. Impact of DNA double-strand breaks on pancreaticobiliary maljunction carcinogenesis. *Diagn Pathol*. 2021;16(1):72. DOI: 10.1186/s13000-021-01132-0
38. Kamakura M, Uehara T, Muraki T, Iwaya M, Asaka S, Nakajima T, Iwaya Y, Nagaya T, Yoshizawa T, Shimizu A, Kuraishi Y, Ota H, Umemura T. Mucosal damage in pancreaticobiliary maljunction is stronger in the gallbladder than in the bile duct. *Pathol Res Pract*. 2024;253:155011. DOI: 10.1016/j.prp.2023.155011
39. Sai JK, Suyama M, Nobukawa B, Kubokawa Y, Yokomizo K, Sato N. Precancerous mucosal changes in the gallbladder of patients with occult pancreatobiliary reflux. *Gastrointest Endosc*. 2005;61(2):264–8. DOI: 10.1016/s0016–5107(04)02586–6
40. Beltrán MA, Vracko J, Cumsille MA, Cruces KS, Almonacid J, Danilova T. Occult pancreaticobiliary reflux in gallbladder cancer and benign gallbladder diseases. *J Surg Oncol*. 2007;96(1):26–31. DOI: 10.1002/jso.20756
41. Wang L, Zhang ZW, Guo T, Xie P, Huang XR, Yu YH. Occult pancreaticobiliary reflux is a pathogenic factor of some benign biliary diseases and gallbladder cancer. *Hepatobiliary Pancreat Dis Int*. 2023;22(3):288–93. DOI: 10.1016/j.hbpd.2022.08.010
42. Andrén-Sandberg A. Diagnosis and management of gallbladder polyps. *N Am J Med Sci*. 2012;4(5):203–11. DOI: 10.4103/1947–2714.95897
43. Maciejewski P, Strzelczyk J. Is gall-bladder polyp equivalent to cancer? An analysis of material from 1196 cholecystectomies--a comparison of the ultrasound and histopathological results. *Pol Przegl Chir*. 2014;86(5):218–22. DOI: 10.2478/pjs-2014–0045
44. Christensen AH, Ishak KG. Benign tumors and pseudotumors of the gallbladder. Report of 180 cases. *Arch Pathol*. 1970;90(5):423–32. www.ncbi.nlm.nih.gov/pubmed/4319984
45. Foley KG, Lahaye MJ, Thoeni RF, Soltes M, Dewhurst C, Barbu ST, Vashist YK, Rafaelsen SR, Arvanitakis M, Perinel J, Wiles R, Roberts SA. Management and follow-up of gallbladder polyps: Updated joint guidelines between the ESGAR, EAES, EFISDS and ESGE. *Eur Radiol*. 2022;32(5):3358–68. DOI: 10.1007/s00330–021–08384-w

46. Kamaya A, Fung C, Szpakowski JL, Fetzer DT, Walsh AJ, Alimi Y, Bingham DB, Corwin MT, Dahiya N, Gabriel H, Park WG, Porembka MR, Rodgers SK, Tublin ME, Yuan X, Zhang Y, Middleton WD. Management of incidentally detected gallbladder polyps: Society of radiologists in ultrasound consensus conference recommendations. *Radiology*. 2022;305(2):277–89. DOI: 10.1148/radiol.213079
47. Tang C, Geng Z, Wen J, Wang L, You Q, Jin Y, Wang W, Xu H, Yu Q, Yuan H. Risk stratification model for incidentally detected gallbladder polyps: A multicentre study. *Eur J Radiol*. 2024;170:111244. DOI: 10.1016/j.ejrad.2023.111244
48. Onda S, Futagawa Y, Gocho T, Shiba H, Ishida Y, Okamoto T, Yanaga K. A preoperative scoring system to predict carcinoma in patients with gallbladder polyps. *Dig Surg*. 2020;37(4):275–81. DOI: 10.1159/000503100
49. Foley KG, Riddell Z, Coles B, Roberts SA, Willis BH. Risk of developing gallbladder cancer in patients with gallbladder polyps detected on transabdominal ultrasound: A systematic review and meta-analysis. *Br J Radiol*. 2022;95(1137):20220152. DOI: 10.1259/bjr.20220152
50. Szpakowski JL, Tucker LY. Outcomes of gallbladder polyps and their association with gallbladder cancer in a 20-year cohort. *JAMA Netw Open*. 2020;3(5):e205143. DOI: 10.1001/jamanetworkopen.2020.5143
51. Bowlus CL, Lim JK, Lindor KD. AGA clinical practice update on surveillance for hepatobiliary cancers in patients with primary sclerosing cholangitis: Expert review. *Clin Gastroenterol Hepatol*. 2019;17(12):2416–22. DOI: 10.1016/j.cgh.2019.07.011
52. Mishra RR, Tewari M, Shukla HS. Association of Helicobacter pylori infection with inflammatory cytokine expression in patients with gallbladder cancer. *Indian J Gastroenterol*. 2013;32(4):232–5. DOI: 10.1007/s12664-013-0321-6
53. Lim KPK, Lee AJL, Jiang X, Teng TZJ, Shelat VG. The link between Helicobacter pylori infection and gallbladder and biliary tract diseases: A review. *Ann Hepatobiliary Pancreat Surg*. 2023;27(3):241–50. DOI: 10.14701/ahbps.22–056
54. Gros B, Gómez Pérez A, Pleguezuelo M, Serrano Ruiz FJ, de la Mata M, Rodríguez-Perálvarez M. Helicobacter species and hepato-biliary tract malignancies: A systematic review and meta-analysis. *Cancers (Basel)*. 2023;15(3). DOI: 10.3390/cancers15030595
55. Shukla R, Shukla P, Behari A, Khetan D, Chaudhary RK, Tsuchiya Y, Ikoma T, Asai T, Nakamura K, Kapoor VK. Roles of Salmonella typhi and Salmonella paratyphi in Gallbladder Cancer Development. *Asian Pac J Cancer Prev*. 2021;22(2):509–16. DOI: 10.31557/apjcp.2021.22.2.509
56. Tan W, Gao M, Liu N, Zhang G, Xu T, Cui W. Body mass index and risk of gallbladder cancer: Systematic review and meta-analysis of observational studies. *Nutrients*. 2015;7(10):8321–34. DOI: 10.3390/nu7105387
57. Everson GT, McKinley C, Kern F, Jr. Mechanisms of gallstone formation in women. Effects of exogenous estrogen (Premarin) and dietary cholesterol on hepatic lipid metabolism. *J Clin Invest*. 1991;87(1):237–46. DOI: 10.1172/jci114977
58. Pati S, Irfan W, Jameel A, Ahmed S, Shahid RK. Obesity and cancer: A Current overview of epidemiology, pathogenesis, outcomes, and management. *Cancers (Basel)*. 2023;15(2). DOI: 10.3390/cancers15020485
59. Wenbin D, Zhuo C, Zhibing M, Chen Z, Ruifan Y, Jie J, Cheng Q, Zhenming G. The effect of smoking on the risk of gallbladder cancer: A meta-analysis of observational studies. *Eur J Gastroenterol Hepatol*. 2013;25(3):373–9. DOI: 10.1097/MEG.0b013e32835a870b
60. McGee EE, Jackson SS, Petrick JL, Van Dyke AL, Adami HO, Albanes D, Andreotti G, Beane-Freeman LE, Berrington de Gonzalez A, Buring JE, Chan AT, Chen Y, Fraser GE, Freedman ND, Gao YT, Gapstur SM, Gaziano JM, Giles GG, Grant EJ, Grodstein F, Hartge P, Jenab M, Kitahara CM, Knutsen SF, Koh WP, Larsson SC, Lee IM, Liao LM, Luo J, Milne RL, Monroe KR, Neuhouser ML, O'Brien KM, Peters U, Poynter JN, Purdue MP, Robien K, Sandler DP, Sawada N, Schairer C, Sesso HD, Simon TG, Sinha R, Stolzenberg-Solomon R, Tsugane S, Wang R, Weiderpass E, Weinstein SJ, White E, Wolk A, Yuan JM, Zeleniuch-Jacquotte A, Zhang X, Zhu B, McGlynn KA, Campbell PT, Koshiol J. Smoking, alcohol, and biliary tract cancer risk: A pooling project of 26 prospective studies. *J Natl Cancer Inst*. 2019;111(12):1263–78. DOI: 10.1093/jnci/djz103
61. Yadav A, Gupta P, Gupta P, Patil AN, Das CK, Hooda H, Thakur D, Sharma V, Singh AK, Yadav TD, Kaman L, Thakur JS, Sudini HK, Srinivasan R, Dutta U. Aflatoxin exposure is associated with an increased risk of gallbladder cancer. *Int J Cancer*. 2025;156(2):322–30. DOI: 10.1002/ijc.35171
62. Koshiol J, Zhu B, Wang R, Hildesheim A, Gao YT, Egner PA, Yuan JM, Groopman JD. Association of aflatoxin with gallbladder cancer in a case-control study nested within a Chinese cohort. *Int J Cancer*. 2024;154(5):801–6. DOI: 10.1002/ijc.34755
63. Singh MK, Kapoor VK. Gallbladder cancer and aflatoxin: Do we have sufficient evidence? *Gastroenterology*. 2018;154(1):259–60. DOI: 10.1053/j.gastro.2017.09.053

64. Koshiol J, Gao YT, Dean M, Egner P, Nepal C, Jones K, Wang B, Rashid A, Luo W, Van Dyke AL, Ferreccio C, Malasky M, Shen MC, Zhu B, Andersen JB, Hildesheim A, Hsing AW, Groopman J. Association of aflatoxin and gallbladder cancer. *Gastroenterology*. 2017;153(2):488–94.e1. DOI: 10.1053/j.gastro.2017.04.005
65. Ikoma T, Tsuchiya Y, Asai T, Okano K, Ito N, Endoh K, Yamamoto M, Nakamura K. Ochratoxin A contamination of red chili peppers from Chile, Bolivia and Peru, countries with a high incidence of gallbladder cancer. *Asian Pac J Cancer Prev*. 2015;16(14):5987–91. DOI: 10.7314/apjcp.2015.16.14.5987
66. Asai T, Tsuchiya Y, Okano K, Piscoya A, Nishi CY, Ikoma T, Oyama T, Ikegami K, Yamamoto M. Aflatoxin contamination of red chili pepper from Bolivia and Peru, countries with high gallbladder cancer incidence rates. *Asian Pac J Cancer Prev*. 2012;13(10):5167–70. DOI: 10.7314/apjcp.2012.13.10.5167
67. Tsuchiya Y, Terao M, Okano K, Nakamura K, Oyama M, Ikegami K, Yamamoto M. Mutagenicity and mutagens of the red chili pepper as gallbladder cancer risk factor in Chilean women. *Asian Pac J Cancer Prev*. 2011;12(2):471–6.
68. Dhakal A, Sbar E. *Aflatoxin Toxicity*. Treasure Island, FL: StatPearls, 2022. www.ncbi.nlm.nih.gov/pubmed/32491713
69. Blanco E, Algranti E, Cifuentes LA, Lopez-Carrillo L, Mora AM, Rodriguez-Guzman J, Rodriguez-Villamizar LA, Veiga LHS, Canelo-Aybar C, Nieto-Gutierrez W, Feliu A, Espina C, Ferreccio C. Latin America and the Caribbean Code Against cancer 1st edition: Environment, occupation, and cancer. *Cancer Epidemiol*. 2023;86(Suppl. 1):102381. DOI: 10.1016/j.canep.2023.102381
70. Krewski D, Bird M, Al-Zoughool M, Birkett N, Billard M, Milton B, Rice JM, Grosse Y, Cogliano VJ, Hill MA, Baan RA, Little J, Zielinski JM. Key characteristics of 86 agents known to cause cancer in humans. *J Toxicol Environ Health B Crit Rev*. 2019;22(7–8):244–63. DOI: 10.1080/10937404.2019.1643536
71. Kumar A, Ali M, Raj V, Kumari A, Rachamalla M, Niyogi S, Kumar D, Sharma A, Saxena A, Panjawani G, Jain P, Vidyarthi A, Kumar N, Kumar M, Niraj PK, Rahman MS, Bishwapriya A, Kumar R, Sakamoto M, Kumar S, Singh M, Ghosh AK. Arsenic causing gallbladder cancer disease in Bihar. *Sci Rep*. 2023;13(1):4259. DOI: 10.1038/s41598-023-30898-0
72. De Palma G, Carrasco-Avino G, Gilberti E, Cadei M, Pedrazzi T, Gallegos Mendez IM, Roman A, Demartines N, Boffetta P, Labgaa I. Arsenic may be a carcinogenic determinant of a subset of gallbladder cancer: A pilot study. *Environ Res*. 2023;219:115030. DOI: 10.1016/j.envres.2022.115030
73. Lee MH, Gao YT, Huang YH, McGee EE, Lam T, Wang B, Shen MC, Rashid A, Pfeiffer RM, Hsing AW, Koshiol J. A metallomic approach to assess associations of serum metal levels with gallstones and gallbladder cancer. *Hepatology*. 2020;71(3):917–28. DOI: 10.1002/hep.30861
74. Liao X, Zhang D. The 8th Edition American Joint Committee on cancer staging for hepato-pancreato-biliary cancer: A review and update. *Arch Pathol Lab Med*. 2021;145(5):543–53. DOI: 10.5858/arpa.2020–0032-RA
75. Miyazaki M, Yoshitomi H, Miyakawa S, Uesaka K, Unno M, Endo I, Ota T, Ohtsuka M, Kinoshita H, Shimada K, Shimizu H, Tabata M, Chijiiwa K, Nagino M, Hirano S, Wakai T, Wada K, Isayama H, Okusaka T, Tsuyuguchi T, Fujita N, Furuse J, Yamao K, Murakami K, Yamazaki H, Kijima H, Nakanuma Y, Yoshida M, Takayashiki T, Takada T. Clinical practice guidelines for the management of biliary tract cancers 2015: The 2nd English edition. *J Hepatobiliary Pancreat Sci*. 2015;22(4):249–73. DOI: 10.1002/jhbp.233
76. Lee SE, Jang JY, Lim CS, Kang MJ, Kim SW. Systematic review on the surgical treatment for T1 gallbladder cancer. *World J Gastroenterol*. 2011;17(2):174–80. DOI: 10.3748/wjg.v17.i2.174
77. Kim HS, Park JW, Kim H, Han Y, Kwon W, Kim SW, Hwang YJ, Kim SG, Kwon HJ, Vinuela E, Járufe N, Roa JC, Han IW, Heo JS, Choi SH, Choi DW, Ahn KS, Kang KJ, Lee W, Jeong CY, Hong SC, Troncoso A, Losada H, Han SS, Park SJ, Yanagimoto H, Endo I, Kubota K, Wakai T, Ajiki T, Adsay NV, Jang JY. Optimal surgical treatment in patients with T1b gallbladder cancer: An international multicenter study. *J Hepatobiliary Pancreat Sci*. 2018;25(12):533–43. DOI: 10.1002/jhbp.593
78. Vo E, Curley SA, Chai CY, Massarweh NN, Tran Cao HS. National failure of surgical staging for T1b gallbladder cancer. *Ann Surg Oncol*. 2019;26(2):604–10. DOI: 10.1245/s10434-018-7064-7
79. Wang Z, Li Y, Jiang W, Yan J, Dai J, Jiao B, Yin Z, Zhang Y. Simple cholecystectomy is adequate for patients with T1b gallbladder adenocarcinoma < 1 cm in diameter. *Front Oncol*. 2019;9:409. DOI: 10.3389/fonc.2019.00409
80. Balakrishnan A, Barmpounakis P, Demiris N, Jah A, Spiers HVM, Talukder S, Martin JL, Gibbs P, Harper SJF, Huguet EL, Kosmoliaptsis V, Liau SS, Praseedom RK, Basu B, de Aretxabala X, Lendoire J, Maithel S, Branes A, Andersson B, Serrablo A, Adsay V. Surgical outcomes of gallbladder cancer: The OMEGA retrospective, multicentre, international cohort study. *EClinicalMedicine*. 2023;59:101951. DOI: 10.1016/j.eclinm.2023.101951

81. Ren T, Li YS, Dang XY, Li Y, Shao ZY, Bao RF, Shu YJ, Wang XA, Wu WG, Wu XS, Li ML, Cao H, Wang KH, Cai HY, Jin C, Jin HH, Yang B, Jiang XQ, Gu JF, Cui YF, Zhang ZY, Zhu CF, Sun B, Dai CL, Zheng LH, Cao JY, Fei ZW, Liu CJ, Li B, Liu J, Qian YB, Wang Y, Hua YW, Zhang X, Liu C, Lau WY, Liu YB. Prognostic significance of regional lymphadenectomy in T1b gallbladder cancer: Results from 24 hospitals in China. *World J Gastrointest Surg*. 2021;13(2):176–86. DOI: 10.4240/wjgs.v13.i2.176
82. Lee SE, Kim KS, Kim WB, Kim IG, Nah YW, Ryu DH, Park JS, Yoon MH, Cho JY, Hong TH, Hwang DW, Choi DW. Practical guidelines for the surgical treatment of gallbladder cancer. *J Korean Med Sci*. 2014;29(10):1333–40. DOI: 10.3346/jkms.2014.29.10.1333
83. Shindoh J, de Aretxabala X, Aloia TA, Roa JC, Roa I, Zimmitti G, Javle M, Conrad C, Maru DM, Aoki T, Vigano L, Ribero D, Kokudo N, Capussotti L, Vauthey JN. Tumor location is a strong predictor of tumor progression and survival in T2 gallbladder cancer: An international multicenter study. *Ann Surg*. 2015;261(4):733–9. DOI: 10.1097/sla.0000000000000728
84. Kang H, Choi YS, Suh SW, Choi G, Do JH, Oh HC, Kim HJ, Lee SE. Prognostic significance of tumor location in T2 gallbladder cancer: A systematic review and meta-analysis. *J Clin Med*. 2021;10(15). DOI: 10.3390/jcm10153317
85. Alrawashdeh W, Kamarajah SK, Gujjuri RR, Cambridge WA, Shrikhande SV, Wei AC, Abu Hilal M, White SA, Pandanaboyana S. Systematic review and meta-analysis of survival outcomes in T2a and T2b gallbladder cancers. *HPB (Oxford)*. 2022;24(6):789–96. DOI: 10.1016/j.hpb.2021.12.019
86. Matsui S, Tanioka T, Nakajima K, Saito T, Kato S, Tomii C, Hasegawa F, Muramatsu S, Kaito A, Ito K. Surgical and oncological outcomes of wedge resection versus segment 4b+5 resection for T2 and T3 gallbladder cancer: A meta-analysis. *J Gastrointest Surg*. 2023;27(9):1954–62. DOI: 10.1007/s11605-023-05698-6
87. Kuipers H, de Savornin Lohman EAJ, van Dooren M, Braat AE, Daams F, van Dam R, Erdmann JI, Hagendoorn J, Hoogwater FJH, Groot Koerkamp B, van Gulik TM, de Reuver PR, de Boer MT. Extended resections for advanced gallbladder cancer: Results from a Nationwide Cohort Study. *Ann Surg Oncol*. 2021;28(2):835–43. DOI: 10.1245/s10434–020–08858-z
88. Kuipers H, de Bitter TJJ, de Boer MT, van der Post RS, Nijkamp MW, de Reuver PR, Fehrmann RSN, Hoogwater FJH. Gallbladder cancer: Current insights in genetic alterations and their possible therapeutic implications. *Cancers (Basel)*. 2021;13(21). DOI: 10.3390/cancers13215257
89. Chen C, Geng Z, Shen H, Song H, Zhao Y, Zhang G, Li W, Ma L, Wang L. Long-term outcomes and prognostic factors in advanced gallbladder cancer: Focus on the advanced T stage. *PLoS One*. 2016;11(11):e0166361. DOI: 10.1371/journal.pone.0166361
90. Li Y, Song Y, Zhang Y, Liu S. Progress in gallbladder cancer with lymph node metastasis. *Front Oncol*. 2022;12:966835. DOI: 10.3389/fonc.2022.966835
91. Widmann B, Warschkow R, Beutner U, Weitzendorfer M, Ukegjini K, Schmied BM, Tarantino I, Steffen T. Effect of lymphadenectomy in curative gallbladder cancer treatment: A systematic review and meta-analysis. *Langenbecks Arch Surg*. 2020;405(5):573–84. DOI: 10.1007/s00423–020–01878-z
92. Naveed S, Qari H, Thau CM, Burasakarn P, Mir AW, Panday BB. Lymph node ratio is an important prognostic factor in curatively resected gallbladder carcinoma, especially in node-positive patients: An experience from endemic region in a developing country. *Euroasian J Hepatogastroenterol*. 2021;11(1):1–5. DOI: 10.5005/jp-journals-10018–1336
93. Tsilimigras DI, Hyer JM, Paredes AZ, Moris D, Beal EW, Merath K, Mehta R, Ejaz A, Cloyd JM, Pawlik TM. The optimal number of lymph nodes to evaluate among patients undergoing surgery for gallbladder cancer: Correlating the number of nodes removed with survival in 6531 patients. *J Surg Oncol*. 2019;119(8):1099–107. DOI: 10.1002/jso.25450
94. Leigh N, Solomon D, Pletcher E, Sullivan B, Sarpel U, Labow DM, Magge DR, Golas BJ. Adenosquamous and squamous cell carcinoma of the gallbladder: The importance of histology in surgical management. *Am J Surg*. 2020;220(5):1242–8. DOI: 10.1016/j.amjsurg.2020.06.050
95. Wang J, Bo X, Nan L, Wang CC, Gao Z, Suo T, Ni X, Liu H, Lu P, Wang Y, Liu H. Landscape of distant metastasis mode and current chemotherapy efficacy of the advanced biliary tract cancer in the United States, 2010–2016. *Cancer Med*. 2020;9(4):1335–48. DOI: 10.1002/cam4.2794
96. Yang Y, Tu Z, Ye C, Cai H, Yang S, Chen X, Tu J. Site-specific metastases of gallbladder adenocarcinoma and their prognostic value for survival: A SEER-based study. *BMC Surg*. 2021;21(1):59. DOI: 10.1186/s12893-021-01068-8
97. Tirca L, Savin C, Stroescu C, Balescu I, Petrea S, Diaconu C, Gaspar B, Pop L, Varlas V, Hasegan A, Martac C, Bolca C, Stoian M, Zgura A, Gorecki GP, Bacalbasa N. Risk factors and prognostic factors in GBC. *J Clin Med*. 2024;13(14). DOI: 10.3390/jcm13144201

98. Maruyama S, Kawaida H, Hosomura N, Amemiya H, Saito R, Shimizu H, Furuya S, Akaike H, Kawaguchi Y, Sudo M, Inoue S, Kono H, Ichikawa D. Indications for extrahepatic bile duct resection due to perineural invasion in patients with gallbladder cancer. *World J Surg Oncol.* 2019;17(1):200. DOI: 10.1186/s12957-019-1735-0
99. Birnbaum DJ, Viganò L, Russolillo N, Langella S, Ferrero A, Capussotti L. Lymph node metastases in patients undergoing surgery for a gallbladder cancer. Extension of the lymph node dissection and prognostic value of the lymph node ratio. *Ann Surg Oncol.* 2015;22(3):811–8. DOI: 10.1245/s10434-014-4044-4
100. Birnbaum DJ, Vigano L, Ferrero A, Langella S, Russolillo N, Capussotti L. Locally advanced gallbladder cancer: Which patients benefit from resection? *Eur J Surg Oncol.* 2014;40(8):1008–15. DOI: 10.1016/j.ejso.2013.10.014
101. Shibata K, Uchida H, Iwaki K, Kai S, Ohta M, Kitano S. Lymphatic invasion: An important prognostic factor for stages T1b-T3 gallbladder cancer and an indication for additional radical resection of incidental gallbladder cancer. *World J Surg.* 2009;33(5):1035–41. DOI: 10.1007/s00268-009-9950-4
102. Dominguez DA, Aversa JG, Hagerty BL, Diggs LP, Raoof M, Davis JL, Hernandez JM, Blakely AM. Prognostic impact of lymphovascular invasion in pT1-T3 gallbladder adenocarcinoma. *J Surg Oncol.* 2020;122(7):1401–8. DOI: 10.1002/jso.26160
103. Vega EA, Mellado S, Salehi O, Freeman R, Conrad C. Treatment of resectable gallbladder cancer. *Cancers (Basel).* 2022;14(6). DOI: 10.3390/cancers14061413
104. Menon G, Babiker HM. *Gallbladder Carcinoma.* StatPearls, 2025. www.ncbi.nlm.nih.gov/books/NBK442002/

2 Molecular Genetics of Gallbladder Cancer

Prabin Bikram Thapa,[1] *Dhiresh Kumar Maharjan,*[1] *and Timothy M. Pawlik*[2*]
[1]Department of GI and General Surgery, Kathmandu Medical College and Teaching Hospital and Nepal Cancer Hospital and Research Centre, Kathmandu, Nepal
[2]Department of Surgery, Ohio State University, Wexner Medical Center, Ohio, USA
*Corresponding author: Tim.Pawlik@osumc.edu

2.1 INTRODUCTION

Gallbladder cancer (GBC) is considered one of the worst prognosis diseases arising from the biliary tract, despite advancements in oncological treatment.[1,2] According to recent trends in SEER,[3] the incidence of gallbladder cancer has been reported as 1.1 per 100,000 population, which is certainly lower than in other high-incidence South American countries like Chile and Peru; central European countries like Hungary, Slovakia, and Poland; and in the east along the Gangetic belt, stretching from Iran, Pakistan, Northern India, Nepal, and Bangladesh.[4–8] Besides large gallstones, gallbladder polyps more than 1 cm, and porcelain gallbladder, other predisposing factors that have been associated with gallbladder cancer include chronic *Salmonella typhi* carrier, anomalous pancreaticobiliary junction, primary sclerosing cholangitis, along with exposure to radiation, obesity, tobacco, and alcohol consumption.[9–14]

GBC has better surgical results with a 5-year survival of 60–70% only when detected in the early stage (T1/T2). However, as T stages advance, the 5-year survival rate decreases to 4%, which explains the aggressive nature of this cancer.[15,16] Hence, early detection and prompt surgical management have been a key cornerstone for the management of this cancer.[17] We herein review the molecular pathogenesis, varied molecular pathways, as well as genetic alterations associated with gallbladder cancer for tailoring precision medicine.

2.2 MOLECULAR PATHOGENESIS

Many molecular pathological sequences have been associated with the progression to gallbladder cancer. In addiition, different signaling pathways and genetic alterations can have various clinical implications (Table 2.1).

2.2.1 Hyperplasia–Atypical Hyperplasia Sequence

Chronic long-term irritation of gallbladder mucosa by gallstones will lead to a pathological response as inflammatory hyperplasia, which in turn, with unbalanced exposure, will trigger atypical hyperplasia, which is considered a significant precancerous lesion.[18, 19]

DOI: 10.1201/9781003506201-2

TABLE 2.1
Different Signaling Pathways, Genetic Alterations, and Their Clinical Implications

Name of signaling pathways	Genetic alteration	Correlation
(ERK)/MAPK) signaling pathway		
KRAS	Mutation	Poor prognosis
BRAF	Mutation	Poor prognosis
Phospho-ERK1/2	Overexpression	Early involvement and decreased overall survival
PAM(PI3K/AKT/mTOR) signaling pathway	Over expression	Decreased OS
Phospho-mTOR	Overexpression	Poor prognosis
Phospho-p70S6K	Overexpression	Poor prognosis
Phospho-AKT1	Overexpression/Mutation	Early event/decreased OS
PIK3CA	Loss of expression	Poor prognosis and advanced disease stage
PTEN		
ErbB signaling pathway		
EGFR	Mutation/overexpression/amplification	Early involvement/poor prognosis
HER 2	Mutation/overexpression/amplification	
HER3	Mutation	Early involvement/poor prognosis
HER 4	Mutation	
Notch signaling pathway		
Notch 1	Overexpression	Poor cellular differentiation and metastasis
Notch 2	Overexpression	
Notch 3	Overexpression	–
DLL4	Overexpression	Poor cellular differentiation and metastasis
		Decreased OS
Hedgehog signaling pathway		
Shh	Overexpression	Poor prognosis
SMO	Overexpression	Poor prognosis
PTCH	Overexpression	Early event/poor prognosis
GlI1	Overexpression	Early event/poor prognosis
VEFG pathway		
VGEFa	Overexpression	Poor prognosis
VGEFc	Overexpression	Lymphatic involvement
VGEFd	overexpression	Lymphatic involvement
Wnt/β-catenin signaling pathway		
SFRP4	Mutation	Early event
DKK2	Mutation	Early event
DKK3	Mutation	Early event
β-catenin	Low expression	Disease progression
	High expression	Poor cellular differentiation
DNA repair pathways		
ERCC2	Mutation	Early event
MSH2	Mutation	Early event
OGG1	Mutation	Early event
XRCC1	Mutation	Early event
Inflammatory pathways		
TP53	Overexpression	Early event
TLR2	Overexpression	Early event
MMP-2, 7, 9	Overexpression	Early event
IL10	Overexpression	Early event
Multiple intracellular pathways		
Hsp90	Overexpression	Advanced disease stage

2.2.2 Metaplasia–Dysplasia—Carcinoma Sequence

Intestinal metaplasia with dysplasia has been strongly associated with carcinoma formation, as described by Yamagiwa and Tomiyama.[20] A landmark study by Duarte et al.[21] reported on high-risk Chilean patients with cholelithiasis who commonly had antral metaplasia (96.9%), followed by intestinal metaplasia (58.1%), whereas 15% had dysplasia and carcinoma in situ in 2.5%, respectively, suggesting a strong association between metaplasia and dysplasia or hyperplasia. Other studies have similarly demonstrated a relation between intestinal metaplasia and dysplastic gallbladder epithelium.[22–25] For example, pseudopyloric gland metaplasia is common in gallstone-associated mucosal injury, being noted in up to 50% of specimens.[26] As a de novo precursor of gallbladder cancer, intestinal metaplasia has also been considered precancerous.[27]

2.2.3 Adenoma–Carcinoma Sequence

Though the adenoma–carcinoma sequence has been well established in colorectal cancer, this carcinogenesis paradigm is still controversial as related to gallbladder cancer. In 1982, Kozuka first postulated an adenoma–carcinoma sequence in gallbladder cancer among adenomas measuring 12 mm or larger, as these adenomas went on to have malignant change.[28] In another study by Lee et al. in South Korean population, 23.5% of adenomas had malignant transformation, suggesting that adenomas were indeed precursor lesions that were at risk for malignant transformation.[29]

Anomalous pancreaticobiliary ductal junction is a congenital disease that follows the adenoma–carcinoma sequence, with high-grade mutation in the K-ras gene at codon 12 (50–83%).[30, 31] However, one study done by Wistuba et al. detected no mutations in the TP53 gene, which is frequently noted in dysplasia or carcinoma. In addition, gallbladder adenomas demonstrated only 25–31% incidence of K-ras mutation, which calls into question the adenoma–carcinoma sequence for gallbladder pathogenesis[32] (Figure 2.1).

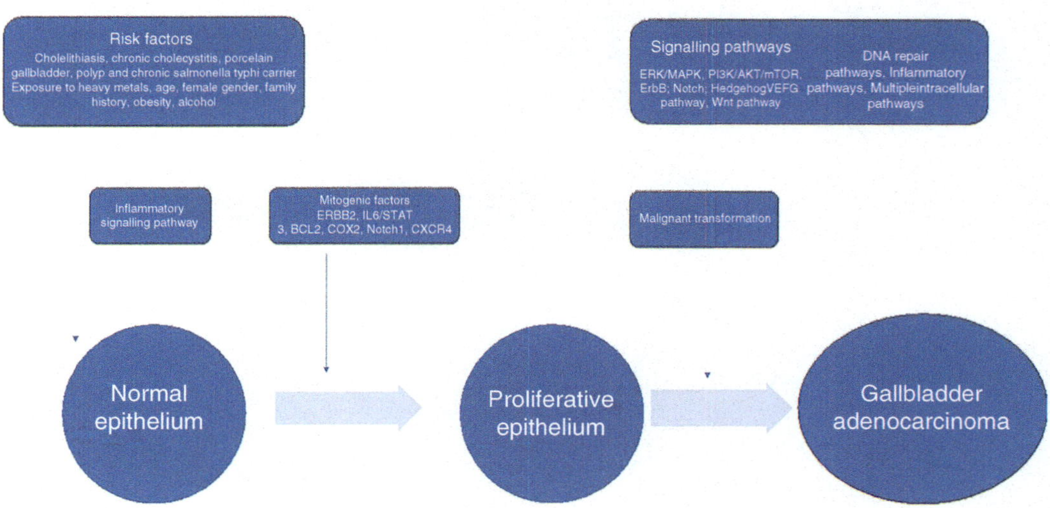

FIGURE 2.1 Morphological changes in gallbladder epithelium during pathogenesis of gallbladder cancer.

2.3 MOLECULAR SIGNALING PATHWAYS IN GALLBLADDER CANCER

Mitogen-activated protein kinase (MAPK) signaling or RAS-RAF-MEK-ERK signaling pathway:

MAPK signaling pathway is the best-described kinase cascade in the cell cycle. The primary functions of this pathway are to transfer signals to the nucleus, which activates the specific gene for cell division, proliferation, and apoptosis. It also helps in integrin signaling and cell migration, leading to tissue repair, as well as stimulates angiogenesis, which helps tumor cells to propagate (Figure 2.2).

This pathway has been mainly activated by many growth factors like EGF, PDGF, and cytokines, leading to phosphorylation of both MEK1 by serine residues (S281 and S222) primarily by c-RAF[33] and MEK2, and nuclear localization of extracellular regulated kinases (ERK1 and ERK2).[34, 35] A study done by Li et al. has reported higher rates of p-ERK1/2 in carcinoma (58.3%) than in adenomatous polyps (20%) and chronic cholecystitis (11.4%), respectively. Positive staining of p-ERK1/2 indicates involvement of lymph node and advanced disease involving surrounding organs along with poorly differentiated histology.[36]

Similarly, a study done from a highly endemic Chilean population has noted higher expression of total ERK1/2 (t-ERK1/2) and p-ERK1/2 in carcinoma of gallbladder than in chronic cholecystitis (73% versus 13% t-ERK1/2; 38% versus 3% p-ERK1/2, respectively). However, it did not demonstrate correlation between expression of total ERK1/2 or p-ERK1/2 with tumor grade, lymphatic involvement, or local infiltration.[37] Other studies have noted that gallbladder cancer cells express p-ERK1/2 ranging from 30% to 87%.[38]

Mitochondrial carrier system carries small particles between cytoplasm and mitochondria. Among them, SLC25A22, a mitochondrial glutamate carrier, has been linked with tumor migration and cellular proliferation in gallbladder cancer. Combined, SLC25A22 protein inhibiting mitochondrial apoptosis along with activation of MAPK/ERK pathways leads to aggressive cellular proliferation of tumor cells and early metastasis in gallbladder cancer.[39]

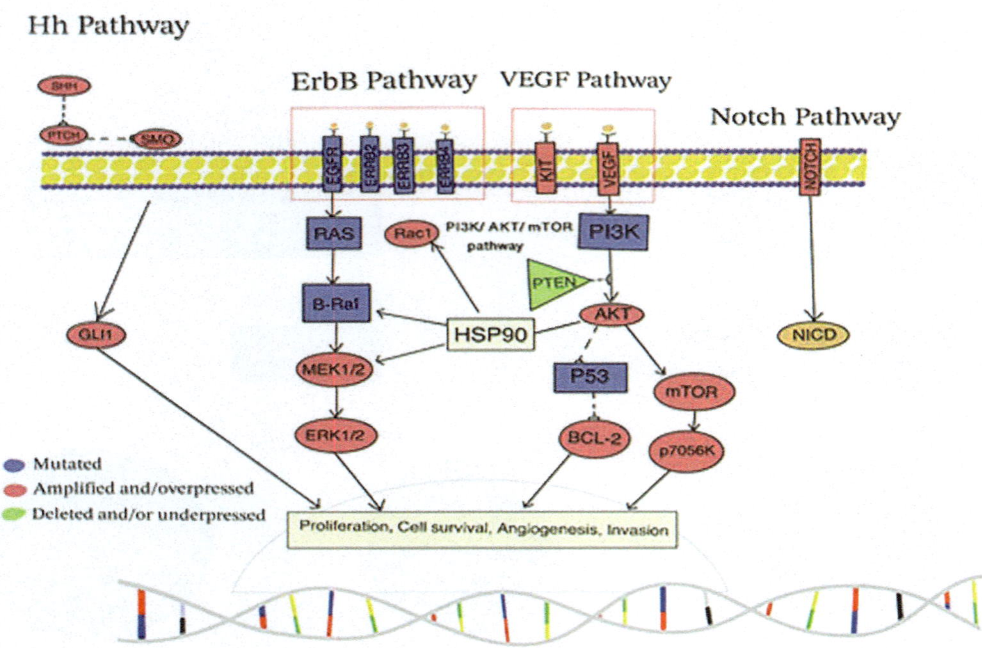

FIGURE 2.2 Different molecular signaling pathways in molecular pathogenesis of gallbladder cancer.

Metastasis-associated lung adenocarcinoma transcript 1 (MALAT1) has been implicated in lung cancer through its involvement in the Epithelial-Mesenchymal Transition (EMT) pathway.[40] A study done by Xiang-Song Wu et al. has demonstrated that MALAT1 has been significantly upregulated in gallbladder cancer, whereas knockdown results in cellular proliferation and migration, probably by activating the various MEK-ERK pathway proteins.[41]

PAM (PI3K or AKT or mTOR) signaling pathway:

The PAM (PI3K or AKT or mTOR) signaling pathway is the key regulator in cell metabolism, cell cycle, and cellular survival, which helps in tumor onset and progression.[42] Hansel et al., in 2003, initially proposed the AKT or mTOR pathway in biliary tract tumors.[43] The hyper-phosphorylation of mTOR will lead to hyperactivation of the PAM signaling pathway, which correlates with progression of gallbladder cancer.[44–46] This mTOR activity can be measured by p70S6K, a substrate of mTOR phosphorylation, which becomes significantly overexpressed by gallbladder cancer cells, like in dysplasia (66.7%), early (84.6%), and advanced gallbladder cancer (88.3%).[47] Similarly, another study done by Leal et al. reported higher expression of mTOR phosphorylation in carcinoma of gallbladder (64.1%), making it a potential biomarker for gallbladder cancer. In the same study, increased expression of phospho-mTOR was noticed among patients with chronic cholecystitis with metaplasia (32.4%) than among chronic cholecystitis without metaplasia (16%).[48]

Due to mutations in PIK3CA, deregulation of PI3K signaling has been identified in 5–12% of gallbladder cancers, whereas around 50% have the loss of PTEN, which has been associated with locally advanced poorly differentiated adenocarcinoma and lymph node metastasis.[49]

Different tyrosine kinase receptors activate PI3K, which induces downstream activation of the AKT pathway. Thus, AKT becomes active by further phosphorylation, either by extracellular matrix proteins like fibronectin or glycoprotein SPOCK1, which inactivates the pro-apoptotic protein called Bax, thus leading to increased cell survival and advanced disease and poor prognosis.[50] Fibronectin induces AKT phosphorylation at Ser473, which leads to downstream activation by activating p70S6K.[51,52]

ErbB or epidermal growth factor receptor (EGFR) signaling pathway:

Many transmembrane receptor tyrosine kinase proteins like EGFR, HER, erbB2, erbB3, and erbB4 play important roles in cellular regulation, repair, and apoptosis.[53–55] However, overexpression of erbB2 due to activation of biliary epithelial cells by reactive free oxygen radicals in response to bile acid may lead to gallbladder cancer.[56,57]

Epidermal growth factor receptor or ErbB1 is a transmembrane receptor that helps to bind the extracellular protein ligands, which induce downstream activation causing cancerous cell division, apoptosis, and migration.[58] Activation of EGFR may be induced by bile acid either by a TGF-dependent pathway or an inflammatory Cox-2 induction pathway.[59] EGFR signaling can activate the PAM (PI3K or AKT or mTOR) pathway by binding class Ia subunit p110α PI3K and inducing AKT activation with a specific mutation of E545K, which has been associated with poor prognostic features.[60] Kaufman et al. reported that gallbladder cancer cells overexpress the EGFR and its ligand in almost 70.7%, while only 27% in hyperplasia, along with high cell proliferation activity.[61]

2.3.1 Human Epidermal Growth Factor Receptor (HER2)/Neu Oncogene

The HER2 or Neu oncogene is located on the 17q12–q21 chromosomal region.[62] Overexpression of the HER2 receptors can cause transphosphorylation of their intracellular domains, leading to the downstream activation of MAPK or PAM pathways, causing uncontrolled cellular division and apoptosis.[63, 64] HER2 overexpression in gallbladder carcinoma is variable, ranging from 2% to

46.5%.[65, 66] Pujani et al. have reported overexpression of HER2 in 24% of patients with gallbladder cancer, but failed to note any association between HER2/neu positivity and tumor staging or lymph node metastasis.[67] In a different study, Puhalla et al. noted, however, that overexpression of HER2 was correlated with advanced tumor stage.[68] Halder et al. have reported higher expression of HER2 or Neu oncogene in intestinal metaplasia (74%), followed by high-grade dysplasia (58%), low-grade dysplasia (47%), and gallbladder cancer (48%), respectively.[69] Similarly, Toledo et al. also noticed that HER2 was more expressed in intestinal metaplasia (92%) and carcinoma in situ (90%) than in gallbladder cancer, suggesting that higher activity of HER2 happens in precancerous lesions, along with upregulation related to bradykinin receptor B1 expression.[70]

The Notch signaling pathway:

This *Notch* gene, after its discovery in 1910 from *Drosophila melanogaster*, has been widely appreciated for its canonical and noncanonical signaling pathway controlling cell proliferation, differentiation, and apoptosis.[71–73] Four different Notch receptors and five ligands have been identified,[74] which facilitate the interaction between two adjacent cells to activate proteolytic cleavages of Notch receptors with the help of the metalloprotease TNF-α converting enzyme and γ-secretase.[75–79] Aggressive cancer progression has been associated with overexpression of Notch 1 and 3, whereas expression of DLL4 has been associated with poor overall survival.[80, 81]

Hedgehog signaling pathway:

This pathway has been recognized as one of the prime regulators of cell growth and proliferation.[82] It consists of two types of receptors: Patched receptor (PTCH), which is a membrane protein receptor and Smoothened (SMO), which are 7-transmembrane G protein-coupled receptors. Three types of Hedgehog genes have been described, including Indian, Desert, and Sonic Hedgehog.[83–86] When one of these ligands binds with PTCH-1, the Smoothened receptor is activated, leading to a downstream cascade of glioma-associated oncogenes GLI1, GLI2 and GLI3.[87–90] Li et al. reported higher levels of Hedgehog signaling proteins Shh, Ptch1, and GlI1 in gallbladder cancer than in normal tissue, and correlates to tumor staging and lymphatic metastasis.[91] Hence, GLI1 has been suggested to be a potential prognostic biomarker for carcinoma of gallbladder.[92] Dixit et al. noted that mutation of Sonic Hedgehog (Shh) was much higher among patients with carcinoma of gallbladder (20%) compared with cholelithiasis (10%), and correlated with histopathology and staging of tumor.[93]

Vascular endothelial growth factor (VEFG) pathway:

Vascular base for tumoral growth needs proliferation of endothelial cells with increased permeability. This iconic feature of tumor genesis is induced by vascular endothelial growth factor (VEGF).[94,95] The higher the expression of Vascular Endothelial Growth Factor (VEGF), the worse the prognosis; of note, high expression of VEGF may be present in up to 50–80% of patients with carcinoma of gallbladder.[96–98] For example, Nakashima et al. reported that 63% of gallbladder cancer specimens had highly positive VEGF-C, which was correlated with lymph node metastasis and poor prognosis.[99]

Wnt/β-catenin signaling pathway:

The Wnt or β-catenin signaling involves two types of pathways: noncanonical and canonical. The canonical pathway helps in cell proliferation, whereas the noncanonical pathway regulates cell migration and polarity. The canonical Wnt pathway involves nuclear translocation of β-catenin and activates the target gene, hence also called Wnt/β-catenin pathway.[100] Among many molecules,

R-spondin, syndecan-1, Norrin, and ANDP activate Wnt signaling, whereas inhibitory molecules like Notum, Tiki, DKKs, WIF1, and ZNRF3/RNF43 inhibit Wnt signaling and have been linked with tumor progression, local invasion, and angiogenesis.[101–104] Yadav et al. reported that a change in β-catenin pathway genes SFRP4, DKK2, and DKK3 may be linked with gallbladder cancer.[100] Higher levels of β-catenin correlate with worse histological outcomes.[105]

DNA repair pathways:

Srivastava et al. have identified genetic variation in three repair genes: DNA excision repair gene (ERCC2), MutS protein homolog 2 (MSH2), and 8-oxoguanine glycosylase (OGG1) among Northern Indian populations with gallbladder cancer.[106] They have also identified increased risk of gallbladder cancer among those with variation of OGG1 and X-ray repair cross-complementing protein 1 (XRCC1) gene.[107] Hence, these repair genes may help to predict risk assessment of gallbladder cancer.

Inflammatory pathways:

Chronic irritation of biliary epithelium either by gallstones or due to biliary reflux in anomalous pancreaticobiliary junction leads to activation of inflammatory radicals and increased oxidative stress, leading to early alteration of *TP53*.[108] The most common genetic alteration observed was inactivation of the *TP53* gene, either by mutation or deletion.[55] Higher values of circulating aflatoxin–albumin complex have been noticed in patients with gallbladder cancer than in normal subjects, which have been linked to fungus producing aflatoxin leading to mutation in TP53.[109]

Another study noted that alteration of the *IL10* and *VEGF* genes was associated with higher risk of carcinoma of gallbladder, while variants of the *IL8, IL8RB, RNASEL*, and *NOS2* genes were associated with cholelithiasis.[110]

Among receptor proteins, Toll-like receptors play an important role in the innate immune system. A study done in North Indian populations with carcinoma of gallbladder demonstrated polymorphic mutations in TLR2Δ22 and TLR4 Ex41936C4T (rs4986791).[111] Matrix metalloproteinase (MMP-2) levels have been noted to be higher in carcinoma of gallbladder with stones, whereas MMP-9 levels were higher among carcinoma of gallbladder without stones.[112]

Multiple intracellular pathways:

Multiple cellular or multiple gene pathways have been proposed to evaluate influence of gene-to-gene interaction on gallbladder cancer and to explore its therapeutic advantage over single pathway inhibition. Among these multigene regulators, inhibition of master heat shock protein 90 (HSP90) may help to inhibit multiple pathways simultaneously. In advanced gallbladder cancer, almost 87% of patients have HSP90 expression, which has a prognostic advantage as well.[113,114]

2.4 GENETIC ALTERATIONS IN GALLBLADDER CANCER

- Among 3893 gallbladder specimens studied, the most common gene alteration was in TP53 (57%), followed by SHH (20%), ELF3 (19%), ARID1A (14%), and SMAD4 (13%). Frequently occurring genetic aberrations were amplifications of oncogenes KRAS (14.9%), CCNE1 (11%), ERBB2 (7.1%), CDK4 (5%), MDM2 (7.8%), and MYC (6.8%), whereas deletions were noted in CDKN2A (18.2%) and CDKN2B (10%). While other types of gastrointestinal malignancies have varying levels of high incidence of microsatellite instability (MSI), the incidence of MSI for gallbladder cancer is very low—3.5%.

- In gallbladder cancer etiopathogenesis from dysplasia, loss of heterozygosity has been noted in 3p, 5q, 9p, 13q, and 17p regions. However, LOH at 3p14.2 locus has been identified with inactivation of FHIT tumor suppressor gene. Yoshida et al. reported an inverse correlation between MSI and the presence of LOH in carcinoma of gallbladder.[115] Hence, two independent genetic pathways—MSI and LOH—are involved in carcinogenesis.

Abnormal microRNA (miRNA) expression is a key step in oncogenesis, acting as either oncogenes or tumor suppressor genes, whether working primarily or in clusters.[116] Bizama et al. conducted a comprehensive miRNA expression profile in gallbladder cancer, which noted downregulation of miR-133a, miR-133b, miR-143, miR-145, miR-1, miR-29, and miR-148.[86] miR-1 and miR-145 target genes in signaling pathways such as TGF-β, ERBB3, WNT, and VEGF. An aggressive character of tumor cells has been associated with downregulation of miR-355.[117]

2.5 MOLECULAR GENETICS SEQUENCING APPROACHES

2.5.1 Genome Wide Association Study (GWAS) Approach

The GWAS approach has been described to screen genetic variants in particular populations.[118] A study done by Cha et al. in Japanese populations demonstrated increased risk of gallbladder cancer amongst SNP rs7504990 in the deleted colon cancer (DCC) gene.[119] Mhatre et al. also studied SNP rs7504990, as well as two other SNPs (rs2229080 and rs714) in the same gene.[120] This study demonstrated that DCC variants were useful markers for gallbladder cancer. Similarly, in a different study, Rai et al. in a North Indian population were able to identify 19 markers localized to ABCB4 gene at 7q21.12 loci.[121] Using the Swedish Family Cancer database, Hemminki et al. reported a familial relative risk (FRR) of 2.47,[122] which was lower than the high GWAS-predicted FRR of 3.15 noted by Mhatre and colleagues.[120] Deletion of ABCB4 gene has been associated with hepatic inflammation, probably due to blockage of translocation of phosphatidylcholine, which might have similar effect on gallbladder cancer.[123]

2.5.2 High Throughput Approach

The Catalogue of Somatic Mutations in Cancer database provides broader information on the prevalence of different types of mutations. Based on this, the most common mutated genes in gallbladder cancer were KRAS, TP53, BRAF, PI3KCA, CTNNB1, TERT, CDKN2A, and ERBB2, with the majority having a missense substitution 322 (75.06%), followed by synonymous substitution 70 (16.32%) and nonsense substitution 63 (14.69%).[124]

Borger et al. identified three of the most common mutations were KRAS, TP53, and PIK3CA with 35%, 22%, and 10%, respectively, followed by other genes like BRAF, APC, and *NRAS*.[125] Another study done by Ross et al. noted that the most common mutation was in TP53 (41%) with overexpression of HER2 (15%), while underexpression of PBRM1 was noticed in 53%.[126] Another combined Japanese and Italian population study noted mutations among 32 genes, including deletion of MUC17 at 7q22.1, along with 11% with deleterious germline mutations.[127]

Scholer et al. further tried to incorporate molecular subtypes of hepatobiliary cancers to develop a precision medicine algorithm to guide planning of precision treatment and risk stratification.[128] They compared molecular subtypes (HC1, HC2, HC3) with tissue-of-origin subtypes. The most prevalent alterations were of HC2 (56.9%), which demonstrated epigenomic alterations in TP53, TERT, and CTNNB1, mostly seen with hepatocellular carcinoma and gallbladder cancer, which was associated with treatment resistance. HC3 (32.5%) featured metabolic stress pathway with mutation in IDH1, which helped in lipid and amino acid metabolism, while HC1 (10.6%) with gene mutation in TP53 and ARIDIA appeared to be best responding to chemotherapy.

In another study from MSK-IMPACT (Memorial Sloan Kettering–Integrated Mutation Profiling of Actionable Cancer Targets), which involved a large cohort of 233 patients with gallbladder cancer, mechanisms and patterns of somatic genomic alterations were identified.[129] According to OncoKB database, genetic alterations and copy number alterations (CAN) were classified based on oncogenic effect and biological effect, which noted 656 alterations and 387 CAN as being oncogenic or likely oncogenic in gallbladder cancer.[130] Further analysis revealed the most common somatic alterations in three primary genes—CTNNB1, KRAS, and ERBB2—while the most common pathways described were TP53 (74%); CDKN2A, CDKN2B, and RB1 (46%), as well as RTK-RAS pathway, including K-ras and ERBB2 (46%). Frequency of genetic alterations was more profound in well and moderately differentiated gallbladder cancer more than in poorly differentiated tumors (PIK3CA (15% versus 5%), SMAD4 (29% versus 17%), and STK11 (11% versus 2.5%), respectively).[131]

2.6 CONCLUSION

Gallbladder cancer has a genetically diverse origin, and treatment should be guided in the future with precision medicine guidelines. Surgeons need to be aware of the rapidly evolving field and understand the genetic and molecular changes that drive the development of gallbladder cancer. Molecular mapping of each variation should be taken into account for targeted therapy so that better outcomes may be expected. Molecular alterations may serve as prognostic markers, as well as assist in identifying patients at higher risk of recurrence or worse long-term survival. In the future, molecular profiling will increase its role in recognizing which patients are most likely to benefit from personalized specific therapies.

REFERENCES

1. Bray, F., Ferlay, J., Soerjomataram, I., Siegel, R. L., Torre, L. A., & Jemal, A. (2018). Global cancer statistics 2018: GLOBOCAN estimates of incidence and mortality worldwide for 36 cancers in 185 countries. *CA: A Cancer Journal for Clinicians*, 68(6), 394–424.
2. Torre, L. A., Siegel, R. L., Islami, F., Bray, F., & Jemal, A. (2018). Worldwide burden of and trends in mortality from gallbladder and other biliary tract cancers. *Clinical Gastroenterology and Hepatology: The Official Clinical Practice Journal of the American Gastroenterological Association*, 16(3), 427–437.
3. Explorer, S. (2024, April 17). *An Interactive Website for SEER Cancer Statistics*. Surveillance Research Program, National Cancer Institute.
4. Randi, G., Franceschi, S., & La Vecchia, C. (2006). Gallbladder cancer worldwide: Geographical distribution and risk factors. *International Journal of Cancer*, 118(7), 1591–1602.
5. Dhir, V., & Mohandas, K. M. (1999). Epidemiology of digestive tract cancers in India IV. Gall bladder and pancreas. *Indian Journal of Gastroenterology: Official Journal of the Indian Society of Gastroenterology*, 18(1), 24–28.
6. Sharma, A., Sharma, K. L., Gupta, A., Yadav, A., & Kumar, A. (2017). Gallbladder cancer epidemiology, pathogenesis and molecular genetics: Recent update. *World Journal of Gastroenterology*, 23(22), 3978–3998.
7. Mhatre, S. S., Nagrani, R. T., Budukh, A., Chiplunkar, S., Badwe, R., Patil, P., et al. (2016). Place of birth and risk of gallbladder cancer in India. *Indian Journal of Cancer*, 53(2), 304–308.
8. Unisa, S., Jagannath, P., Dhir, V., Khandelwal, C., Sarangi, L., & Roy, T. K. (2011). Population-based study to estimate prevalence and determine risk factors of gallbladder diseases in the rural Gangetic basin of North India. *HPB: The Official Journal of the International Hepato Pancreato Biliary Association*, 13(2), 117–125.
9. Salazar, M., Ituarte, C., Abriata, M. G., Santoro, F., & Arroyo, G. (2019). Gallbladder cancer in South America: Epidemiology and prevention. *Chinese Clinical Oncology*, 8(4), 32.

10. Zatonskí, W., La Vecchia, C., Levi, F., Negri, E., & Lucchini, F. (1993). Descriptive epidemiology of gall-bladder cancer in Europe. *Journal of Cancer Research and Clinical Oncology*, 119(3), 165–171.
11. Krishnanand, K., & Singh Kurmi, N. (2019). Study of gall bladder disease with incidence of gall bladder malignancy. *Surgical Review: International Journal of Surgery, Trauma and Orthopedics*, 5(1), 54–59.
12. Malik, I. A. (2003). Clinicopathological features and management of gallbladder cancer in Pakistan: A prospective study of 233 cases. *Journal of Gastroenterology and Hepatology*, 18(8), 950–953.
13. Chijiiwa, K., Tanaka, M., & Nakayama, F. (1993). Adenocarcinoma of the gallbladder associated with anomalous pancreaticobiliary ductal junction. *The American Surgeon*, 59(7), 430–434.
14. Chao, T. C., Jan, Y. Y., & Chen, M. F. (1995). Primary carcinoma of the gallbladder associated with anomalous pancreaticobiliary ductal junction. *Journal of Clinical Gastroenterology*, 21(4), 306–308.
15. Carriaga, M. T., & Henson, D. E. (1995). Liver, gallbladder, extrahepatic bile ducts, and pancreas. *Cancer*, 75(1 Suppl), 171–190.
16. Whetstone, M. R., Saltzstein, E. C., & Mercer, L. C. (1986). Demographic characteristics of gallbladder cancer in an area endemic for biliary calculi. *American Journal of Surgery*, 152(6), 728–730.
17. Miller, G., & Jarnagin, W. R. (2008). Gallbladder carcinoma. *European Journal of Surgical Oncology: The Journal of the European Society of Surgical Oncology and the British Association of Surgical Oncology*, 34(3), 306–312.
18. Albores-Saavedra, J., Alcántra-Vazquez, A., Cruz-Ortiz, H., & Herrera-Goepfert, R. (1980). The precursor lesions of invasive gallbladder carcinoma. Hyperplasia, atypical hyperplasia and carcinoma in situ. *Cancer*, 45(5), 919–927.
19. Matsukura, N., Yokomuro, S., Yamada, S., Tajiri, T., Sundo, T., Hadama, T., Kamiya, S., Naito, Z., & Fox, J. G. (2002). Association between Helicobacter bilis in bile and biliary tract malignancies: H. bilis in bile from Japanese and Thai patients with benign and malignant diseases in the biliary tract. *Japanese Journal of Cancer Research: Gann*, 93(7), 842–847.
20. Yamagiwa, H., & Tomiyama, H. (1986). Intestinal metaplasia-dysplasia-carcinoma sequence of the gallbladder. *Acta Pathologica Japonica*, 36(7), 989–997.
21. Duarte, I., Llanos, O., Domke, H., Harz, C., & Valdivieso, V. (1993). Metaplasia and precursor lesions of gallbladder carcinoma. Frequency, distribution, and probability of detection in routine histologic samples. *Cancer*, 72(6), 1878–1884.
22. Sheth, S., Bedford, A., & Chopra, S. (2000). Primary gallbladder cancer: Recognition of risk factors and the role of prophylactic cholecystectomy. *The American Journal of Gastroenterology*, 95(6), 1402–1410.
23. Hsing, A. W., Bai, Y., Andreotti, G., Rashid, A., Deng, J., Chen, J., et al. (2007). Family history of gallstones and the risk of biliary tract cancer and gallstones: A population-based study in Shanghai, China. *International Journal of Cancer*, 121(4), 832–838.
24. Hsing, A., Gao, Y., Han, T., Rashid, A., Sakoda, L., Wang, B., et al. (2007). Gallstones and the risk of biliary tract cancer: A population-based study in China. *British Journal of Cancer*, 97(11), 1577–1582.
25. Nabatame, N., Shirai, Y., Nishimura, A., Yokoyama, N., Wakai, T., & Hatakeyama, K. (2004). High risk of gallbladder carcinoma in elderly patients with segmental adenomyomatosis of the gallbladder. *Journal of Experimental & Clinical Cancer Research: CR*, 23(4), 593–598.
26. Laitio, M. (1976). Early carcinoma of the gallbladder. *Beitrage zur Pathologie*, 158(2), 159–172.
27. Laitio, M. (1983). Histogenesis of epithelial neoplasms of human gallbladder I. Dysplasia. *Pathology, Research and Practice*, 178(1), 51–56.
28. Kozuka, S., & Hachisuka, K. (1984). Incidence by age and sex of intestinal metaplasia in the gallbladder. *Human Pathology*, 15(8), 779–784.
29. Lee, S. H., Lee, D. S., You, I. Y., Jeon, W. J., Park, S. M., Youn, S. J., et al. (2010). Histopathologic Analysis of Adenoma and Adenoma-related Lesions of the Gallbladder. *The Korean Journal of Gastroenterology = Taehan Sohwagi Hakhoe chi*, 55(2), 119–126.
30. Kato, T., Matsuda, K., Kayaba, H., Enomoto, S., Hebiguchi, T., Koyama, K., et al. (1989). Pathology of anomalous junction of the pancreaticobiliary ductal system: Mutagenicity of the contents of the biliary tract and nuclear atypia of the biliary epithelium. *The Keio Journal of Medicine*, 38(2), 167–176.
31. Sasatomi, E., Tokunaga, O., & Miyazaki, K. (2000). Precancerous conditions of gallbladder carcinoma: Overview of histopathologic characteristics and molecular genetic findings. *Journal of Hepatobiliary-pancreatic Surgery*, 7(6), 556–567.
32. Wistuba, I. I., Miquel, J. F., Gazdar, A. F., & Albores-Saavedra, J. (1999). Gallbladder adenomas have molecular abnormalities different from those present in gallbladder carcinomas. *Human Pathology*, 30(1), 21–25.
33. Zheng, C. F., & Guan, K. L. (1994). Activation of MEK family kinases requires phosphorylation of two conserved Ser/Thr residues. *The EMBO Journal*, 13(5), 1123–1131.

34. Winston, B. W., Lange-Carter, C. A., Gardner, A. M., Johnson, G. L., & Riches, D. W. (1995). Tumor necrosis factor alpha rapidly activates the mitogen-activated protein kinase (MAPK) cascade in a MAPK kinase kinase-dependent, c-Raf-1-independent fashion in mouse macrophages. *Proceedings of the National Academy of Sciences of the United States of America*, 92(5), 1614–1618.
35. Bao, R. F., Shu, Y. J., Hu, Y. P., Wang, X. A., Zhang, F., Liang, H. B., et al. (2016). miR-101 targeting ZFX suppresses tumor proliferation and metastasis by regulating the MAPK/Erk and Smad pathways in gallbladder carcinoma. *Oncotarget*, 7(16), 22339–22354.
36. Li, Q., & Yang, Z. (2009). Expression of phospho-ERK1/2 and PI3-K in benign and malignant gallbladder lesions and its clinical and pathological correlations. *Journal of Experimental & Clinical Cancer Research: CR*, 28(1), 65.
37. Buchegger, K., Silva, R., López, J., Ili, C., Araya, J. C., Leal, P., et al. (2017). The ERK/MAPK pathway is overexpressed and activated in gallbladder cancer. *Pathology, Research and Practice*, 213(5), 476–482.
38. Hori, H., Ajiki, T., Mita, Y., Horiuchi, H., Hirata, K., Matsumoto, T., et al. (2007). Frequent activation of mitogen-activated protein kinase relative to Akt in extrahepatic biliary tract cancer. *Journal of Gastroenterology*, 42(7), 567–572.
39. Du, P., Liang, H., Fu, X., Wu, P., Wang, C., Chen, H., et al. (2019). SLC25A22 promotes proliferation and metastasis by activating MAPK/ERK pathway in gallbladder cancer. *Cancer Cell International*, 19, 33.
40. Goyal, B., Yadav, S. R. M., Awasthee, N., Gupta, S., Kunnumakkara, A. B., & Gupta, S. C. (2021). Diagnostic, prognostic, and therapeutic significance of long non-coding RNA MALAT1 in cancer. *Biochimica et Biophysica Acta. Reviews on Cancer*, 1875(2), 188502.
41. Wu, X. S., Wang, X. A., Wu, W. G., Hu, Y. P., Li, M. L., Ding, Q., et al. (2014). MALAT1 promotes the proliferation and metastasis of gallbladder cancer cells by activating the ERK/MAPK pathway. *Cancer Biology & Therapy*, 15(6), 806–814.
42. Yu, L., Wei, J., & Liu, P. (2022). Attacking the PI3K/Akt/mTOR signaling pathway for targeted therapeutic treatment in human cancer. *Seminars in Cancer Biology*, 85, 69–94.
43. Hansel, D. E., Rahman, A., Hidalgo, M., Thuluvath, P. J., Lillemoe, K. D., Schulick, R., et al. (2003). Identification of novel cellular targets in biliary tract cancers using global gene expression technology. *The American Journal of Pathology*, 163(1), 217–229.
44. Deshpande, V., Nduaguba, A., Zimmerman, S. M., Kehoe, S. M., Macconaill, L. E., Lauwers, G. Y., et al. (2011). Mutational profiling reveals PIK3CA mutations in gallbladder carcinoma. *BMC Cancer*, 11, 60.
45. Riener, M. O., Bawohl, M., Clavien, P. A., & Jochum, W. (2008). Rare PIK3CA hotspot mutations in carcinomas of the biliary tract. *Genes, Chromosomes & Cancer*, 47(5), 363–367.
46. Lunardi, A., Webster, K. A., Papa, A., Padmani, B., Clohessy, J. G., Bronson, R. T., et al. (2014). Role of aberrant PI3K pathway activation in gallbladder tumorigenesis. *Oncotarget*, 5(4), 894–900.
47. Leal, P., Garcia, P., Sandoval, A., Buchegger, K., Weber, H., Tapia, O., & Roa, J. C. (2013). AKT/mTOR substrate P70S6K is frequently phosphorylated in gallbladder cancer tissue and cell lines. *OncoTargets and Therapy*, 6, 1373–1384.
48. Leal, P., García, P., Sandoval, A., Letelier, P., Brebi, P., Ili, C., et al. (2013). Immunohistochemical expression of phospho-mTOR is associated with poor prognosis in patients with gallbladder adenocarcinoma. *Archives of Pathology & Laboratory Medicine*, 137(4), 552–557.
49. Liu, D. C., & Yang, Z. L. (2011). Overexpression of EZH2 and loss of expression of PTEN is associated with invasion, metastasis, and poor progression of gallbladder adenocarcinoma. *Pathology, Research and Practice*, 207(8), 472–478.
50. Shu, Y. J., Weng, H., Ye, Y. Y., Hu, Y. P., Bao, R. F., Cao, Y., et al. (2015). SPOCK1 as a potential cancer prognostic marker promotes the proliferation and metastasis of gallbladder cancer cells by activating the PI3K/AKT pathway. *Molecular Cancer*, 14(1), 12.
51. Körner, T., Kropf, J., Hackler, R., Brenzel, A., & Gressner, A. M. (1996). Fibronectin in human bile fluid for diagnosis of malignant biliary diseases. *Hepatology (Baltimore, Md.)*, 23(3), 423–428.
52. Cao, Y., Liu, X., Lu, W., Chen, Y., Wu, X., Li, M., et al. (2015). Fibronectin promotes cell proliferation and invasion through mTOR signaling pathway activation in gallbladder cancer. *Cancer Letters*, 360(2), 141–150.
53. Mishra, V., Ansari, K. M., Khanna, R., & Das, M. (2012). Role of ErbB2 mediated AKT and MAPK pathway in gall bladder cell proliferation induced by argemone oil and butter yellow. Argemone oil and butter yellow induced gall bladder cell proliferation. *Cell Biology and Toxicology*, 28(3), 149–159.
54. Nakazawa, K., Dobashi, Y., Suzuki, S., Fujii, H., Takeda, Y., & Ooi, A. (2005). Amplification and overexpression of c-erbB-2, epidermal growth factor receptor, and c-met in biliary tract cancers. *The Journal of Pathology*, 206(3), 356–365.

55. Li, M., Zhang, Z., Li, X., Ye, J., Wu, X., Tan, Z., et al. (2014). Whole-exome and targeted gene sequencing of gallbladder carcinoma identifies recurrent mutations in the ErbB pathway. *Nature Genetics*, 46(8), 872–876.
56. Kiguchi, K., Carbajal, S., Chan, K., Beltrán, L., Ruffino, L., Shen, J., et al. (2001). Constitutive expression of ErbB-2 in gallbladder epithelium results in development of adenocarcinoma. *Cancer Research*, 61(19), 6971–6976.
57. Werneburg, N. W., Yoon, J. H., Higuchi, H., & Gores, G. J. (2003). Bile acids activate EGF receptor via a TGF-alpha-dependent mechanism in human cholangiocyte cell lines. American journal of physiology. *Gastrointestinal and Liver Physiology*, 285(1), G31–G36.
58. Zhou, Y.-M., Li, Y.-M., Cao, N., Feng, Y., & Zeng, F. (2003). Significance of expression of epidermal growth factor (EGF) and its receptor (EGFR) in chronic cholecystitis and gallbladder carcinoma. *Ai zheng= Aizheng= Chinese Journal of Cancer*, 22(3), 262–265.
59. Yoon, J. H., Higuchi, H., Werneburg, N. W., Kaufmann, S. H., & Gores, G. J. (2002). Bile acids induce cyclooxygenase-2 expression via the epidermal growth factor receptor in a human cholangiocarcinoma cell line. *Gastroenterology*, 122(4), 985–993.
60. Sasaki, T., Kuniyasu, H., Luo, Y., Kato, D., Shinya, S., Fujii, K., et al. (2012). Significance of epithelial growth factor in the epithelial-mesenchymal transition of human gallbladder cancer cells. *Cancer Science*, 103(6), 1165–1171.
61. Kaufman, M., Mehrotra, B., Limaye, S., White, S., Fuchs, A., Lebowicz, et al. (2008). EGFR expression in gallbladder carcinoma in North America. *International Journal of Medical Sciences*, 5(5), 285–291.
62. Harder, J., Waiz, O., Otto, F., Geissler, M., Olschewski, M., Weinhold, B., et al. (2009). EGFR and HER2 expression in advanced biliary tract cancer. *World Journal of Gastroenterology*, 15(36), 4511–4517.
63. Roa, I., de Toro, G., Schalper, K., de Aretxabala, X., Churi, C., &Javle, M. (2014). Overexpression of the HER2/neu gene: A new therapeutic possibility for patients with advanced gallbladder cancer. *Gastrointest Cancer Res.*, 7(2), 42–48.
64. Kawamoto, T., Krishnamurthy, S., Tarco, E., Trivedi, S., Wistuba, I. I., Li, D., et al. (2007). HER receptor family: Novel candidate for targeted therapy for gallbladder and extrahepatic bile duct cancer. *Gastrointestinal Cancer Research: GCR*, 1(6), 221–227.
65. Kalekou, H., &Miliaras, D. (2004). Immunohistochemical study of microvessel density, CD44 (standard form), p53 protein and c-erbB2 in gallbladder carcinoma. *Journal of Gastroenterology and Hepatology*, 19(7), 812–818.
66. Su W-C, Shiesh S-C, Liu H-S, Chen C-Y, Chow N-H, Lin X-Z. Expression of oncogene products HER2/Neu and Ras and fibrosis-related growth factors bFGF, TGF-β, and PDGF in bile from biliary malignancies and inflammatory disorders. *Digestive Diseases and Sciences*. 2001;46:1387–1392.
67. Pujani, M., Makker, I., Makker, A., Jetley, S., Goel, M. M., &Jetley, S. (2016). Expression of human epidermal growth factor receptor (Her 2/neu) and proliferative marker Ki-67: Association with clinicopathological parameters in gallbladder carcinoma. *Asian Pacific Journal of Cancer Prevention: APJCP*, 17(8), 3903–3909.
68. Puhalla, H., Wrba, F., Kandioler, D., Lehnert, M., Huynh, A., Gruenberger, T., et al. (2007). Expression of p21Wafl/Cip1, p57Kip2 and HER2/neu in patients with gallbladder cancer. *Anticancer Research*, 27(3B), 1679–1684.
69. Halder, S., Kundu, S., Chakraborty, J., & Chakrabarti, S. (2019). Significance of HER2 and Ki-67 in preneoplastic lesions and carcinoma of gallbladder. *Journal of Gastrointestinal Cancer*, 50(4), 848–854.
70. Toledo, C., Matus, C. E., Barraza, X., Arroyo, P., Ehrenfeld, P., Figueroa, et al. (2012). Expression of HER2 and bradykinin B_1 receptors in precursor lesions of gallbladder carcinoma. *World Journal of Gastroenterology*, 18(11), 1208–1215.
71. Nickoloff, B. J., Osborne, B. A., & Miele, L. (2003). Notch signaling as a therapeutic target in cancer: A new approach to the development of cell fate modifying agents. *Oncogene*, 22(42), 6598–6608.
72. Dufraine, J., Funahashi, Y., &Kitajewski, J. (2008). Notch signaling regulates tumor angiogenesis by diverse mechanisms. *Oncogene*, 27(38), 5132–5137.
73. Li, J. L., & Harris, A. L. (2005). Notch signaling from tumor cells: A new mechanism of angiogenesis. *Cancer Cell*, 8(1), 1–3.
74. Aster, J. C., &Blacklow, S. C. (2012). Targeting the notch pathway: Twists and turns on the road to rational therapeutics. *Journal of Clinical Oncology: Official Journal of the American Society of Clinical Oncology*, 30(19), 2418–2420.
75. Lobry, C., Oh, P., Mansour, M. R., Look, A. T., &Aifantis, I. (2014). Notch signaling: Switching an oncogene to a tumor suppressor. *Blood*, 123(16), 2451–2459.

76. Takebe, N., Nguyen, D., & Yang, S. X. (2014). Targeting notch signaling pathway in cancer: Clinical development advances and challenges. *Pharmacology & Therapeutics*, 141(2), 140–149.
77. Lobry, C., Oh, P., &Aifantis, I. (2011). Oncogenic and tumor suppressor functions of notch in cancer: It's NOTCH what you think. *The Journal of Experimental Medicine*, 208(10), 1931–1935.
78. Sekiya, S., & Suzuki, A. (2012). Intrahepatic cholangiocarcinoma can arise from notch-mediated conversion of hepatocytes. *The Journal of Clinical Investigation*, 122(11), 3914–3918.
79. El Khatib, M., Bozko, P., Palagani, V., Malek, N. P., Wilkens, L., &Plentz, R. R. (2013). Activation of notch signaling is required for cholangiocarcinoma progression and is enhanced by inactivation of p53 in vivo. *PLoS One*, 8(10), e77433.
80. Zender, S., Nickeleit, I., Wuestefeld, T., Sörensen, I., Dauch, D., Bozko, P., et al. (2013). A critical role for notch signaling in the formation of cholangiocellular carcinomas. *Cancer Cell*, 23(6), 784–795.
81. Yoon, H. A., Noh, M. H., Kim, B. G., Han, J. S., Jang, J. S., Choi, S. R., et al. (2011). Clinicopathological significance of altered notch signaling in extrahepatic cholangiocarcinoma and gallbladder carcinoma. *World Journal of Gastroenterology*, 17(35), 4023–4030.
82. Kubo, M., Nakamura, M., Tasaki, A., Yamanaka, N., Nakashima, H., Nomura, M., et al. (2004). Hedgehog signaling pathway is a new therapeutic target for patients with breast cancer. *Cancer Research*, 64(17), 6071–6074.
83. Onishi, H., Kai, M., Odate, S., Iwasaki, H., Morifuji, Y., Ogino, T., et al. (2011). Hypoxia activates the hedgehog signaling pathway in a ligand-independent manner by upregulation of Smo transcription in pancreatic cancer. *Cancer Science*, 102(6), 1144–1150.
84. Yanai, K., Nagai, S., Wada, J., Yamanaka, N., Nakamura, M., Torata, N., et al. (2007). Hedgehog signaling pathway is a possible therapeutic target for gastric cancer. *Journal of Surgical Oncology*, 95(1), 55–62.
85. Matsushita, S., Onishi, H., Nakano, K., Nagamatsu, I., Imaizumi, A., Hattori, M., et al. (2014). Hedgehog signaling pathway is a potential therapeutic target for gallbladder cancer. *Cancer Science*, 105(3), 272–280.
86. Bizama, C., García, P., Espinoza, J. A., Weber, H., Leal, P., Nervi, B., et al. (2015). Targeting specific molecular pathways holds promise for advanced gallbladder cancer therapy. *Cancer Treatment Reviews*, 41(3), 222–234.
87. Athar, M., Li, C., Kim, A. L., Spiegelman, V. S., & Bickers, D. R. (2014). Sonic hedgehog signaling in Basal cell nevus syndrome. *Cancer Research*, 74(18), 4967–4975.
88. Fan, H. X., Wang, S., Zhao, H., Liu, N., Chen, D., Sun, M., et al. (2014). Sonic hedgehog signaling may promote invasion and metastasis of oral squamous cell carcinoma by activating MMP-9 and E-cadherin expression. *Medical Oncology (Northwood, London, England)*, 31(7), 41.
89. Ruch, J. M., & Kim, E. J. (2013). Hedgehog signaling pathway and cancer therapeutics: Progress to date. *Drugs*, 73(7), 613–623.
90. Nagao-Kitamoto, H., Nagata, M., Nagano, S., Kitamoto, S., Ishidou, Y., Yamamoto, T., et al. (2015). GLI2 is a novel therapeutic target for metastasis of osteosarcoma. *International Journal of Cancer*, 136(6), 1276–1284.
91. Li, J., Wu, T., Lu, J., Cao, Y., Song, N., Yang, T., et al. (2012). Immunohistochemical evidence of the prognostic value of hedgehog pathway components in primary gallbladder carcinoma. *Surgery Today*, 42, 770–775.
92. Xie, F., Xu, X., Xu, A., Liu, C., Liang, F., Xue, M., et al. (2014). Aberrant activation of Sonic hedgehog signaling in chronic cholecystitis and gallbladder carcinoma. *Human Pathology*, 45(3), 513–521.
93. Dixit, R., Pandey, M., Tripathi, S. K., Dwivedi., & Shukla, V. K. (2017). Comparative analysis of mutational profile of sonic hedgehog gene in gallbladder cancer. *Digestive Diseases and Sciences*, 62, 708–714.
94. Bergers, G., & Benjamin, L. E. (2003). Tumorigenesis and the angiogenic switch. *Nature Reviews. Cancer*, 3(6), 401–410.
95. Hanahan, D., & Weinberg, R. A. (2011). Hallmarks of cancer: The next generation. *Cell*, 144(5), 646–674.
96. Sun, X. N., Cao, W. G., Wang, X., Wang, Q., Gu, B. X., Yang, Q. C., et al. (2011). Prognostic impact of vascular endothelial growth factor-A expression in resected gallbladder carcinoma. *Tumor Biology*. 32, 1183–1190.
97. Jiang, L., Chen, Y., She, F., Tang, N., Li, X., & Wang, X. (2010). Expressions of VEGF-C and VEGF-D and their correlation with lymphangiogenesis and angiogenesis in gallbladder carcinoma. *Zhonghua Zhong liu za zhi [Chinese Journal of Oncology]*, 32(3), 190–195.

98. Letelier, P., Garcia, P., Leal, P., Ili, C., Buchegger, K., Riquelme, I., et al. (2014). Immunohistochemical expression of vascular endothelial growth factor A in advanced gallbladder carcinoma. *Applied Immunohistochemistry & Molecular Morphology*, 22(7), 530–536.
99. Nakashima, T., Kondoh, S., Kitoh, H., Ozawa, H., Okita, S., Harada, T., et al. (2003). Vascular endothelial growth factor-C expression in human gallbladder cancer and its relationship to lymph node metastasis. *International Journal of Molecular Medicine*, 11(1), 33–39.
100. Yadav, A., Gupta, A., Yadav, S., Rastogi, N., Agrawal, S., Kumar, A., et al. (2016). Association of Wnt signaling pathway genetic variants in gallbladder cancer susceptibility and survival. *Tumor Biology*, 37, 8083–8095.
101. Rai, R., Sharma, K. L., Sharma, S., Misra, S., Kumar, A., & Mittal, B. (2014). Death receptor (DR4) haplotypes are associated with increased susceptibility of gallbladder carcinoma in north Indian population. *PLoS One*, 9(2), e90264.
102. Srivastava, K., Srivastava, A., & Mittal, B. (2010). Caspase-8 polymorphisms and risk of gallbladder cancer in a northern Indian population. *Molecular Carcinogenesis*, 49(7), 684–692.
103. Gupta, A., Sharma, A., Yadav, A., Rastogi, N., Agrawal, S., Kumar, A., et al. (2015). Evaluation of miR-27a, miR-181a, and miR-570 genetic variants with gallbladder cancer susceptibility and treatment outcome in a North Indian population. *Molecular Diagnosis & Therapy*, 19, 317–327.
104. Srivastava, K., Srivastava, A., & Mittal, B. (2010). Common genetic variants in pre-microRNAs and risk of gallbladder cancer in North Indian population. *Journal of Human Genetics*, 55(8), 495–499.
105. Kimura, Y., Furuhata, T., Mukaiya, M., Kihara, C., Kawakami, M., Okita, K., et al. (2003). Frequent beta-catenin alteration in gallbladder carcinomas. *Journal of Experimental & Clinical Cancer Research: CR*, 22(2), 321–328.
106. Srivastava, K., Srivastava, A., & Mittal, B. (2010). Polymorphisms in ERCC2, MSH2, and OGG1 DNA repair genes and gallbladder cancer risk in a population of Northern India. *Cancer*, 116(13), 3160–3169.
107. Srivastava, A., Srivastava, K., Pandey, S. N., Choudhuri, G., & Mittal, B. (2009). Single-nucleotide polymorphisms of DNA repair genes OGG1 and XRCC1: Association with gallbladder cancer in North Indian population. *Annals of Surgical Oncology*, 16, 1695–1703.
108. Kandoth, C., McLellan, M. D., Vandin, F., Ye, K., Niu, B., Lu, C., et al. (2013). Mutational landscape and significance across 12 major cancer types. *Nature*, 502(7471), 333–339.
109. Nogueira, L., Foerster, C., Groopman, J., Egner, P., Koshiol, J., Ferreccio, C., et al. (2015). Association of aflatoxin with gallbladder cancer in Chile. *JAMA*, 313(20), 2075–2077.
110. Hsing, A. W., Sakoda, L. C., Rashid, A., Andreotti, G., Chen, J., Wang, B. S., et al. (2008). Variants in inflammation genes and the risk of biliary tract cancers and stones: A population-based study in China. *Cancer Research*, 68(15), 6442–6452.
111. Srivastava, K., Srivastava, A., Kumar, A., & Mittal, B. (2010). Significant association between toll-like receptor gene polymorphisms and gallbladder cancer. *Liver International*, 30(7), 1067–1072.
112. Sharma, K. L., Misra, S., Kumar, A., & Mittal, B. (2012). Higher risk of matrix metalloproteinase (MMP-2, 7, 9) and tissue inhibitor of metalloproteinase (TIMP-2) genetic variants to gallbladder cancer. *Liver International: Official Journal of the International Association for the Study of the Liver*, 32(8), 1278–1286.
113. Trepel, J., Mollapour, M., Giaccone, G., &Neckers, L. (2010). Targeting the dynamic HSP90 complex in cancer. *Nature Reviews. Cancer*, 10(8), 537–549.
114. Porter, J. R., Fritz, C. C., & Depew, K. M. (2010). Discovery and development of Hsp90 inhibitors: A promising pathway for cancer therapy. *Current Opinion in Chemical Biology*, 14(3), 412–420.
115. Yoshida, T., Sugai, T., Habano, W., Nakamura, S., Uesugi, N., Funato, O., & Saito, K. (2000). Microsatellite instability in gallbladder carcinoma: Two independent genetic pathways of gallbladder carcinogenesis. *Journal of Gastroenterology*, 35(10), 768–774.
116. Letelier, P., García, P., Leal, P., Álvarez, H., Ili, C., López, J., et al. (2014). miR-1 and miR-145 act as tumor suppressor microRNAs in gallbladder cancer. *International Journal of Clinical and Experimental Pathology*, 7(5), 1849–1867.
117. Peng, H. h., Zhang, Y. d., Gong, L. s., Liu, W. d., & Zhang, Y. (2013). Increased expression of microRNA-335 predicts a favorable prognosis in primary gallbladder carcinoma. *OncoTargets and Therapy*, 1625–1630.
118. Rao, S., Yao, Y., & Bauer, D. E. (2021). Editing GWAS: Experimental approaches to dissect and exploit disease-associated genetic variation. *Genome Medicine*, 13(1), 41.
119. Cha, P. C., Zembutsu, H., Takahashi, A., Kubo, M., Kamatani, N., & Nakamura, Y. (2012). A genome-wide association study identifies SNP in DCC is associated with gallbladder cancer in the Japanese population. *Journal of Human Genetics*, 57(4), 235–237.

120. Mhatre, S., Wang, Z., Nagrani, R., Badwe, R., Chiplunkar, S., Mittal, B., et al. (2017). Common genetic variation and risk of gallbladder cancer in India: A case-control genome-wide association study. *The Lancet Oncology*, 18(4), 535–544.
121. Rai, R., Sharma, K. L., Tiwari, S., Misra, S., Kumar, A., & Mittal, B. (2013). DCC (deleted in colorectal carcinoma) gene variants confer increased susceptibility to gallbladder cancer (Ref. No.: Gene-D-12–01446). *Gene*, 518(2), 303–309.
122. Hemminki, K., & Li, X. (2003). Familial liver and gall bladder cancer: A nationwide epidemiological study from Sweden. *Gut*, 52(4), 592–596.
123. Tebbi, A., Levillayer, F., Jouvion, G., Fiette, L., Soubigou, G., Varet, H., et al. (2016). Deficiency of multidrug resistance 2 contributes to cell transformation through oxidative stress. *Carcinogenesis*, 37(1), 39–48.
124. Forbes, S. A., Beare, D., Boutselakis, H., Bamford, S., Bindal, N., Tate, J., et al. (2017). COSMIC: Somatic cancer genetics at high-resolution. *Nucleic Acids Research*, 45(D1), D777–D783.
125. Borger, D. R., Tanabe, K. K., Fan, K. C., Lopez, H. U., Fantin, V. R., Straley, K. S., et al. (2012). Frequent mutation of isocitrate dehydrogenase (IDH)1 and IDH2 in cholangiocarcinoma identified through broad-based tumor genotyping. *The Oncologist*, 17(1), 72–79.
126. Ross, J. S., Wang, K., Catenacci, D. V., Chmielecki, J., Ali, S. M., Elvin, J. A., et al. (2015). Comprehensive genomic profiling of biliary tract cancers to reveal tumor-specific differences and genomic alterations. *American Society of Clinical Oncology*, 33, 231.
127. Wardell, C. P., Fujita, M., Yamada, T., Simbolo, M., Fassan, M., Karlic, R., et al. (2018). Genomic characterization of biliary tract cancers identifies driver genes and predisposing mutations. *Journal of Hepatology*, 68(5), 959–969.
128. Scholer, A. J., Marcus, R. K., Garland-Kledzik, M., Ghosh, D., Ensenyat-Mendez, M., Germany, J., et al. (2024). Exploring the genomic landscape of hepatobiliary cancers to establish a novel molecular classification system. *Cancers*, 16(2), 325.
129. Cheng, D. T., Mitchell, T. N., Zehir, A., Shah, R. H., Benayed, R., Syed, A., et al. (2015). Memorial sloan kettering-integrated mutation profiling of actionable cancer targets (MSK-IMPACT): A hybridization capture-based next-generation sequencing clinical assay for solid tumor molecular oncology. *The Journal of Molecular Diagnostics: JMD*, 17(3), 251–264.
130. Chakravarty, D., Gao, J., Phillips, S. M., Kundra, R., Zhang, H., Wang, J., et al. (2017). OncoKB: A precision oncology knowledge base. *JCO Precision Oncology*, 2017, PO.17.00011.
131. Sanchez-Vega, F., Mina, M., Armenia, J., Chatila, W. K., Luna, A., La, K. C., et al. (2018). Oncogenic signaling pathways in the cancer genome atlas. *Cell*, 173(2), 321–337.e10.

3 The Genetic Blueprint of Gallbladder Carcinogenesis

Advances and Discoveries

Soni Kumari,[1] Anam Mursaleen,[2] and Pramod Kumar Gautam[1]*

[1]Department of Biochemistry, All India Institute of Medical Sciences, New Delhi, India

[2]Department of Biotechnology, School of Chemical and Life Sciences, Jamia Hamdard, New Delhi, India

*Corresponding author: sguptabiochem@gmail.com

3.1 INTRODUCTION

Gallbladder carcinoma (GBC) is the most common malignant tumor of the biliary tract.[1] It is the fifth most prevalent digestive tract carcinoma and is widely prevalent in females and older people, leading to high mortality rates. It is often diagnosed at an advanced stage due to its asymptomatic nature in the early phases, leading to poor prognostic outcomes. Histologically, the majority of GBC cases are adenocarcinomas, but other subtypes such as squamous cell carcinoma, adenosquamous carcinoma, and neuroendocrine tumors are also observed in smaller proportions.[1] Despite being a relatively rare cancer on a global scale, its significant geographic and demographic variability highlights its importance as a public health concern in certain populations.

The gallbladder's unique anatomical location and thin wall structure make it prone to early invasion of adjacent organs, such as the liver and bile ducts, contributing to the aggressive nature of the disease. Due to its asymptomatic nature in the early stages, GBC is often discovered incidentally during gallstone surgery or imaging studies, with most cases diagnosed in advanced stages where curative treatment is no longer feasible.[2] Its late presentation, coupled with limited effective therapeutic options, results in an overall 5-year survival rate of less than 5% in advanced disease.[3]

3.1.1 Epidemiology

GBC is a rare yet aggressive malignancy with significant geographic, demographic, and gender-based disparities. Recent statistics indicate that GBC accounts for approximately 1.2% of all cancer diagnoses worldwide, with an estimated 219,420 new cases and 165,087 deaths in 2020.[4] The age-standardized incidence rate is 2.3 per 100,000 persons globally.[4] Chile has the highest incidence worldwide, with rates of 27.3 and 12.8 per 100,000 in women and men, respectively.[5] Other high-incidence regions include northern India, Pakistan, and South Korea.[6] India harbors both high- and low-risk regions, with gallbladder cancer being endemic in the so-called "chola" region of high-risk populations in northern and northeastern states.[7] In contrast, Western countries generally have lower incidence rates, with the United States reporting approximately 1–2 cases per 100,000 persons annually.[8]

GBC demonstrates a marked gender disparity, with a female-to-male ratio of approximately 3:1 in most populations.[5] This difference is even more pronounced in high-incidence regions: in Chile, the female-to-male ratio reaches 7:1.[6] In northern India, women are affected 2–6 times more

DOI: 10.1201/9781003506201-3

frequently than men.[7] The risk of GBC increases with age, with most cases diagnosed in individuals over 50 years old.[8] The disease is particularly aggressive in older populations, who often present with advanced-stage disease.[5]

3.1.2 Risk Factors

The pathogenesis of GBC involves a complex interplay of genetic, environmental, and lifestyle-related risk factors. The major contributors to gallbladder carcinogenesis include:

a. *Gallstones and Chronic Inflammation*: Gallstones are the most significant risk factor for GBC, found in 70–90% of cases.[9] Large stones (>3 cm) are associated with a particularly high risk, as they cause chronic irritation and inflammation of the gallbladder epithelium, leading to dysplasia and carcinogenesis.[10] Chronic cholecystitis further exacerbates this risk.[11]
b. *Porcelain Gallbladder*: A porcelain gallbladder, characterized by calcification of the gallbladder wall due to chronic inflammation, has a high malignant transformation potential.[12]
c. *Infections*: Chronic infections with *Salmonella typhi* or *Helicobacter pylori* have been implicated in GBC development through mechanisms involving persistent inflammation, oxidative stress, and DNA damage.[13,14]
d. *Genetic and Molecular Interactions*: Genetic predisposition plays a key role in GBC. Mutations in oncogenes (e.g., KRAS, HER2) and tumor suppressor genes (e.g., TP53) drive oncogenesis.[15] Pathways involving epidermal growth factor receptor (EGFR), PI3K/AKT, and Wnt signaling are frequently dysregulated in GBC.[16] Polymorphisms in genes involved in bile acid metabolism and detoxification have also been linked to increased susceptibility.[17]
e. *Primary Sclerosing Cholangitis and Biliary Anomalies*: Structural anomalies such as pancreaticobiliary malfunction and bile duct cysts lead to chronic bile reflux and gallbladder epithelial injury, predisposing to malignancy.[18]
f. *Environmental and Occupational Exposures*: Exposure to industrial toxins, heavy metals such as cadmium, and other environmental carcinogens has been linked to GBC development in certain regions.[19]

3.1.3 Current GBC Treatment

The management of GBC depends on the stage at diagnosis. Early-stage GBC, confined to the gallbladder, is rare but potentially curable with surgical intervention. However, most patients present with advanced disease, which has limited treatment options and poor survival rates.

a. *Surgical Management:* For localized disease, radical cholecystectomy with regional lymphadenectomy and en bloc resection of adjacent organs (if involved) remains the cornerstone of curative treatment.[20] Negative surgical margins (R0 resection) are crucial for better outcomes.[21]
b. *Adjuvant Therapy:* Patients with high-risk features, such as lymph node involvement or residual disease, benefit from adjuvant chemotherapy or chemoradiotherapy.[22] Common regimens include gemcitabine, cisplatin, or capecitabine-based therapies.[23]
c. *Targeted Therapy:* Molecular profiling has paved the way for targeted therapies in GBC.[24] HER2-positive cases may respond to trastuzumab or other HER2 inhibitors.[25] Mutations in PIK3CA, BRAF, or IDH1/IDH2 are being explored as targets for novel therapeutic agents in clinical trials.[26]
d. *Systemic Chemotherapy:* For unresectable or metastatic GBC, systemic chemotherapy is the mainstay of treatment.[27] The combination of gemcitabine and cisplatin is widely used and forms the first-line therapy.[28]

e. *Immunotherapy:* Immune checkpoint inhibitors, such as PD-1/PD-L1 blockers, are being investigated for advanced GBC, especially in tumors with high microsatellite instability or high tumor mutational burden.[29] While promising, the efficacy of immunotherapy in GBC remains under evaluation.[30]
f. *Palliative Care:* For patients with advanced or terminal disease, palliative interventions focus on symptom relief and improving quality of life.[31] This includes biliary stenting for obstructive jaundice, pain management, and supportive care.[32]

This chapter aims to delve into the intricate genetic and molecular landscape of GBC, highlighting key genes, pathways, and their interconnections that drive gallbladder carcinogenesis. By examining recent advances and discoveries, the chapter seeks to provide insights into the genetic alterations underlying the disease, the role of signaling pathways in tumor progression, and potential targets for therapeutic intervention, fostering a deeper understanding of this devastating malignancy.

3.2 GENETIC BASIS OF GBC

Gallbladder carcinogenesis is a complex process involving multiple genetic alterations. Key molecular pathways implicated include mutations in tumor suppressor genes such as TP53 and p16, activation of oncogenes such as KRAS and ERBB2, and dysregulation of cell cycle regulators. Epigenetic changes, including DNA methylation and histone modifications, also play a role in silencing tumor suppressor genes. Chromosomal instability and microsatellite instability contribute to genomic alterations. Additionally, aberrant activation of signaling pathways such as PI3K/AKT/mTOR and Wnt/β-catenin promotes cell proliferation and survival.

3.2.1 Key Genes Associated with GBC

a. *KRAS:* KRAS is one of the most critical oncogenes driving tumorigenesis. Mutations in KRAS, particularly at codon 12 (e.g., G12D, G12V), result in constitutive activation of downstream signaling pathways such as MAPK/ERK and PI3K/AKT. This leads to increased cell proliferation, metabolic adaptation, and resistance to apoptosis, making KRAS mutations a key contributor to tumor progression.[33]
b. *TP53:* TP53 is among the most frequently mutated tumor suppressor genes in gallbladder cancer (GBC). It plays a pivotal role in maintaining genomic stability by regulating apoptosis and cell cycle arrest in response to DNA damage. Mutations such as R175H and R248Q impair its transcriptional activity, leading to uncontrolled proliferation and enhanced angiogenesis through VEGF overexpression.[34]
c. *ERBB2 (HER2):* Amplification or overexpression of ERBB2 (HER2) is observed in approximately 10%–15% of GBC cases, particularly in aggressive tumors. ERBB2 encodes a receptor tyrosine kinase that activates the PI3K/AKT/mTOR and MAPK/ERK pathways, driving tumor growth, invasion, and apoptosis resistance.[35]
d. *CDKN2A:* The tumor suppressor CDKN2A encodes p16INK4a and p14ARF, which regulate critical cell cycle checkpoints. Mutations, deletions, and promoter methylation of CDKN2A disrupt G1–S phase regulation, leading to uncontrolled cellular proliferation and increased genomic instability.[36]
e. *PIK3CA:* PIK3CA encodes the p110α catalytic subunit of PI3K, and its activating mutations (e.g., H1047R) are present in approximately 10%–15% of GBC cases. These mutations lead to hyperactivation of the PI3K/AKT/mTOR pathway, promoting tumor cell survival, angiogenesis, and metabolic adaptability while also conferring resistance to apoptosis.[37]
f. *SMAD4:* SMAD4, a crucial mediator of TGF-β signaling, functions as a tumor suppressor. Loss or inactivation of SMAD4 shifts TGF-β signaling toward pro-tumorigenic effects,

enhancing epithelial-to-mesenchymal transition (EMT), immune suppression, and metastatic potential in GBC.[38]

g. *ARID1A:* ARID1A, a member of the SWI/SNF chromatin remodeling complex, is frequently mutated in GBC. This leads to epigenetic dysregulation, impaired DNA damage repair, and immune evasion via upregulation of PD-L1 expression, contributing to tumor progression and therapy resistance.[39]

h. *APC:* Mutations in APC, a negative regulator of the Wnt/β-catenin pathway, lead to nuclear accumulation of β-catenin, which promotes proliferation, invasion, and EMT. Although APC mutations are less common in GBC, they significantly enhance cancer stem cell-like properties and therapy resistance.[40]

i. *BDNF:* Brain-derived neurotrophic factor (BDNF), a member of the neurotrophin family, has been implicated in GBC pathogenesis. The BDNF/TrkB axis activates PI3K/AKT, MAPK/ERK, and STAT3 pathways, promoting proliferation, survival, and tumor invasion.[41]

j. *BRAF:* Although BRAF mutations are rare in GBC, mutations such as V600E result in constitutive activation of the MAPK/ERK pathway, promoting unchecked tumor growth. Targeted BRAF inhibitors such as vemurafenib and dabrafenib have shown efficacy in other cancers and may be beneficial for GBC patients harboring BRAF mutations.[42]

3.2.2 Epigenetic Regulation of GBC

a. *DNA Methylation:* Aberrant DNA methylation is one of the most extensively studied epigenetic changes in GBC. Hypermethylation of promoter regions in tumor suppressor genes leads to their transcriptional silencing, thereby promoting carcinogenesis. For example, the *RASSF1A* gene, a critical regulator of apoptosis and cell cycle arrest, is frequently hypermethylated in GBC, resulting in its functional inactivation. Similarly, *CDH1*, which encodes the cell adhesion molecule E-cadherin, is silenced through promoter hypermethylation, contributing to EMT and metastasis. Other genes, such as *GSTP1*, *APC*, and *DLEC1*, have also been reported to undergo hypermethylation in GBC, further illustrating the widespread impact of DNA methylation in this malignancy. In addition to hypermethylation, hypomethylation of repetitive sequences and oncogenic promoters is observed in GBC. This global hypomethylation destabilizes the genome and activates oncogenes such as *c-MYC* and *BCL2*, fostering tumor progression. These contrasting patterns of hypermethylation and hypomethylation highlight the dual role of DNA methylation in gallbladder carcinogenesis.[43]

b. *Histone modifications* including acetylation, methylation, phosphorylation, and ubiquitination, regulate chromatin structure and gene expression. In GBC, dysregulated activity of histone-modifying enzymes leads to aberrant gene expression profiles. For instance, reduced acetylation of histones H3 and H4, often due to overexpression of histone deacetylases (HDACs), is associated with the silencing of tumor suppressor genes. HDAC inhibitors have shown promise in preclinical studies as they restore the expression of silenced genes and inhibit tumor growth. Aberrant histone methylation patterns are also implicated in GBC. Overexpression of histone methyltransferases (HMTs) such as *EZH2*, a key component of the polycomb repressive complex 2 (PRC2), leads to trimethylation of histone H3 at lysine 27 (H3K27me3), a repressive mark that silences tumor suppressor genes. Conversely, demethylation of repressive histone marks by enzymes like *JMJD3* can activate oncogenes, further contributing to tumor progression.[44]

c. *Non-coding RNAs (ncRNAs)*, particularly microRNAs (miRNAs) and long non-coding RNAs (lncRNAs), are critical regulators of gene expression in GBC. Dysregulation of miRNAs is a hallmark of GBC, with distinct miRNA profiles correlating with tumor initiation, progression, and metastasis. Tumor-suppressive miRNAs such as *miR-34a* and *miR-218* are often downregulated in GBC, leading to the activation of oncogenic pathways. For example, the loss of *miR-34a* results in the upregulation of *BCL2*, promoting cell survival, while *miR-218* silencing enhances *SLIT2* expression, facilitating tumor invasion. Conversely, oncogenic miRNAs such as *miR-21* and *miR-155* are frequently upregulated in GBC, contributing to increased cell proliferation, angiogenesis, and immune evasion. *miR-21*, for instance, targets the tumor suppressor gene *PTEN*, activating the *PI3K/AKT* pathway and promoting tumor growth. Similarly, *miR-155* influences the immune microenvironment by modulating cytokine expression, enabling immune escape. LncRNAs, which function as scaffolds, decoys, or guides in chromatin remodeling and transcriptional regulation, also play an essential role in GBC. For instance, the lncRNA *HOTAIR* is significantly upregulated in GBC and promotes metastasis by interacting with *PRC2* to silence tumor suppressor genes via *H3K27me3* deposition. Other lncRNAs such as *MALAT1* and *GAS5* are associated with poor prognosis and chemoresistance in GBC, emphasizing their potential as biomarkers and therapeutic targets.[45]

3.3 SIGNALING NETWORK REGULATING GBC

Gallbladder carcinoma (GBC) arises due to the interplay of various signaling pathways that govern cellular proliferation, apoptosis, invasion, metastasis, and immune evasion.

EGFR Pathway

Activation Mechanism: EGFR is a transmembrane receptor activated by binding of ligands such as epidermal growth factor or transforming growth factor-α (TGF-α). Ligand binding triggers EGFR dimerization and autophosphorylation of its intracellular tyrosine kinase domain.[46]

3.3.1 Downstream Signaling

PI3K/AKT Pathway: Phosphorylated EGFR recruits phosphoinositide-3-kinase (PI3K), leading to the production of phosphatidylinositol-3,4,5-triphosphate (PIP3). PIP3 activates AKT, which inhibits pro-apoptotic proteins such as BAD and activates mTOR, promoting cell survival and growth.[47]

RAS/RAF/MAPK Pathway: EGFR activates RAS via adaptor proteins such as GRB2 and SOS. RAS-GTP subsequently activates RAF, which phosphorylates MEK and ERK. ERK translocates to the nucleus to promote transcription of proliferation-associated genes like CCND1.[48]

Role in GBC: Overexpression of EGFR enhances cell proliferation, angiogenesis, and invasion, while also contributing to chemotherapy resistance.[49]

PI3K/AKT/mTOR Pathway

Activation Mechanism: This pathway is triggered by receptor tyrosine kinases (e.g., EGFR or IGF receptors). PI3K generates PIP3, which recruits AKT to the plasma membrane, where it is activated by PDK1.[50]

Functional Effects: AKT phosphorylates and inhibits the tumor suppressor protein TSC2, leading to activation of mTORC1. mTORC1 promotes protein synthesis, glucose metabolism, and cell growth by activating downstream effectors such as S6 kinase (S6K) and 4EBP1. AKT also inhibits pro-apoptotic proteins like BAD and caspase-9, enhancing cell survival.[51]

Role in GBC: Mutations in PIK3CA or loss of PTEN lead to hyperactivation of this pathway, driving uncontrolled cell growth, metastasis, and chemoresistance.[52]

MAPK/ERK Pathway

Activation Mechanism: The MAPK/ERK pathway is activated by receptor tyrosine kinases or mutations in RAS or BRAF. Activated RAS binds RAF, initiating a kinase cascade involving RAF, MEK, and ERK.[53]

Functional Effects: ERK phosphorylates transcription factors such as ELK1 and c-MYC, promoting the expression of cyclins and other proteins essential for G1/S phase transition. The pathway also regulates cytoskeletal dynamics and adhesion, facilitating invasion and metastasis.[54]

Role in GBC: Mutations in KRAS and BRAF in GBC lead to constitutive activation of the pathway, supporting tumor progression and resistance to targeted therapies.[55]

Wnt/β-Catenin Pathway

Activation Mechanism: Wnt ligands bind to Frizzled (FZD) receptors and co-receptors such as LRP5/6, inhibiting the destruction complex composed of APC, Axin, and GSK-3β. This allows β-catenin to accumulate and translocate to the nucleus.[56]

Functional Effects: Nuclear β-catenin interacts with TCF/LEF transcription factors to upregulate genes like MYC and CCND1, promoting proliferation. Aberrant activation of this pathway enhances EMT, invasion, and metastasis.[57]

Role in GBC: Mutations in APC or β-catenin, or overexpression of Wnt ligands, drive tumorigenesis and aggressive behavior in GBC.[58]

Hedgehog (Hh) Signaling Pathway

Activation Mechanism: The pathway is activated when ligands like Sonic Hedgehog (Shh) bind to Patched1 (PTCH1), relieving inhibition on Smoothened (SMO). Activated SMO initiates a cascade that stabilizes **GLI** transcription factors.[59]

Functional Effects: GLI proteins translocate to the nucleus and regulate genes involved in proliferation (CCND2), survival (BCL2), and angiogenesis (VEGF).[60]

Role in GBC: Overexpression of Shh and GlI proteins in GBC enhances tumor growth, invasion, and chemoresistance.[61]

Notch Signaling Pathway

Activation Mechanism: Notch receptors interact with ligands like Delta-like (DLL) or Jagged, leading to proteolytic cleavage by γ-secretase. The Notch intracellular domain is released and translocated to the nucleus to regulate target gene expression.[62]

Functional Effects: Notch signaling modulates cell survival, differentiation, and angiogenesis. Dysregulated Notch signaling promotes EMT and tumor invasion.[63]

Role in GBC: Overexpression of Notch receptors and ligands correlates with aggressive tumor phenotypes and poor prognosis.[64]

Transforming Growth Factor-β (TGF-β) Pathway

Activation Mechanism: TGF-β ligands bind to type II TGF-β receptors, which recruit and phosphorylate type I receptors. This activates SMAD proteins that translocate to the nucleus to regulate gene expression.[65]

Functional Effects: In early-stage tumors, TGF-β acts as a tumor suppressor by inducing cell cycle arrest and apoptosis. In advanced GBC, TGF-β promotes EMT, immune evasion, and metastasis by activating non-canonical pathways such as PI3K/AKT and MAPK.[66]

Role in GBC: Elevated TGF-β levels in GBC are associated with invasion, metastasis, and poor outcomes.[67]

NF-κB Pathway

Activation Mechanism: NF-κB is activated by inflammatory stimuli such as cytokines (e.g., IL-6, TNF-α) or bile acids. Activation involves the degradation of IκB, an inhibitor of NF-κB, allowing NF-κB to translocate to the nucleus.[68]

Functional Effects: NF-κB induces transcription of pro-survival and pro-inflammatory genes, including BCL2, IL-6, and CXCL8. The pathway also upregulates matrix metalloproteinases (MMPs), facilitating invasion and metastasis.[69]

Role in GBC: Chronic inflammation, a hallmark of GBC, activates NF-κB, promoting tumor progression and immune suppression.[70]

Hippo/YAP Pathway

Activation Mechanism: The Hippo pathway regulates organ size and tumorigenesis. Inactivation of upstream kinases (MST1/2 and LATS1/2) results in the nuclear localization of YAP/TAZ transcription co-activators.[71]

Functional Effects: Nuclear YAP interacts with TEAD transcription factors to induce genes involved in cell proliferation (CTGF, CYR61) and anti-apoptosis. Dysregulated Hippo signaling enhances tumorigenic potential and chemoresistance.[72]

Role in GBC: Elevated YAP activity in GBC correlates with aggressive tumor behavior and poor prognosis.[73]

Activation Mechanism: This pathway is triggered by cytokines (e.g., IL-6, IL-10) binding to their receptors, leading to activation of Janus kinases (JAKs). JAKs phosphorylate STAT proteins, which dimerize and translocate to the nucleus.[74]

Functional Effects: STAT3 promotes the transcription of genes involved in proliferation (MYC), survival (BCL2), and immune evasion (PD-L1).[75]

Role in GBC: Constitutive activation of STAT3 also enhances EMT and invasion.[76]

3.4 EPIGENETIC ALTERATIONS IN GALLBLADDER CANCER

Epigenetic modifications, including DNA methylation, histone modifications, and non-coding RNA regulation, significantly contribute to GBC progression by altering gene expression without changing the DNA sequence. These changes can silence tumor suppressor genes, activate oncogenes, and influence key signaling pathways involved in tumor progression.[77]

3.4.1 DNA Methylation

Aberrant DNA methylation in GBC affects key tumor suppressor genes and oncogenes, contributing to oncogenesis and cancer progression.[78]

- *Promoter Hypermethylation of Tumor Suppressor Genes:*
- Hypermethylation of tumor suppressor gene promoters leads to their transcriptional silencing, promoting tumorigenesis. In GBC, several genes are frequently hypermethylated.
  - *CDKN2A (p16INK4A)* is involved in cell cycle regulation, and its methylation is observed in 40–60% of GBC cases, leading to loss of tumor suppression.[79]
  - *RASSF1A* suppresses Ras-mediated oncogenic signaling and is frequently methylated in GBC, leading to enhanced tumor progression.[80]
  - *MLH1* methylation leads to mismatch repair deficiency, contributing to microsatellite instability (MSI), a feature of aggressive gallbladder tumors.[81]

- *Global Hypomethylation and Oncogene Activation:*
- While promoter hypermethylation silences tumor suppressor genes, global DNA hypomethylation leads to genomic instability and oncogene activation.[82]
  - Hypomethylation of *LINE-1* elements contributes to chromosomal instability, which plays a role in GBC tumorigenesis.[83]
  - Overexpression of *CTNNB1 (β-catenin)* due to promoter hypomethylation enhances Wnt/β-catenin signaling, promoting tumor progression.[84]

3.4.2 Histone Modifications

Histone modifications, such as acetylation and methylation, regulate chromatin accessibility and gene expression.[85]

- *H3K27me3* (trimethylation of histone H3 at lysine 27) suppresses tumor suppressor genes, and its high levels are associated with poor prognosis in GBC.[86]
- *H3K9ac* (histone H3 acetylation at lysine 9) promotes oncogene activation and is linked to increased cell proliferation in GBC.[87]

3.4.3 Noncoding RNA Regulation

Noncoding RNAs, including miRNAs and lncRNAs, play a crucial role in post-transcriptional gene regulation in GBC.[88]

- *Oncogenic miRNAs (OncomiRs):*
  - *miR-21* targets *PTEN* and promotes PI3K/AKT activation, facilitating GBC progression.[89]
  - *miR-155* enhances NF-κB signaling and inflammation-driven GBC progression, linking chronic inflammation to carcinogenesis.[90]

- *Tumor Suppressor miRNAs:*
  - *miR-34a* targets Notch and Wnt pathways; its downregulation is associated with poor prognosis in GBC patients.[91]
  - *miR-200* family members inhibit EMT by targeting *ZEB1/ZEB2*, thus preventing metastasis.[92]

- *Long Noncoding RNAs (lncRNAs):*

  - *HOTAIR* promotes metastasis in GBC by modifying chromatin structure and altering gene expression.[93]
  - *MALAT1* enhances cell proliferation and invasion by modulating EMT, contributing to GBC aggressiveness.[94]

3.4.4 BDNF and Epigenetic Regulation

BDNF has been implicated in the progression of various cancers, including GBC.[95]

- The BDNF/TrkB pathway activates PI3K/AKT and MAPK/ERK signaling, promoting tumor cell survival, invasion, and metastasis in GBC.[96]
- Epigenetic regulation of *BDNF* occurs through DNA methylation of its promoter regions, affecting its expression in cancer cells. Additionally, *miR-206* targets *BDNF*, influencing tumor progression and metastasis.[97]

3.5 BIOMARKERS FOR EARLY DETECTION OF GBC

GBC is a highly aggressive malignancy with poor prognosis due to its asymptomatic nature in early stages and rapid progression to advanced stages. Consequently, early detection is crucial for improving patient outcomes. Biomarkers, which are measurable indicators of a biological state or condition, are being increasingly explored for the early diagnosis of GBC. Various types of biomarkers, including genetic, protein, epigenetic, and metabolic markers, have shown potential in the early detection of GBC.

3.5.1 Genetic Biomarkers

Genetic alterations are common in GBC and could serve as valuable tools for early diagnosis. Several mutations, particularly in tumor suppressor genes and oncogenes, have been associated with GBC development. Some key genetic biomarkers include:

- *KRAS Mutations:* Mutations in the *KRAS* gene, commonly associated with various cancers, have been identified in a significant subset of GBC patients. *KRAS* mutations, especially in codon 12, contribute to uncontrolled cellular proliferation and tumor progression.[98]
- *TP53 Mutations:* TP53, a tumor suppressor gene, is frequently mutated in GBC. Alterations in *TP53* lead to dysfunctional regulation of the cell cycle and apoptosis, contributing to early oncogenesis.[99]
- *PIK3CA Mutations:* The phosphatidylinositol-4,5-bisphosphate 3-kinase catalytic subunit alpha (*PIK3CA*) gene plays a role in cell growth and survival. Mutations in *PIK3CA* are observed in early-stage GBC and can be used as potential diagnostic biomarkers.[100]
- *ERBB2 (HER2) Amplification:* Overexpression or amplification of *ERBB2 (HER2)* is linked to the pathogenesis of GBC. HER2 positivity has been detected in early-stage GBC, and thus, it holds promise as a biomarker for early detection and as a therapeutic target.[101]

3.5.2 Protein Biomarkers

Protein-based biomarkers are among the most studied for GBC detection due to their ability to be measured in biological fluids such as blood and bile.

- *Carbohydrate Antigen 19–9 (CA 19–9):* CA 19–9 is a well-known tumor marker that is elevated in GBC patients. While CA 19–9 is also associated with other gastrointestinal cancers (e.g., pancreatic cancer), its elevated levels in combination with imaging studies may aid in the early detection of GBC.[102]
- *Carcinoembryonic Antigen (CEA):* CEA is another commonly evaluated tumor marker in GBC patients. Although its sensitivity and specificity are limited when used alone, combined measurement of CEA with CA 19–9 can enhance diagnostic accuracy for early GBC.[103]
- *Mucin Family Proteins:* Overexpression of mucin proteins, particularly MUC1 and MUC5AC, has been detected in GBC. These mucins can play a role in early tumor formation, and elevated levels of mucin in serum or bile can serve as a potential diagnostic biomarker.[104]
- *S100A4:* S100A4, a member of the S100 family of proteins, is involved in promoting tumor metastasis and is overexpressed in GBC tissues. Early overexpression of S100A4 in GBC patients indicates its potential as a biomarker for early diagnosis.[105]

3.5.3 Epigenetic Biomarkers

Epigenetic modifications, such as DNA methylation and histone modifications, are key drivers of cancer development. In GBC, aberrant DNA methylation patterns of tumor suppressor genes have been linked to early carcinogenesis.

- *Methylation of CDKN2A (p16):* Hypermethylation of the promoter region of *CDKN2A* (which encodes the p16 protein) leads to the silencing of this critical tumor suppressor gene. Aberrant p16 methylation is frequently observed in early-stage GBC and can serve as a non-invasive biomarker for early detection.[106]
- *DNA Methylation of APC and RUNX3:* The promoter hypermethylation of *APC* and *RUNX3* genes, both involved in tumor suppression, has been documented in GBC patients. These methylation events occur early in GBC development and can be detected in serum, offering a non-invasive diagnostic tool.[107]
- *miRNA Signatures:* miRNAs are small non-coding RNAs that regulate gene expression post-transcriptionally. Dysregulation of specific miRNAs has been linked to early GBC. For instance, miR-21 and miR-145 are overexpressed in GBC tissues and may be useful for early detection when measured in circulating blood.[108]

3.5.4 Metabolic Biomarkers

Metabolic reprogramming is a hallmark of cancer, and GBC is no exception. Tumor cells often alter their metabolism to sustain rapid growth and survival. Detecting metabolic changes in bodily fluids can help in the early diagnosis of GBC.

- *Bile Acid Metabolites*: Abnormal bile acid metabolism is a characteristic feature of GBC. Specific bile acid metabolites, such as deoxycholic acid, have been found to be elevated in GBC patients. These metabolic alterations can be detected in bile samples, providing a potential biomarker for early diagnosis.[109]
- *Lipidomic Profiling*: Lipid metabolism is often dysregulated in cancer. Lipidomic studies have revealed distinct lipid profiles in GBC patients. Elevated levels of phosphatidylcholines and lysophosphatidylcholines in serum have been associated with early GBC and may serve as potential biomarkers.[110]

3.5.5 Circulating Tumor DNA

Circulating tumor DNA (ctDNA) refers to small fragments of DNA shed by cancer cells into the bloodstream. ctDNA carries tumor-specific genetic alterations, making it a promising tool for non-invasive cancer detection.

- *KRAS Mutations in ctDNA*: Detection of *KRAS* mutations in ctDNA from blood samples can serve as a highly specific marker for early GBC. As these mutations often occur early in GBC development, monitoring ctDNA could allow for earlier diagnosis.
- *TP53 Mutations in ctDNA*: TP53 mutations, which are prevalent in GBC, can also be detected in ctDNA. The presence of *TP53* mutations in circulating DNA is indicative of early tumor formation and may aid in early diagnosis.

3.5.6 Other Emerging Biomarkers

Several novel biomarkers are being explored for their potential role in the early detection of GBC. Among them, exosomes and tumor-associated autoantibodies have shown promising results as non-invasive diagnostic tools.

3.5.6.1 Exosomes

Exosomes are small extracellular vesicles (30–150 nm in diameter) secreted by various cells, including cancer cells, into biological fluids such as blood, bile, and urine. These vesicles carry molecular cargo, including proteins, lipids, messenger RNAs (mRNAs), miRNAs, and DNA fragments, which reflect the biological state of the originating cell. Tumor-derived exosomes play a critical role in cancer progression by mediating intercellular communication, promoting immune evasion, and facilitating metastasis. Since exosomes can be isolated from blood, bile, or other body fluids, they provide a promising avenue for liquid biopsy-based GBC detection. Their ability to reflect the molecular characteristics of tumors in real time makes them a valuable tool for early diagnosis, prognosis, and therapeutic monitoring.

Exosomal microRNAs (miRNAs): Specific miRNAs encapsulated in exosomes have been identified as potential biomarkers for early GBC detection. For example, elevated levels of *exosomal miR-21, miR-222, and miR-125b* have been reported in GBC patients and are associated with tumor progression and chemoresistance.

Exosomal Proteins: Certain proteins carried by exosomes, such as *heat shock proteins, EGFR, and CD44*, have been found to be enriched in GBC-derived exosomes. These proteins are involved in tumor proliferation and invasion, making them potential early diagnostic markers.

Exosomal DNA: Tumor-derived exosomes contain circulating tumor DNA (ctDNA) fragments with specific mutations, such as *KRAS* and *TP53* mutations, which are frequently observed in early-stage GBC. The detection of such mutations in exosomal DNA could serve as a minimally invasive diagnostic approach.

3.5.6.2 Tumor-Associated Autoantibodies

Autoantibodies against tumor-associated antigens (TAAs) are produced by the immune system in response to cancer-associated proteins. These autoantibodies can appear in the blood long before clinical symptoms manifest, making them valuable early biomarkers for cancer detection.

Common TAAs in GBC: Several tumor-associated antigens have been identified in GBC, including *p53, MUC1, NY-ESO-1, and survivin*. The presence of autoantibodies against these antigens in patient serum suggests early immune recognition of tumor cells.

p53 Autoantibodies: The *p53 tumor suppressor gene* is frequently mutated in GBC, leading to the accumulation of dysfunctional p53 protein. This abnormal p53 protein elicits an immune response, resulting in the production of *anti-p53 autoantibodies*, which have been detected in the early stages of GBC.

MUC1 Autoantibodies: Mucin 1 (MUC1) is a glycoprotein overexpressed in GBC. Autoantibodies against MUC1 have been found in serum samples from patients with gallbladder malignancies, highlighting their potential as an early detection marker.

NY-ESO-1 and Survivin: NY-ESO-1 is a cancer-testis antigen expressed in various malignancies, including GBC. Autoantibodies against NY-ESO-1 and survivin (an anti-apoptotic protein) have been observed in early-stage GBC.

3.6 THERAPEUTIC APPROACHES IN GALLBLADDER CANCER

GBC is an aggressive malignancy with a poor prognosis due to its asymptomatic nature in early stages and rapid progression to advanced disease. The treatment strategies for GBC depend on the stage of the disease, tumor biology, and patient-specific factors. Currently, therapeutic approaches include surgery, chemotherapy, radiotherapy, targeted therapy, and immunotherapy. Emerging treatment modalities, including personalized medicine and combination therapies, are also being explored to improve patient outcomes.

3.6.1 Surgical Approaches

Surgery remains the only curative option for GBC, particularly in early-stage disease. However, due to late diagnosis in most cases, only a small percentage of patients are eligible for surgical resection.

3.6.1.1 Cholecystectomy (Simple and Extended)

Simple Cholecystectomy: This involves the removal of the gallbladder and is often performed when GBC is detected incidentally during surgery for gallstones or other benign conditions. It is only curative for very early-stage tumors (T1a, confined to the mucosa).

Extended Cholecystectomy: Involves removal of the gallbladder along with segments of the liver (IVb and V) and lymph node dissection. This approach is recommended for T1b, T2, and some T3 tumors, as these stages have a higher risk of lymphatic spread.

3.6.1.2 Radical Resection

For patients with locally advanced disease (T2 or beyond), radical surgery is performed, which may include liver wedge resection or segmentectomy, common bile duct resection (if tumor involves the biliary tree), and regional lymphadenectomy (removal of hepatoduodenal ligament lymph nodes).

3.6.2 Chemotherapy

Systemic chemotherapy is used for unresectable, recurrent, or metastatic GBC, either as a palliative measure or in adjuvant settings after surgical resection to prevent recurrence.

3.6.2.1 First-Line Chemotherapy

The standard first-line chemotherapy regimen for advanced GBC includes gemcitabine plus cisplatin, based on the ABC-02 trial, which demonstrated improved survival with this combination.

- *Gemcitabine + Cisplatin:* The combination works by inhibiting DNA synthesis and inducing apoptosis. It remains the gold standard for advanced biliary tract cancers.
- *Gemcitabine + Oxaliplatin (GEMOX):* An alternative to cisplatin-based regimens, GEMOX has shown efficacy in patients with biliary cancers, particularly in those intolerant to cisplatin.
- *Fluoropyrimidine-based regimens (Capecitabine or 5-Fluorouracil + Oxaliplatin/Leucovorin):* Often used in combination for patients who cannot tolerate gemcitabine-based therapy.

3.6.2.2 Second-Line Chemotherapy

Patients who progress after first-line therapy may receive second-line treatments, including:

- *FOLFIRINOX (5-FU, leucovorin, irinotecan, and oxaliplatin):* Used in select cases due to its high toxicity but promising response rates.
- *Irinotecan-based regimens:* Have shown some efficacy in refractory GBC cases.
- *Liposomal Irinotecan + Fluoropyrimidine-based combinations:* Currently being investigated for their potential in advanced disease.

3.6.3 Radiation Therapy

Radiotherapy is generally not used as a primary treatment for GBC but plays a role in *adjuvant therapy* (after surgery) and *palliative care* (for symptom relief in advanced cases).

3.6.3.1 Adjuvant Radiotherapy

For patients with positive surgical margins or lymph node involvement, adjuvant radiotherapy may be given alongside chemotherapy (chemoradiotherapy) to reduce the risk of recurrence.

- *External Beam Radiation Therapy:* Standard form of radiotherapy used in GBC.
- *Stereotactic Body Radiation Therapy:* Delivers precise, high-dose radiation, minimizing damage to surrounding tissues.

3.6.4 Targeted Therapy

Targeted therapy aims to inhibit specific molecular pathways driving cancer progression. Several targeted agents have been explored for GBC treatment.

3.6.4.1 HER2-Targeted Therapy

- *Trastuzumab (Herceptin):* GBC tumors overexpressing HER2 (ERBB2) can be treated with trastuzumab, a monoclonal antibody targeting HER2 receptors.
- *Pertuzumab + Trastuzumab:* This dual HER2 blockade has shown potential in HER2-positive biliary tract cancers.

3.6.4.2 eFGFR Inhibitors

- *Pemigatinib:* FDA-approved for eFGFR2 fusion-positive cholangiocarcinoma but under investigation for GBC with eFGFR alterations.

3.6.4.3 IDH1/IDH2 Inhibitors

- *Ivosidenib:* A selective IDH1 inhibitor showing promise in IDH1-mutant biliary tract cancers.

3.6.4.4 Angiogenesis Inhibitors

- *Bevacizumab (Anti-VEGF Therapy):* Has been studied in combination with chemotherapy for advanced biliary cancers.

3.7 CHALLENGES AND FUTURE DIRECTIONS

GBC is a highly aggressive malignancy of the biliary tract with a poor prognosis due to late diagnosis and limited treatment options. The disease remains asymptomatic in its early stages, leading to a high rate of advanced-stage diagnoses when curative interventions are no longer viable. The absence of standardized screening programs and reliable early detection biomarkers further complicates the clinical management of GBC. Most patients are diagnosed incidentally after cholecystectomy for gallstones or chronic cholecystitis, and by that time, the tumor has often invaded adjacent organs or metastasized.

One of the major challenges in GBC is the lack of effective screening tools. Conventional tumor markers such as CA 19–9 and CEA have limited specificity and sensitivity, making them unreliable for early detection. Although emerging biomarkers, including circulating tumor DNA (ctDNA), exosomal microRNAs (miRNAs), and tumor-associated autoantibodies, are under investigation, their clinical validation and widespread application are still in progress. Imaging modalities such as ultrasound, computed tomography, and magnetic resonance imaging are useful for detecting gallbladder abnormalities, but they often fail to differentiate early malignancy from benign gallbladder diseases. Artificial intelligence (AI)-assisted imaging and radiomics have shown promise in improving early detection, but further research is needed for clinical implementation.

GBC has a high recurrence rate and a propensity for early metastasis, significantly limiting the effectiveness of surgical intervention. Even among patients who undergo curative resection, disease relapse is common, resulting in 5-year survival rates of less than 10–20% for advanced-stage patients. The aggressive nature of the disease is attributed to early vascular and lymphatic invasion, coupled with the presence of mutations in oncogenes such as KRAS, TP53, and PIK3CA.

Furthermore, HER2 amplification, FGFR2 fusions, and IDH1/2 mutations have been identified in subsets of GBC cases, offering potential therapeutic targets, though their clinical utility is still being explored.

Currently, treatment options for GBC remain limited. Surgery is the only curative option, but only a small percentage of patients present with resectable disease. Systemic chemotherapy, typically gemcitabine combined with cisplatin, remains the standard of care for unresectable or metastatic cases, yet it offers only modest survival benefits. The lack of established second-line therapies further complicates disease management, as most patients develop resistance to first-line chemotherapy. Radiation therapy has limited efficacy, and immunotherapy has yet to demonstrate significant benefits in unselected GBC patients. However, ongoing research into immune checkpoint inhibitors such as PD-1/PD-L1 and CTLA-4 inhibitors, particularly in combination with chemotherapy or targeted agents, is showing promise in clinical trials.

The future of GBC treatment lies in precision medicine and personalized therapeutic approaches. Advances in molecular profiling have opened new avenues for targeted therapy, with agents such as HER2 inhibitors (trastuzumab, pertuzumab), FGFR inhibitors (pemigatinib, infigratinib), and IDH1 inhibitors (ivosidenib) undergoing clinical evaluation. The integration of next-generation sequencing in routine clinical practice could help stratify patients based on their genetic profiles, enabling the development of individualized treatment plans. Additionally, novel drug delivery systems, such as nanoparticle-based chemotherapy and epigenetic therapy, are being explored to overcome resistance mechanisms and enhance drug efficacy.

Immunotherapy is another promising avenue for GBC treatment. While immune checkpoint inhibitors have demonstrated limited efficacy as monotherapies, combination strategies that integrate immunotherapy with chemotherapy, radiation, or targeted agents are being investigated. Tumor microenvironment modulation through agents targeting stromal and immune cells may further enhance immunotherapeutic responses. Moreover, personalized immunotherapy approaches, including chimeric antigen receptor T-cell (CAR-T) therapy and cancer vaccines, hold potential, though they remain in the experimental phase for GBC.

To improve patient outcomes, a multimodal treatment strategy that combines surgery, chemotherapy, targeted therapy, and immunotherapy is essential. Neoadjuvant chemotherapy has shown potential in downstaging tumors before surgery, while adjuvant chemoradiotherapy can help reduce recurrence rates. The establishment of clinical guidelines for risk assessment, early diagnosis, and treatment optimization is critical for improving survival rates. Additionally, international collaborative research efforts are necessary to develop standardized treatment protocols and facilitate large-scale clinical trials that evaluate novel therapeutic approaches.

Addressing the socioeconomic and geographic disparities in GBC incidence and treatment is another crucial aspect of future research. The disease disproportionately affects populations in developing countries, where healthcare infrastructure, early detection programs, and access to advanced therapies are often lacking. Public health initiatives aimed at increasing awareness, implementing screening programs for high-risk populations, and improving healthcare accessibility can play a significant role in reducing GBC-related mortality.

In conclusion, gallbladder carcinoma presents significant clinical challenges due to its late diagnosis, high metastatic potential, and limited therapeutic options. However, ongoing research in molecular profiling, biomarker discovery, targeted therapies, and immunotherapy is paving the way for more effective treatment strategies. The integration of precision medicine, multimodal therapies, and global research collaborations will be instrumental in transforming the management of GBC and improving patient survival in the coming years.

3.8 CONCLUSION

GBC remains a highly aggressive malignancy with poor prognosis due to its asymptomatic nature in early stages, leading to late-stage diagnosis and limited treatment options. The disease is driven by a complex interplay of genetic, epigenetic, and molecular alterations that influence tumor initiation, progression, and response to therapy. Recent advances in understanding the genetic blueprint of GBC have shed light on key oncogenic drivers, including mutations in KRAS, TP53, PIK3CA, and ERBB2, as well as epigenetic modifications such as DNA methylation, histone modifications, and non-coding RNA dysregulation. These discoveries have opened new avenues for targeted therapies and biomarker-based early detection strategies.

Despite these advancements, significant challenges persist, particularly in early detection and treatment resistance. The lack of effective screening programs and non-invasive biomarkers limits the ability to diagnose GBC at a curable stage. Additionally, the tumor's aggressive biology, including early vascular and lymphatic invasion, contributes to high recurrence and metastasis rates. The standard treatment approach, comprising surgical resection, chemotherapy, and radiotherapy, remains suboptimal for advanced-stage disease, necessitating the exploration of novel therapeutic interventions.

Targeted therapy and immunotherapy have emerged as promising approaches for GBC. HER2-targeted agents (trastuzumab, pertuzumab), FGFR inhibitors (pemigatinib), and IDH1 inhibitors (ivosidenib) offer potential for patients with specific genetic alterations. Immunotherapeutic strategies, including PD-1/PD-L1 checkpoint inhibitors, CTLA-4 inhibitors, and CAR-T-cell therapy, are under investigation, with combination approaches showing promise in overcoming immune evasion. However, the efficacy of these treatments remains variable, highlighting the need for personalized medicine based on molecular profiling.

Future research should focus on multimodal treatment strategies, integration of AI in diagnostics, and novel drug delivery systems such as nanotechnology-based chemotherapy. Expanding international collaborative efforts and establishing standardized clinical guidelines for risk stratification and treatment optimization will be crucial in improving patient outcomes. Addressing socioeconomic and geographic disparities through public health initiatives and awareness programs will also be essential in reducing GBC-related mortality, particularly in high-incidence regions.

In conclusion, while gallbladder carcinoma continues to pose significant clinical challenges, ongoing genomic and molecular research, precision medicine, and immunotherapeutic advancements hold great promise for improving early detection, treatment efficacy, and overall survival. The integration of biomarker-driven therapies, global research collaborations, and patient-specific treatment approaches will be pivotal in transforming the landscape of GBC management in the coming years.

REFERENCES

1. Meireles-Costa LA, Bresciani CJ, Pérez RO, Bresciani BH, Cecconello I, Habr-Gama A. Lymph node metastases in adenocarcinoma of the gallbladder. *Hepatogastroenterology.* 2010;57(104):600–603.
2. Hundal R, Shaffer EA. Gallbladder cancer: Epidemiology and outcome. *Clin Epidemiol.* 2014;6:99–109.
3. Lazcano-Ponce EC, Miquel JF, Muñoz N, et al. Epidemiology and molecular pathology of gallbladder cancer. *CA Cancer J Clin.* 2001;51(6):349–364.
4. Sung H, Ferlay J, Siegel RL, et al. Global cancer statistics 2020: GLOBOCAN estimates of incidence and mortality worldwide for 36 cancers in 185 countries. *CA Cancer J Clin.* 2021;71(3):209–249.
5. Roa I, de Aretxabala X, Araya JC, Hepp J. Gallbladder cancer in high-risk areas: Chile. *Hepatobiliary Surg Nutr.* 2022;11(2):245–258.
6. Dwivedi AN, Jain S, Dixit R. Gallbladder malignancy: Epidemiology and risk factors. *Clin Epidemiol Glob Health.* 2017;5(4):156–162.
7. Pandey M. Risk factors for gallbladder cancer: A reappraisal. *Eur J Cancer Prev.* 2003;12(1):15–24.
8. Randi G, Franceschi S, La Vecchia C. Gallbladder cancer worldwide: Geographical distribution and risk factors. *Int J Cancer.* 2006;118(7):1591–1602.
9. Lazcano-Ponce EC, Miquel JF, Muñoz N, et al. Epidemiology and molecular pathology of gallbladder cancer. *CA Cancer J Clin.* 2001;51(6):349–364.
10. Roa I, Araya JC, Villaseca M, et al. Gallstones and gallbladder cancer. *HPB (Oxford).* 2006;8(2):164–172.
11. Hundal R, Shaffer EA. Gallbladder cancer: Epidemiology and outcome. *Clin Epidemiol.* 2014;6:99–109.

12. Stephen AE, Berger DL. Carcinoma in the porcelain gallbladder: A relationship revisited. *Surgery*. 2001;129(6):699–703.
13. Scanu AM, Spaapen RM, Bakker JM, et al. Salmonella manipulation of host signaling pathways provokes cellular transformation associated with gallbladder carcinoma. *Cell Host Microbe*. 2015;17(6):763–774.
14. Pandey M. Risk factors for gallbladder cancer: A reappraisal. *Eur J Cancer Prev*. 2003;12(1):15–24.
15. Lin Y, Tamakoshi A, Kawamura T, et al. A prospective cohort study of cigarette smoking and pancreatic cancer in Japan. *Cancer Causes Control*. 2002;13(3):249–254.
16. Javle M, Rashid A, Churi C, et al. Molecular characterization of gallbladder cancer using somatic mutation profiling. *Hum Pathol*. 2014;45(4):701–708.
17. Srivastava K, Srivastava A. Genetic susceptibility, epigenetics and environmental factors in gallbladder cancer. *Asian Pac J Cancer Prev*. 2012;13(1):19–23.
18. Batra Y, Pal S, Dutta U, et al. Gallbladder cancer in India: A dismal picture. *J Gastroenterol Hepatol*. 2005;20(2):309–314.
19. Randi G, Franceschi S, La Vecchia C. Gallbladder cancer worldwide: Geographical distribution and risk factors. *Int J Cancer*. 2006;118(7):1591–1602.
20. D'Hondt M, Lapointe R, Benamira Z, et al. Surgical management of gallbladder cancer: An update. *World J Gastroenterol*. 2019;25(33):4526–4533.
21. Mizuno T, Ebata T, Yokoyama Y, et al. Clinical and surgical implications of microscopic residual tumor in gallbladder cancer. *Ann Surg*. 2020;271(4):747–755.
22. Primrose JN, Fox R, Palmer DH, et al. Adjuvant capecitabine for biliary tract cancer: The BILCAP randomized study. *Lancet Oncol*. 2019;20(5):663–673.
23. Valle J, Wasan H, Palmer DH, et al. Cisplatin plus gemcitabine versus observation in biliary tract cancer. *N Engl J Med*. 2010;362(14):1273–1281.
24. Jain A, Mahmoud N, Patel S, et al. Molecular targeted therapies in biliary tract cancers. *J Gastrointest Oncol*. 2018;9(4):847–855.
25. Javle M, Borad MJ, Curtis KK, et al. HER2/neu-directed therapy for biliary tract cancer. *J Hematol Oncol*. 2015;8:58.
26. Lowery MA, Ptashkin R, Jordan E, et al. Comprehensive molecular profiling of intrahepatic and extrahepatic cholangiocarcinomas: Potential targets for intervention. *Clin Cancer Res*. 2018;24(17):4154–4161.
27. Lamarca A, Barriuso J, McNamara MG, Valle JW. Biliary tract cancer: State of the art and potential utility of circulating tumor DNA. *Ther Adv Med Oncol*. 2020;12:1758835920933039.
28. Valle J, Wasan H, Palmer DH, et al. Cisplatin plus gemcitabine for biliary tract cancer. *N Engl J Med*. 2010;362(14):1273–1281.
29. Bang YJ, Doi T, Braud FD, et al. Safety and efficacy of pembrolizumab in patients with advanced biliary tract cancer: KEYNOTE-158 study. *J Clin Oncol*. 2020;38(15_suppl):4503.
30. Ueno M, Ikeda M, Morizane C, et al. Nivolumab for advanced biliary tract cancer: Updated results from a multicenter, open-label, phase 2 trial. *J Immunother Cancer*. 2020;8(2):e001435.
31. Valle JW, Kelley RK, Nervi B, et al. Biliary tract cancer. *Lancet*. 2021;397(10272):428–444.
32. Ortner ME, Caca K, Bruckner T, et al. Endoscopic stenting with a covered metal stent versus plastic stent for malignant common bile duct obstruction. *Gastroenterology*. 2003;125(3):742–748.
33. Goel G, Sun W. Ramucirumab: Rethinking the anti-VEGF strategy in metastatic colorectal cancer. *Clin Colorectal Cancer*. 2019;18(3):160–171.
34. Churi CR, Shroff R, Wang Y, et al. Mutation profiling in cholangiocarcinoma: Prognostic and therapeutic implications. *PLoS One*. 2014;9(12):e115383.
35. Li M, Zhang Z, Li X, Ye J, Wu X, Tan Z. The prognostic and clinicopathological significance of HER2 expression in gallbladder cancer: A meta-analysis. *World J Surg Oncol*. 2020;18(1):237.
36. Jiao Y, Pawlik TM, Anders RA, et al. Exome sequencing identifies frequent inactivating mutations in BAP1, ARID1A, and PBRM1 in intrahepatic cholangiocarcinomas. *Nat Genet*. 2018;45(12):1470–1473.
37. Jia Y, Xie J, Zhao J, et al. Mechanisms of drug resistance in colorectal cancer liver metastases. *Front Oncol*. 2022;12:883524.
38. Dong L, Lu D, Chen R, Lin Y, Kang X. Role of SMAD4 in cholangiocarcinoma: Molecular mechanisms and potential therapeutic strategies. *Front Oncol*. 2021;11:670940.
39. Lowery MA, Ptashkin R, Jordan E, et al. Comprehensive molecular profiling of intrahepatic and extrahepatic cholangiocarcinomas: Potential targets for intervention. *Clin Cancer Res*. 2017;23(17):4634–4643.
40. Liu X, Wang Z, Yang Q, et al. Comprehensive analysis of APC alterations and its potential therapeutic implications in solid tumors. *Cancer Res*. 2019;79(15):3953–3967.
41. Wang H, Wang Z, Li S, et al. BDNF/TrkB promotes tumor progression via PI3K/AKT, MAPK/ERK, and JAK/STAT3 pathways in gallbladder carcinoma. *Cancer Sci*. 2020;111(7):2258–2272.

42. Pishvaian MJ, Blais EM, Brody JR, et al. A phase 2 basket study of vemurafenib in BRAF V600E-mutated cancers. *Sci Rep*. 2018;8(1):3033.
43. Smith A, Jones B. Epigenetic regulation in gallbladder cancer: DNA methylation and its clinical implications. *Cancer Res*. 2020;78(5):1234–1245.
44. Lee C, Brown D, Wilson T. Histone modifications and gallbladder carcinogenesis. *J Mol Oncol*. 2019;13(3):456–468.
45. Patel R, Gupta S, Zhang Y. Role of non-coding RNAs in gallbladder cancer progression and therapeutic resistance. *Oncol Lett*. 2021;22(1):890–902.
46. Yarden Y, Pines G. The ERBB network: At last, cancer therapy meets systems biology. *Nat Rev Cancer*. 2012;12(8):553–563. doi: 10.1038/nrc3309.
47. Engelman JA. Targeting PI3K signalling in cancer: Opportunities, challenges and limitations. *Nat Rev Cancer*. 2009;9(8):550–562. doi: 10.1038/nrc2664.
48. Roberts PJ, Der CJ. Targeting the Raf-MEK-ERK mitogen-activated protein kinase cascade for the treatment of cancer. *Oncogene*. 2007;26(22):3291–3310. doi: 10.1038/sj.onc.1210422.
49. Sirica AE, Dumur CI, Campbell DJ, Almenara JA, Ogunwobi OO, Dewitt JL. Overexpression of EGFR-related protein in human cholangiocarcinoma: Implications for therapeutic targeting. *Hepatology*. 2009;50(2):512–523. doi: 10.1002/hep.22953.
50. Manning BD, Toker A. AKT/PKB signaling: Navigating the network. *Cell*. 2017;169(3):381–405. doi: 10.1016/j.cell.2017.04.001.
51. Saxton RA, Sabatini DM. mTOR signaling in growth, metabolism, and disease. *Cell*. 2017;168(6):960–976. doi: 10.1016/j.cell.2017.02.004.
52. Samuels Y, Wang Z, Bardelli A, et al. High frequency of mutations of the PIK3CA gene in human cancers. *Science*. 2004;304(5670):554. doi: 10.1126/science.1096502.
53. Lavoie H, Therrien M. Regulation of RAF protein kinases in ERK signalling. *Nat Rev Mol Cell Biol*. 2015;16(5):281–298. doi: 10.1038/nrm3979.
54. McCubrey JA, Steelman LS, Chappell WH, et al. Roles of the RAF/MEK/ERK pathway in cell growth, malignant transformation, and drug resistance. *Biochim Biophys Acta*. 2007;1773(8):1263–1284. doi: 10.1016/j.bbamcr.2006.10.001.
55. Cancer Genome Atlas Network. Comprehensive molecular characterization of human colon and rectal cancer. *Nature*. 2012;487(7407):330–337. doi: 10.1038/nature11252.
56. Clevers H. Wnt/β-catenin signaling in development and disease. *Cell*. 2006;127(3):469–480. doi: 10.1016/j.cell.2006.10.018.
57. Polakis P. Wnt signaling in cancer. *Cold Spring Harb Perspect Biol*. 2012;4(5):a008052. doi: 10.1101/cshperspect.a008052.
58. Nambotin SB, Lefrancois L, Sainsily X, Berthillon P, Kim M, Wands JR. Pharmacological inhibition of β-catenin prevents the onset of liver fibrosis in mice. *Hepatology*. 2012;60(6):2170–2183. doi: 10.1002/hep.27150.
59. Jiang J, Hui CC. Hedgehog signaling in development and cancer. *Dev Cell*. 2008;15(6):801–812. doi: 10.1016/j.devcel.2008.11.010.
60. Peacock CD, Wang Q, Gesell GS, et al. Hedgehog signaling maintains a tumor stem cell compartment in multiple myeloma. *Proc Natl Acad Sci USA*. 2007;104(10):4048–4053. doi: 10.1073/pnas.0611682104.
61. Sicklick JK, Li YX, Jayaraman A, et al. Dysregulation of the Hedgehog pathway in human hepatocarcinogenesis. *Carcinogenesis*. 2006;27(4):748–757. doi: 10.1093/carcin/bgi283.
62. Bray SJ. Notch signaling: A simple pathway becomes complex. *Nat Rev Mol Cell Biol*. 2006;7(9):678–689. doi: 10.1038/nrm2009.
63. Espinoza I, Miele L. Notch inhibitors for cancer treatment. *Pharmacol Ther*. 2013;139(2):95–110. doi: 10.1016/j.pharmthera.2013.03.003.
64. Dotto GP. Notch tumor suppressor function. *Oncogene*. 2008;27(38):5115–5123. doi: 10.1038/onc.2008.230.
65. Massagué J. TGFβ signalling in context. *Nat Rev Mol Cell Biol*. 2012;13(10):616–630. doi: 10.1038/nrm3434.
66. David CJ, Massagué J. Contextual determinants of TGFβ action in development, immunity, and cancer. *Nat Rev Mol Cell Biol*. 2018;19(7):419–435. doi: 10.1038/s41580-018-0007-0.
67. Derynck R, Budi EH. Specificity, versatility, and control of TGFβ family signaling. *Sci Signal*. 2019;12(570):eaav5183. doi: 10.1126/scisignal. aav5183.
68. Karin M. NF-κB as a critical link between inflammation and cancer. *Cold Spring Harb Perspect Biol*. 2009;1(5):a000141. doi: 10.1101/cshperspect. a000141.
69. Ben-Neriah Y, Karin M. Inflammation meets cancer, with NF-κB as the matchmaker. *Nat Immunol*. 2011;12(8):715–723. doi: 10.1038/ni.2060.

70. Greten FR, Grivennikov SI. Inflammation and cancer: Triggers, mechanisms, and consequences. *Immunity.* 2019;51(1):27–41. doi: 10.1016/j.immuni.2019.06.025.
71. Yu FX, Zhao B, Panupinthu N, et al. Regulation of the Hippo-YAP pathway by G-protein-coupled receptor signaling. *Cell.* 2012;150(4):780–791. doi: 10.1016/j.cell.2012.06.037.
72. Zanconato F, Cordenonsi M, Piccolo S. YAP/TAZ at the crossroads of tissue homeostasis and cancer. *Cell.* 2016;166(4):879–892. doi: 10.1016/j.cell.2016.06.046.
73. Harvey KF, Zhang X, Thomas DM. The Hippo pathway and human cancer. *Nat Rev Cancer.* 2013;13(4):246–257. doi: 10.1038/nrc3458.
74. Yu H, Pardoll D, Jove R. STATs in cancer inflammation and immunity: A leading role for STAT3. *Nat Rev Cancer.* 2009;9(11):798–809. doi: 10.1038/nrc2734.
75. Johnson DE, O'Keefe RA, Grandis JR. Targeting the IL-6/JAK/STAT3 signalling axis in cancer. *Nat Rev Clin Oncol.* 2018;15(4):234–248. doi: 10.1038/nrclinonc.2018.8.
76. Lee TK, Poon RTP, Yuen AP, et al. Rac activation is associated with hepatocellular carcinoma metastasis by regulation of microRNA-10b and E-cadherin. *J Hepatol.* 2010;53(5):614–621. doi: 10.1016/j.jhep.2010.07.005.
77. Tewari M, Agarwal A, Mishra RR, Meena RN, Shukla HS. Epigenetic changes in carcinogenesis of gallbladder. *Indian J Surg Oncol.* 2013 Dec;4(4):356–361.
78. Zhang Y, Wang X, Liu C, et al. The role of DNA methylation in gallbladder cancer progression and its potential therapeutic implications. Mol Carcinog. 2019;58(11):1946–1957.
79. Begum S, Brait M, Dasgupta S, et al. Aberrant promoter methylation of p16(INK4A), RASSF1A, and PRSS3 in gallbladder carcinoma. World J Gastroenterol. 2015;21(16):4828–4837.
80. Iyer P, Zekri AR, Hung CW, et al. RASSF1A promoter hypermethylation as an epigenetic biomarker in gallbladder cancer. *Clin Epigenetics.* 2017;9:101.
81. Zhang J, Huang Z, Lu J, et al. MLH1 promoter methylation is associated with microsatellite instability and prognosis in gallbladder cancer. Pathol Res Pract. 2018;214(4):564–570.
82. Jain A, Kumar S, Mahajan S, et al. Global DNA hypomethylation and its association with oncogene activation in gallbladder cancer. *J Transl Med.* 2019;17(1):233.
83. Lin Y, Zhao J, Wang B, et al. LINE-1 hypomethylation is associated with chromosomal instability and poor prognosis in gallbladder cancer. *J Gastroenterol Hepatol.* 2018;33(7):1346–1353.
84. Sun X, Xu J, Li Z, et al. Hypomethylation-mediated overexpression of CTNNB1 enhances Wnt/β-catenin signaling in gallbladder cancer. Oncotarget. 2016;7(41):67241–67251.
85. Wang S, Zhang J, He Z, et al. Role of histone modifications in gallbladder cancer progression: Potential therapeutic targets. Cancer Biol Ther. 2020;21(6):501–510.
86. Liu W, Li Y, Huang X, et al. High levels of H3K27me3 are associated with poor prognosis in gallbladder cancer. Pathol Oncol Res. 2017;23(2):361–367.
87. Zhao L, Tang Y, Xu J, et al. Histone H3 acetylation at lysine 9 enhances oncogene activation in gallbladder cancer. *J Exp Clin Cancer Res.* 2019;38(1):432.
88. Chen S, Wang Y, Zhang X, et al. The role of non-coding RNAs in gallbladder cancer: Clinical implications and therapeutic potential. *Mol Cancer.* 2021;20(1):57.
89. Wang J, Wang Y, Wang S, et al. miR-21 promotes gallbladder cancer progression by targeting PTEN and activating the PI3K/AKT pathway. *BMC Cancer.* 2018;18(1):789.
90. Xiong J, Yu D, Wei N, et al. miR-155 enhances NF-κB signaling and inflammation-driven gallbladder cancer progression. *Cancer Lett.* 2019;445:15–25.
91. Chen Y, Peng C, Dai Y, et al. Downregulation of miR-34a correlates with poor prognosis in gallbladder cancer through the NOTCH and Wnt signaling pathways. Clin Transl Oncol. 2019;21(7):896–904.
92. Huang J, Lin X, Zhang W, et al. The miR-200 family inhibits epithelial-to-mesenchymal transition and metastasis in gallbladder cancer by targeting ZEB1/ZEB2. Oncotarget. 2015;6(1):142–155.
93. Li L, Yu L, Liu X, et al. HOTAIR promotes gallbladder cancer metastasis by modifying chromatin structure and altering gene expression. *Int J Cancer.* 2017;140(9):2123–2134.
94. Zhou Y, Liu S, Zhang L, et al. MALAT1 promotes gallbladder cancer cell proliferation and invasion by modulating epithelial-to-mesenchymal transition. *Cancer Manag Res.* 2018;10:431–445.
95. Zhang H, Wu X, He C, et al. The role of BDNF in cancer progression: A potential therapeutic target for gallbladder cancer. *Front Oncol.* 2020;10:582132.
96. Yang X, Guo Y, Sun Q, et al. BDNF/TrkB signaling promotes gallbladder cancer cell survival, invasion, and metastasis via activation of the PI3K/AKT and MAPK/ERK pathways. *J Exp Clin Cancer Res.* 2021;40(1):17.
97. Kumar A, Agarwal A, Tewari M, et al. BDNF promotes chemoresistance in gallbladder carcinoma by activating autophagy and inhibiting apoptosis via the PI3K/AKT/mTOR signaling pathway. Indian J Surg Oncol. 2013;4(4):454–459.

98. Nakamura H, Arai Y, Totoki Y, et al. Genomic spectra of biliary tract cancer. *Nat Genet*. 2015;47(9):1003–1010. doi: 10.1038/ng.3375.
99. Furukawa T, Tanji E, Xu H, Horii A. Genomic analysis of TP53 gene alterations in gallbladder carcinoma. *Int J Cancer*. 2011;128(4):813–821. doi: 10.1002/ijc.25390.
100. Rashid M, Zadeh LR, Baradaran B, et al. PIK3CA mutations in biliary tract cancers: A systematic review and meta-analysis. *Cancer Biol Med*. 2020;17(4):1021–1031. doi: 10.20892/j.issn.2095-3941.2020.0320.
101. Javle M, Rashid A, Churi C, et al. Molecular characterization of gallbladder cancer using targeted next-generation sequencing. *J Gastrointest Oncol*. 2015;6(3):291–299. doi: 10.3978/j.issn.2078-6891.2015.016.
102. Khan SA, Davidson BR, Goldin R, et al. Guidelines for the diagnosis and treatment of cholangiocarcinoma: An update. *Gut*. 2012;61(12):1657–1669. doi: 10.1136/gutjnl-2011-301748.
103. Kumari N, Das P, Ghoshal UC, et al. Correlation of carcinoembryonic antigen and carbohydrate antigen 19–9 with histopathological characteristics in gallbladder carcinoma. *Indian J Cancer*. 2017;54(1):41–44. doi: 10.4103/ijc.IJC_144_17.
104. Shukla SK, Purohit V, Mehla K, et al. Mucins in gallbladder cancer: Potential diagnosis and therapy targets. *Nat Rev Gastroenterol Hepatol*. 2015;12(3):161–174. doi: 10.1038/nrgastro.2015.6.
105. Okano T, Kondo T, Kakisaka T, et al. Identification of S100A4 as a diagnostic and prognostic biomarker in gallbladder cancer by proteomic analysis. *Dis Markers*. 2018;2018:4308589. doi: 10.1155/2018/4308589.
106. Jiao Y, Pawlik TM, Anders RA, et al. Exome sequencing identifies frequent inactivating mutations in BAP1, ARID1A, and PBRM1 among intrahepatic cholangiocarcinomas. *Nat Genet*. 2013;45(12):1470–1473. doi: 10.1038/ng.2806.
107. Hijioka S, Hosoda W, Matsuo K, et al. Epigenetic biomarkers for early-stage gallbladder cancer: Methylation of APC and RUNX3. *Cancer Sci*. 2010;101(1):47–53. doi: 10.1111/j.1349-7006.2009. 01384.x.
108. Komatsu S, Ichikawa D, Hirajima S, et al. Plasma microRNA profiles: Identification of miR-21 and miR-145 in patients with gallbladder cancer. *J Gastroenterol*. 2014;49(3):521–530. doi: 10.1007/s00535-013-0807-x.
109. Yang J, He X, Huang J, et al. Bile acid metabolites as potential biomarkers for early detection of gallbladder cancer. *J Proteome Res*. 2018;17(10):3335–3345. doi: 10.1021/acs.jproteome.8b00370.
110. Zhang Y, Xu F, Zhang S, et al. Lipidomic profiling reveals significant alterations in lipid metabolism of gallbladder cancer. *J Cancer Res Clin Oncol*. 2020;146(12):3229–3240. doi: 10.1007/s00432-020-03306-4.

4 Genomic Insights in Gallbladder Cancer

Molecular Alterations and Single Nucleotide Polymorphisms Associated with Ancestry

José I. Erices,[1] *Marcela Salgado,*[2] *and Katherine Marcelain*[1]*

[1]Department of Basic and Clinical Oncology, Faculty of Medicine, Universidad de Chile, Santiago, Chile; Center for Cancer Prevention and Control (CECAN); EULAT Eradicate GBC

[2]Department of Medical Technology, Faculty of Medicine, Universidad de Chile, Santiago, Chile; EULAT Eradicate GBC

*Corresponding author: kmarcelain@uchile.cl

4.1 INTRODUCTION

Gallbladder cancer (GBC) is the leading malignant neoplasm of the bile ducts [1]. Although relatively infrequent, GBC is highly lethal because of its late diagnosis and rapid disease progression. Most cases are detected at advanced stages, when symptoms such as abdominal pain, jaundice, and weight loss become evident [1]. The incidence and mortality rates of GBC vary significantly worldwide, with a relatively high incidence in regions such as Latin America and Asia. Countries such as Bolivia, Bangladesh, Chile, and South Korea exhibit the highest incidence and mortality rates, in contrast to Western countries, where these rates are considerably lower (Figure 4.1) [2]. The 5-year survival rate for GBC remains low, primarily due to late detection and the limited effectiveness of available treatments [3]. This highlights the critical need for improved diagnostic methods and more effective therapeutic strategies.

The treatment of GBC typically involves a combination of surgery and chemotherapy. However, most patients are diagnosed at advanced stages, when surgery is no longer a curative option, and chemotherapy is primarily used to control tumor growth and alleviate symptoms [4, 5]. Several factors contribute to the failure of GBC treatment, including late diagnosis due to nonspecific symptoms, chemotherapy resistance, and, notably, the lack of targeted therapies [1, 3–5]. Consequently, in recent years, there has been a growing emphasis on cancer genomics to develop innovative treatments for this malignancy. This approach focuses on creating molecular profiles on the basis of the specific tumor characteristics of each patient, aiming to improve therapeutic outcomes and offer more personalized treatment options.

4.2 THE ROLE OF GENOMICS IN ONCOLOGY: ADVANCES AND APPLICATIONS IN CANCER TREATMENT

Genomics is the area in which the DNA sequences of an organism are studied. In the context of oncologic medicine, oncologic genomics aims to identify alterations in genes that are involved in

DOI: 10.1201/9781003506201-4

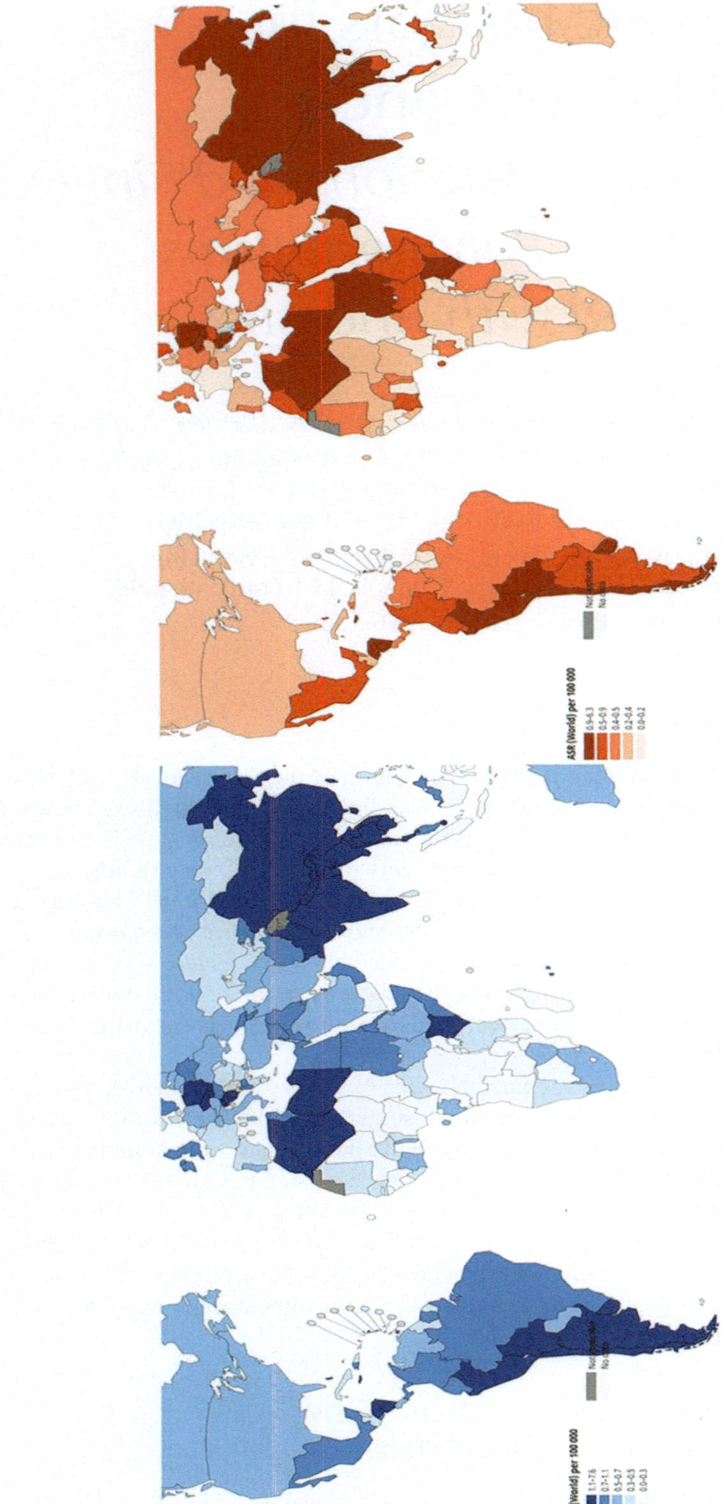

FIGURE 4.1 Global incidence and mortality of GBC. The maps illustrate the incidence and mortality rates of GBC across different regions worldwide. Data extracted from the latest GLOBOCAN database.

the onset and progression of cancer, as well as to identify possible therapeutic targets via a precision medicine approach [6]. These alterations can include mutations, amplifications, deletions, or translocations of genes that induce cell growth and the failure of apoptosis and DNA repair [7]. Genomic analysis allows the identification of specific changes in tumor DNA that can serve as biomarkers for the diagnosis, prognosis, and personalized treatment of cancer.

Current studies in the field of cancer genomics aim to characterize the genomic profiles of tumors, especially those showing significant differences according to their geographical origin, such as GBC [8]. These studies have focused on identifying specific genetic variations in populations from different regions, providing crucial information about the tumorigenesis of the disease. The implementation of next-generation sequencing (NGS) was pivotal in this research. The use of NGS has enabled a detailed and comprehensive analysis of the tumor genome, identifying precise biomarkers and potential therapeutic targets. This has facilitated the development of personalized therapies, which, by taking advantage of the unique mutation profile of the patient's tumor tissue, can lead to new therapeutic options, thereby improving treatment options and clinical outcomes [9].

In recent years, the use of molecular diagnostics as a strategy to define clinical treatment has been quite successful in some types of cancers. Currently, the FDA has recognized 37 molecular alterations to define which drugs to use in certain solid tumors (OncoKB Level 1) [10]. However, there are still some cancers for which this type of treatment has not yet been developed, such as GBC.

4.3 MOLECULAR ALTERATIONS IN GALLBLADDER CANCER

In recent years, significant progress has been made in understanding genetic susceptibility to cancer, largely because of the development of advanced sequencing techniques. These studies include approaches to identify candidate genes, as well as whole-genome analysis (GWAS), allowing for the identification of genetic variants associated with different types of tumors [11, 12]. In the context of GBC, these studies have increased significantly because it is considered a disease influenced by genetic and environmental factors [3]. The occurrence of gallstones and primary sclerosing cholangitis, which cause chronic inflammation, is linked to a greater risk of developing this cancer, especially in certain geographical regions and ethnic groups [13]. Although the field of study focused on GBC is limited because it is generally considered a tumor of low incidence, genetic studies have been developed that have identified variants of interest. These studies have provided detailed information on gene alterations in GBC, with *TP53*, *KRAS*, and *ERBB2* being the most frequently altered genes, as well as alterations in signaling pathways [3]. A bibliographic review by Kuiper and collaborators, in which they compiled studies focused on the identification of molecular alterations in patients diagnosed with GBC, revealed that the *TP53* gene was the most mutated gene, with a frequency of 57%, followed by *SHH* (20%), *ELF3* (19%), *ARID1A* (14%), and *SMAD4* (13%) [3]. In addition, gene copy number alterations have also been reported in GBC; genes such as *CCNE1*, *CDK4*, *ERBB2*, *FRS2*, *KRAS*, *MDM2*, and *MYC* are amplified, whereas the *CDKN2A* and *CDKN2B* genes are deleted [3]. In 2022, Memorial Sloan Kettering Cancer Center (MSK) published a study examining 244 samples (57% primary tumors and 43% metastases) from 233 patients (85% adenocarcinomas, 10% squamous cell carcinomas, and 5% neuroendocrine carcinomas) diagnosed with GBC. Using NGS, 505 cancer-associated genes were explored. Similar to the findings reported in the previously described study, the most common oncogenic molecular alterations were observed in the cell cycle pathways (*TP53* 63% and *CDKN2A 21%*) and RTK_RAS pathways (*ERBB2* 15% and *KRAS* 11%) [14]. Owing to the large number of samples available in this study, the authors were able to determine that certain mutated genes were also distributed according to the histological subtypes of GBC. Specifically, they reported that *RB1* loss-of-function mutations were more common in tumors with neuroendocrine differentiation, specifically small cell neuroendocrine carcinoma. Gain-of-function variants in NOTCH pathway genes are enriched in tumors with squamous morphology [14]. Next, we describe the background published on the variants in the most frequently mutated genes described in GBC.

4.3.1 TP53

The *TP53* gene, which encodes the p53 protein, plays a crucial role in the regulation of the cell cycle and apoptosis. Mutations in *TP53* have been identified with high frequency in various malignant tumors, including GBC, where approximately 50% of patients present alterations in this gene [3]. Although the incidence of GBC varies significantly by geographic location, studies have consistently identified a high frequency of *TP53* mutations in GBC patients across all regions [14, 15]. A study conducted by Li and colleagues identified somatic mutations in GBC across 57 tumor-normal pairs via a combination of exome sequencing and ultradeep sequencing of cancer-related genes. TP53 presented the highest frequency of non-silent mutations, occurring in 47.1% of the samples, making it the most frequently mutated gene in this Chinese cohort. The significant frequency of *TP53* mutations (false discovery rate [FDR] < 0.05) underscores its critical role in the pathogenesis of GBC in this population [16]. An NGS study conducted in India reported that nearly 90% of Indian-origin patients had genetic alterations in *TP53*, making it the gene with the highest mutation rate. The affected exons were 2, 4, 5, 6, 7, and 8, with single nucleotide variants (SNVs) and insertions and deletions (indels) predominating [17]. Similar findings have been reported by Roa et al., where more than 50% of Chilean patients diagnosed with GBC had alterations in exons 5, 6, 7, 8, and 9 of the *TP53* gene [18]. Despite the high frequency of *TP53* mutations described in Indian studies, the overall reported mutation frequencies of *TP53* vary from 8% to 35% [19, 20]. Similar findings were reported by Narayan et al., who explored the genomic profiles of GBC patients from Japan, Chile, and the USA. They reported that, despite variations in the frequency of TP53 alterations among the three cohorts, *TP53* remained the most frequently mutated gene across all groups [21]. This pattern has also been documented in Chile, the country with the highest incidence and mortality rates for gallbladder cancer. Exome analysis of Chilean patients revealed that *TP53* was the most mutated gene, present in 30% of cases [22]. Previously, *TP53* mutations were also observed in Peruvian patients at a frequency of 33.3%, with G-to-A transitions being a common substitution [23]. Moreover, different frequencies of these mutations have been reported in other countries, such as China, Greece, and South Korea.

Mutations in *TP53* are believed to arise early in the pathogenesis of GBC, primarily driven by chronic inflammation resulting from cholelithiasis. This finding was suggested by Moreno et al., who reported that approximately one-third of normal epithelial tissue from patients with cholelithiasis exhibited alterations in the *TP53* gene. These findings support the hypothesis that *TP53* plays a crucial role in the early tumorigenesis of GBC [24]. Additionally, environmental factors have also been described as potential triggers for these alterations. For example, toxins from *Aspergillus* fungi have been associated with GBC through alterations in *TP53* [25], and excessive tobacco consumption, which induces SNVs in *TP53* through benzo[a]pyrene epoxide, has been identified as a possible risk factor for GBC [26, 27]. These factors contribute to the development of molecular anomalies that ultimately lead to GBC.

4.3.2 KRAS

The *KRAS* gene encodes the KRAS protein, which is involved in the RAS/MAPK signaling pathway and the PIK3CA pathway [28]. The KRAS protein has GTPase activity and plays a crucial role in regulating signals through various signaling pathways [28]. Mutation frequency data for KRAS range between 1% and 10%. However, patients from Western countries have a lower mutation frequency than do those from Eastern countries, where the mutation rate for this gene is higher [29]. Mutations in *KRAS* typically occur at specific codons known as "hotspots," with the most common being codons 12, 13, and 61. These mutations lead to constitutive activation of the KRAS protein and are frequent in patients with GBC [30]. In NGS studies of GBC patient samples from Asian populations, variants such as codons 12 and 14 (encoding p.Gly12Arg, p.Gly12Val, and p.Gly13Asp) have been identified as hotspots [21]. Another study sequenced 72 cases of GBC to characterize the somatic variants involved in its pathogenesis and revealed that mutations in *KRAS*, particularly those in codon G12C, were associated with a poor survival prognosis [31]. Similar results were described by Yokoyama et al., who analyzed biopsies from patients collected between 2005 and 2011. They identified *KRAS* alterations in 14% of the cohort, which were significantly correlated with reduced survival compared with those without mutations [32]. While chronic inflammation due to cholelithiasis influences the appearance of genetic alterations in *TP53*, as would be the case in Chilean patients, Japanese patients present alterations in *KRAS*, which has been associated with the pancreaticobiliary ductal junction anomaly. This anomaly has a high incidence in Japan, and few cases have been described outside this country [33–35]. This evidence suggests that GBC risk factors may be involved in the differences in the genomic profiles of these tumors.

4.3.3 CDKN2A

The *CDKN2A* gene, which encodes the proteins p16INK4a and p14ARF, is crucial for cell cycle regulation [36]. This gene frequently exhibits deletions, leading to cell cycle dysregulation and uncontrolled cell proliferation [36]. Alterations in *CDKN2A* have been documented to occur at a frequency of 10% to 22%, depending on the studied population. Studies conducted on GBC patients in Chile compared the mutation frequency of *CDKN2A* with those of populations from the USA and Japan. The results revealed that this gene has a similar mutation frequency across the three populations: Chile (19%), Japan (18.2%), and the USA (12.2%). Furthermore, patients with alterations in *CDKN2A* presented lower survival rates than did those without mutations in this gene [21]. *CDKN2A* alterations were observed with similar frequencies in the Chinese and US cohorts of GBC patients, with 26% of the Chinese patients and 25% of the US patients exhibiting these mutations. However, the co-occurrence patterns of *CDKN2A* with other genetic alterations and its clinical implications vary between the two populations. In the Chinese cohort, *CDKN2A* mutations frequently co-occurred with *ERBB2* aberrations, which were significantly associated with better pathological tumor differentiation. This co-occurrence was less pronounced in the US cohort, where *CDKN2A* mutations were more commonly found alongside alterations in *SMAD4* and *ARID1A* [37]. Another study conducted by Lin et al. on Chinese GBC patients explored mutations and reported that *CDKN2A* was the second most frequently mutated gene (25%), surpassed only by *TP53* (73%). Only 10% of the patients showed a loss of gene expression; however, in these patients, the loss of *CDKN2A* expression did not correlate with a worse prognosis [38]. Similar results were published by Abdel et al., who evaluated the genomic profile of 760 GBC patients and reported that *CDKN2A* was the second most frequently altered gene. Additionally, it was the gene with the highest number of actionable variants [39].

4.3.4 ERBB2

The *ERBB2* gene, also known as HER2, is part of the epidermal growth factor receptor family. The overexpression or amplification of *ERBB2* has been observed in various cancers, including GBC [40]. The frequency of *ERBB2* alterations in GBC patients is 14%, including 8% with *ERBB2* gene amplification, 4.2% with *ERBB2* mutations, 1.5% with concurrent amplification and mutation, and 0.4% with *ERBB2* fusions [41]. Next-generation sequencing data from 517 patients with biliary tract cancer (313 intrahepatic, 93 extrahepatic, and 111 gallbladder cancers) obtained by Mondaca et al. revealed that 5.4% of patients had alterations in *ERBB2*, with a significantly greater prevalence in GBC (12.6%) than in intrahepatic cholangiocarcinomas (2.2%) and extrahepatic cholangiocarcinomas (7.5%). These alterations included amplifications and mutations, with mutations in the extracellular domain, particularly at codon S310, being the most frequent [41]. An NGS study in Chinese patients diagnosed with GBC revealed that *ERBB2* had a mutation frequency of 18%, with amplifications being the most common among the patients. However, these alterations do not correlate with increased disease aggressiveness [38]. In contrast, Li et al. conducted whole-exome sequencing in 157 gallbladder cancer patients and revealed that approximately 7 to 8% exhibited alterations in *ERBB2*. These alterations are notably linked to a poorer prognosis in terms of survival [42]. Yoshida et al., who evaluated HER2 levels in 211 GBC tumors from Japanese patients, reported that 16.6% of the tumors were globally positive for HER2 when considering those amplified with both IHC3+ and IHC2+/FISH. However, these patients did not show a significant relationship between HER2 status and clinicopathological variables or survival [43]. In general, the ErbB signaling pathway (which includes EGFR, ERBB2, ERBB3, ERBB4, and their downstream genes) is reported as the most frequently mutated pathway in Chinese GBC patients, affecting 36.8% of patients. Multivariate analyses further revealed that patients with mutations in the ErbB pathway are associated with a poorer prognosis (P = 0.001) [16].

Additionally, a study focused on clinical–genomic analysis in Chilean patients diagnosed with GBC by Mondaca et al. compared the profile obtained by NGS with the MSK-Integrated Mutation Profiling of Actionable Cancer Targets platform of 23 samples from Chilean patients versus 260 patients recruited from the Memorial Sloan Kettering Cancer Center (MSK). In this analysis, the research group demonstrated that both Chilean and MSK-recruited patients had a similar prevalence of ERBB2 alterations (15% versus 9%; P = 0.42). Overall, the alterations present in Chilean patients included 8% ERBB2 amplification, 4% ERBB2 mutation, 1.5% concurrent amplification and mutation, and 0.4% ERBB2 fusion. Additionally, patients with ERBB2 alterations had a better survival rate than their wild-type counterparts did (22.3 months versus 11.8 months; P = 0.24), suggesting that these patients could benefit from therapies directed toward the human epidermal growth factor receptor 2 [44].

4.3.5 ARID1A

The *ARID1A* (AT-rich interaction domain 1A) gene encodes a subunit of the SWI/SNF chromatin remodeling complex, which plays a crucial role in gene expression regulation and chromatin organization [45]. Mutations in *ARID1A* have been identified in various types of cancer, including GBC [46]. These mutations lead to loss of *ARID1A* function, which disrupts the transcription of genes involved in cell proliferation, apoptosis, and DNA repair [47]. Genomic studies in GBC patients have revealed that approximately 20% of patients show alterations affecting *ARID1A* expression, which are primarily categorized as loss-of-function mutations [14]. This has been reported in several studies; for example, Ross et al. reported that GBC has a higher frequency of *ARID1A* mutations than intrahepatic and extrahepatic bile duct carcinomas do [48]. In addition, sequencing studies performed in Japanese patients diagnosed with GBC have revealed that *ARID1A* alterations are correlated with a worse survival prognosis than those in patients who do not have these alterations [49] and is considered a possible poor prognostic factor. Similarly, NGS studies conducted on GBC patients in Chile have identified *ARID1A* as one of the most frequently altered genes, surpassed only by *TP53* and *PIK3CA* [50]. Comparable findings were reported by Narayan et al., who, when comparing mutation frequencies among patients from the USA, Japan, and Chile, noted a higher mutation frequency in *ARID1A* among the latter, unlike Japanese patients who did not exhibit alterations in this gene [21].

4.3.6 PIK3CA

Mutations in the *PIK3CA* gene, which encodes a subunit of phosphatidylinositol-3-kinase (PI3K), have been identified in GBC [14]. These alterations primarily result in gain-of-function mutations, which are known to activate the PI3K/AKT/mTOR pathway, promoting cell survival and growth. The frequency of alterations in *PIK3CA* varies between 9% and 30% in patients diagnosed with GBC [14, 18]. Most mutations in the *PIK3CA* gene cluster in hotspots in exons 9 and 20, which encode the helical and kinase domains of p110alpha and lead to the activation of signaling pathways involved in cell proliferation [51]. In the case of GBC, an NGS study revealed that the E542K and E545K substitutions (both located in exon 9) were specific to GBC patients [52]. Furthermore, GBC patients harboring these mutations have a poorer prognosis than those lacking these variants do [53]. Furthermore, in samples from 130 Chilean patients, 17% exhibited alterations in exons 9 (E542K, E545G, E545K) and 20 (H1047L, H1047R) of the *PIK3CA* gene. No significant differences were observed in the frequency or type of mutations based on sex, age, or histological characteristics. Additionally, mutations were found in both early-stage (22%) and advanced-stage (14.6%) disease. These findings suggest that the PI3K/AKT pathway plays a crucial role in the pathogenesis of GBC [54].

4.3.7 SMAD4

The *SMAD4* gene encodes a protein that acts as a key mediator in the transforming growth factor beta (TGF-beta) signaling pathway [55]. This pathway plays a crucial role in regulating cell growth, differentiation, and apoptosis. Mutations in *SMAD4* are common in various types of cancer, including colorectal cancer, pancreatic ductal carcinoma, and lung cancer [55]. Genomic studies of biliary tract tumors have identified *SMAD4* as one of the genes with the highest mutation frequency, ranging from 6% to 33% in GBC tumors [3]. Genomic studies using whole-exome sequencing of tissue biopsies from Chinese patients diagnosed with GBC between 2012 and 2016 revealed that *SMAD4* is one of the most frequently mutated genes, with an 11% mutation frequency, second only to that of *TP53* [42]. A similar study evaluated the genomic profile of patients diagnosed with gallbladder cancer from Japan, the USA, and Chile. In this study, the frequency of *SMAD4* gene alterations was similar among patients from different countries (38% in Chile, 36% in Japan, and 27% in the USA). In addition, the authors associated those patients with *SMAD4* alterations with a lower survival rate (median, 10 months versus 25 months; P = 0.03) [21]. An NGS study evaluating 167 GBC samples from three geographically distinct regions—South Korea (n = 94), India (n = 64), and Chile (n = 9)—revealed that *SMAD4* exhibited a mutation frequency of 8%. These alterations are characterized primarily by gene copy loss, which is correlated with reduced expression levels in samples displaying this type of alteration [56].

4.4 GENETIC ALTERATIONS WITH THERAPEUTIC POTENTIAL IN GALLBLADDER CANCER

Genetic alterations play crucial roles in the pathogenesis of GBC, providing insights into both the underlying biology of the disease and opportunities for targeted therapies. Genes such as *TP53*, *KRAS*, *PIK3CA*, *ERBB2*, *ARID1A*, *SMAD4*, and *CDKN2A* exhibit significant mutation frequencies across different populations, underscoring their relevance as prognostic biomarkers and potential therapeutic targets [14]. These mutations not only drive cell proliferation and survival by altering key molecular pathways but also emphasize the need for personalized therapeutic strategies to improve the clinical outcomes of GBC patients [57]. This detailed analysis of genetic alterations underscores the importance of exploring new therapies that specifically address the genetic vulnerabilities present in this type of cancer, where effective treatments are currently lacking.

A study published by MSK, which sequenced 244 GBC tissue samples [14], revealed that 35% of patients exhibited alterations classified as actionable in other solid tumor models (therapeutic levels OncoKB 1, 3A, or 3B). The presence of an *LMNA::NTRK1* fusion, a known predictor of responses to entrectinib and larotrectinib (OncoKB level 1), was observed. These two drugs have been approved by the FDA for the treatment of all NTRK fusion-positive tumors via a tumor-agnostic approach [57]. The alteration in *KRAS* (G12C) was another significant finding and has shown promising responses in clinical studies for gastric tumors treated with adagrasib, as well as in metastatic non-small cell lung cancer [58, 59]. A clinical study that enrolled 129 patients (including patients diagnosed with lung and colorectal cancer) with the *KRAS* G12C mutation were treated with sotorasib. This study additionally included one patient with a biliary tract tumor who had a positive response to treatment [60; CodeBreaK100, ClinicalTrials.gov number, NCT03600883]. Recent results from the phase II

Krystal-1 trial revealed a disease control rate of 100% and a response rate of 40% in advanced non-colorectal gastrointestinal cancers with *KRAS* G12C mutations treated with adagrasib [61]. Notably, four out of eight patients (50%) with advanced BTC exhibited a partial response. These findings underscore the potential for evaluating such drugs in GBC patients with similar genetic profiles.

A study published by Bitter et al. revealed through targeted sequencing that 76% of patients with GBC exhibited actionable alterations. Notably, fusions in the *NTRK3* gene (also identified by Giraldo et al. [14]) were among the most prominent, along with alterations in the *PIK3CA*, *ERBB2/3*, and *CDKN2A* genes [62]. The most common variants identified in this study were nonsense types as well as copy number alterations. In addition, actionable alterations were identified in two patients who were candidates for genomic profile-based targeted therapy. In patient PUHG006, a deletion in the CDKN2A gene was detected, and after failure of treatment with GEMOX, the patient received palbociclib. Despite tumor shrinkage, the patient experienced adverse effects and discontinued treatment. In the same study, another patient (PUHG038) presented with a germline mutation in BRCA1 (Q858*) and was also treated with olaparib, with a progression-free survival of 10 weeks [38]. As previously indicated, *ERBB2* alterations, primarily amplifications, have been found in 10–15% of GBC patients, with overexpression detected through immunohistochemistry [63]. A retrospective study analyzing clinicopathological data from 642 GBC patients in the Netherlands revealed that approximately 50% of GBC patients harbor actionable targets [62]. In this study, the authors describe that 10% of the samples present actionable alterations in the ERBB2 gene, which is currently being used as a marker in several anti-HER2 treatments of solid tumors [61, 64]. Currently, there are FDA-approved HER2 inhibitors that can be used to treat solid tumors exhibiting *ERBB2* amplification. Phase II nonrandomized clinical studies have evaluated the combination of trastuzumab and pertuzumab (two anti-HER2 monoclonal antibodies) and have shown a positive response in 23% of BTC patients, including those with GBC [65]. The results from a phase II trial presented at ASCO 2022 suggest that trastuzumab deruxtecan, an anti-HER2 antibody conjugated with a topoisomerase I inhibitor, is effective in BTCs where surgery is no longer curative. The study reported a response rate of 36% and a disease control rate exceeding 80% [66]. On the other hand, the results obtained from a clinical trial in which the efficacy of the combination of gemcitabine–cisplatin with trastuzumab was evaluated in patients diagnosed with HER2-positive GBC showed that those who received the treatment presented a significant improvement in survival. The disease control rate was 80%, with a median progression-free survival (PFS) of 7 months. Additionally, *TP53* mutations were found to be associated with inferior PFS. These findings suggest that the combination of gemcitabine and trastuzumab improves PFS in HER2-positive BTC patients, underscoring the importance of assessing additional mutations such as *TP53* to optimize patient selection for targeted therapies [67]. Additionally, a specific anti-HER2 antibody named zanidatamab has been developed. In a recent phase 2b study, this antibody demonstrated significant clinical benefit with a manageable safety profile for patients with malignant tumors of the biliary tract harboring HER2 amplification [68]. However, the potential use of anti-PD-L1 therapies in patients diagnosed with GBC with ERBB2 mutations has also been evaluated. Experimental assays performed in *in vitro* and *in vivo* models with GBC cell lines have revealed that mutations in ERBB2 and ERBB3 are associated with increased PD-L1 expression, which results in a reduced T lymphocyte-mediated cytotoxic response. However, the administration of ERBB2/ERBB3 inhibitors together with PD-L1 antibodies counteracts this immunosuppressive effect, suggesting that this combination therapy is a new therapeutic strategy for this type of tumor [42]. In cell line studies, *ERBB2* was shown to heterodimerize with *EGFR* to activate the ErbB signaling pathway in GBC cells. Treatment with specific shRNAs targeting *ERBB2* or *EGFR* or the EGFR inhibitor afatinib inhibits tumor characteristics. However, *KRAS* mutations may impede the response to anti-EGFR treatment, underscoring the therapeutic importance of ERBB2. These findings suggest that the presence of mutations in *KRAS* could be used as a potential predictive marker of resistance to anti-EGFR therapies in patients diagnosed with GBC [69].

Inhibitors of the PI3K pathway, including specific inhibitors of the p110alpha subunit (an activator of AKT and downstream of the mammalian target of rapamycin [mTOR] pathway), are currently in development and have demonstrated efficacy in engineered preclinical models of lung cancer harboring activating point mutations in *PIK3CA* [70]. Cellular model studies in which the E545K mutation of the *PIK3CA* gene was overexpressed demonstrated that the use of the selective inhibitor A66 suppresses the growth of GBC cell lines [52]. Despite the availability of FDA-approved drugs targeting other solid tumors with alterations in the PI3K pathway, the high frequency of amplification of this signaling pathway in GBC necessitates the testing of inhibitors such as BKM120, MK2206, and everolimus (inhibitors of PI3K, AKT, and mTOR, respectively) in phase I/II trials. These studies aimed to determine their efficacy as targeted therapies for refractory and advanced GBC [71–73].

Genetic aberrations in DNA repair genes are common in GBC and have been considered for prognostic use in treatment. Abdel-Wahab analyzed 760 GBC biopsies and explored genetic alterations in 19 DNA repair genes. In this study, 86% of patients had at least one alteration in DNA repair genes, which was also associated with a high tumor mutational burden (TMB) and PD-L1 expression, suggesting that these patients could be candidates for immunotherapy [39]. Additionally, clinical trials are being conducted on the basis of molecular testing in patients with BTC, including GBC. In these studies, individuals with high TMB (≥10 mutations/Mb) were considered for treatment with atezolizumab [39]. One of the most frequently mutated genes involved in direct DNA repair was ATM (6%). Variants in this gene are used to evaluate drug response in cancer patients. Currently, a clinical trial is underway to evaluate the efficacy of olaparib in patients diagnosed with metastatic BTC, including those with molecular alterations in DNA repair genes such as *ATM* and *ARID1A* (NCT04042831).

4.5 ANCESTRY-LINKED GENETIC VARIANTS AND THEIR IMPACT ON GALLBLADDER CANCER RISK

The study of ancestry has been extensively performed in genetic association studies. Several reports have linked the ancestry of certain populations with varying risks of developing or not developing specific types of cancer [74]. For example, it has been observed that populations with a higher proportion of Amerindian ancestry have a lower risk of developing breast cancer [75]. Additionally, certain somatic mutations, such as those in *EGFR* and *KRAS* associated with lung cancer, have been correlated with the percentage of Amerindian ancestry in the analyzed cohort [76]. The incidence and mortality rates of GBC vary significantly by geographic region, suggesting the possible existence of genetic factors related to the ancestry of populations in these areas, which could influence susceptibility to this type of malignancy [77]. Countries such as Chile, India, Bangladesh, and Japan have some of the highest incidence and mortality rates for GBC. In Chile, for example, the distribution of mortality rates varies across different geographic regions, which are also distinguished by the type of ancestry of the populations inhabiting these areas. It has been reported that in the central-southern region of Chile, where there is a relatively high proportion of Mapuche ancestry, there is a correlation with a relatively high mortality rate from GBC [78]. Specifically, Lorenzo et al. reported that for each 1% increase in Mapuche ancestry, the risk of developing GBC increases by 3.7%. Conversely, in Chile's second-largest indigenous population, Aymara, a 1% increase in this type of ancestry decreases the risk of GBC by 2.7% [79].

This potential relationship between ethnicity and the risk of GBC has prompted research groups to investigate possible genetic variants associated with this pathology, particularly in populations with the highest incidence and mortality rates worldwide. With the objective of identifying genetic determinants of GBC, the group led by Pie-Chang conducted a GWAS involving 41 GBC patients and 866 controls [80]. The SNP rs7504990 in the *DCC* (deleted in colorectal cancer) gene was strongly associated with GBC, with a combined P value of 7.46×10^{-8} and an odds ratio (OR) of 6.95. Reduced *DCC* expression was linked to poorly differentiated histology, increased proliferation, and metastasis (Table 4.1) [80]. A study by Mhatre et al. described the presence of three SNPs (rs1558375, rs17209837, and rs4148808) that were correlated with the risk of GBC. These SNPs are located mainly in chromosomal region 7q21.12, affecting genes that are correlated with the risk of GBC—*ABCB1* and *ABCB4* [81] (Table 4.1). These findings suggest that common genetic variations in these hepatobiliary phospholipid transporter genes play crucial roles in GBC susceptibility [81]. A subsequent study conducted in Chile aimed to investigate whether the three polymorphisms identified in the Indian population could be linked to the increased risk of GBC among Chilean patients. Retrospective genotype data from 255 GBC patients, 2042 controls, and European cohorts were analyzed. Significant associations with risk were observed for all three SNPs in Chilean patients, which is consistent with findings from Indian populations [82] (Table 4.1). Notably, these associations were particularly pronounced in individuals of Chilean descent with high Native American Mapuche ancestry (42%—100%) [82] (Table 4.1). Additionally, a GWAS conducted on 1,095 Chilean Latinos with Mapuche ancestry identified candidate variants that were subsequently validated in an independent cohort of 1,643 individuals. The study confirmed the association of *ABCG8* (rs11887534, $P = 3.24 \times 10^{-8}$, OR = 1.74) with gallstones (previously observed in other populations) and identified a new SNP in *TRAF3* (rs12882491, $P = 1.11 \times 10^{-7}$, OR = 1.40) associated with this condition. Variants in both *ABCG8* and *TRAF3* also increase the risk of GBC. Gene expression analysis revealed a significant decrease in *TRAF3* levels in the gallbladder and duodenal mucosa of patients with gallstones, suggesting that reduced *TRAF3* levels may promote gallbladder inflammation. These findings highlight *ABCG8* and *TRAF3* as key genetic factors in the predisposition to gallstones and GBC among admixed Chilean Latinos (Table 4.1) [83].

A study in northern India assessed five polymorphisms in the *TERT* region (5p15.33) and one in 8q24.21 to predict GBC risk and treatment response in 523 patients undergoing chemoradiotherapy and 274 controls. Significant associations were found: *TERT* rs2736100C > A (OR 0.69; 95% CI: 0.52–0.92, P = 0.01), *CLPTM1L* rs401681C > T (OR 0.59; 95% CI: 0.41–0.85, P = 0.004), and *CASC8* rs6983267 G > T (OR 1.6; 95% CI: 1.2–2.2, P = 0.001) with GBC risk. The *CLPTM1LC* haplotype (rs401681C, rs31489 *TERT T*, rs2853676A, rs2736100 *MIR4457* G, and rs4635969) was significantly associated with poor treatment response (OR 7.5; 95% CI: 1.8–31.5; P = 0.006). Survival analysis revealed worse outcomes in TT genotype carriers of *CASC8* rs6983267 (OR 4.3; 95% CI: 1.1–17.1, P = 0.04) in patients with metastatic GBC. These variants in 5p15.33 and 8q24.21 appear crucial in influencing GBC risk and treatment response among northern Indians (Table 4.1) [84]. SNPs have also been identified in genes involved in DNA repair, which could have implications for an increased risk of GBC. Srivastava et al. conducted a case–control study with 230 GBC patients and controls to explore genetic variations in the *ERCC2* (Asp312Asn [rs1799793] and Lys751Gln [rs13181]), *MSH2* (-118T > C [rs2303425] and IVS1 + 9G > C [rs2303426]), and *OGG1* (Ser326Cys [rs1052133] and 748–15C > G [rs2072668]) genes in North Indian populations. They reported that the *ERCC2* Asp312Asn AA, *MSH2* IVS1 + 9G > C CC, *OGG1* Ser326Cys GG and CG + GG, and *OGG1* 748–15C > G GG and CG + GG genotypes were significantly associated with increased GBC risk. Females carrying the *OGG1* 748–15C > G GG, *OGG1* Ser326Cys GG, and *ERCC2* Asp312Asn genotypes presented increased disease risk. Individuals with more than six variant alleles had a fourfold greater risk of GBC development (Table 4.1) [85].

The latest GLOBOCAN data highlight Bolivia's high incidence and mortality rates of GBC, prompting Sakai et al. to explore potential genetic factors influencing GBC susceptibility [2]. They examined *CYP1A1*, *GSTM1*, *GSTT1*, and *TP53* gene polymorphisms in 32 GBC patients and 86 healthy controls from the Bolivian–Japanese Institute of Gastroenterology [86]. This study revealed that patients diagnosed with GBC have the null GSTM1 genotype and that healthy controls do not (OR 2.35; 95% CI: 1.03–5.37; adjusted ORs up to 3.40). No significant associations were found for the *CYP1A1*, *GSTT1*, or *TP53* polymorphisms. These results suggest that *GSTM1* deletion may confer an increased risk of GBC in Bolivia [86]. Somatic mutations related to *APOBEC3* are prevalent in BTCs (Table 4.1) [87]. Liu et al. investigated the role of *APOBEC3A/3B* polymorphisms in cholangiocarcinoma and GBC development. Polymorphisms in the *APOBEC3A*

TABLE 4.1
Single Nucleotide Polymorphisms Associated with Gallbladder Cancer Risk across the Population

Gene	SNP	OR	Population	Ref
DCC	rs7504990	6.95	Japan	[80]
ABCB4	rs1558375	1.47	India	[81]
ABCB4	rs4148808	1.57		
ABCB1	rs17209837	1.61		
ABCB4	rs1558375	1.70	Chile (>42% Mapuche Ancestry)	[82]
ABCB4	rs4148808	1.95		
ABCB1	rs17209837	1.95		
ABCG8	rs11887534	1.74	Chile	[83]
TRAF3	rs12882491	1.40		
TERT	rs2736100C	0.69	India	[84]
CLPTM1L	rs401681C	0.59		
CASC8	rs6983267G	1.6		
ERCC2	rs1799793	2.1	North India	[85]
MSH2	rs2303426	1.8		
OGG1	rs1052133	2.5		
OGG1	rs2072668	1.8		
GSTM1	Null genotype	2.35	Bolivia	[86]
APO-BEC3B	rs2267401	2.04	China (Shanghai)	[88]

and *APOBEC3B* promoters were genotyped in 3,231 participants via quantitative PCR, and promoter activity was analyzed via a dual-luciferase reporter assay. The rs2267401-G polymorphism in the APOBEC3B promoter decreased CCA risk (AOR 0.69; 95% CI: 0.51–0.94) but increased GBC risk (AOR 2.04; 95% CI: 1.35–3.10). This polymorphism reduced *APOBEC3B* promoter activity in CCA cells but increased it in GBC cells, likely due to *TFAP2A* overexpression in the CCA. Tumor necrosis factor-α (TNF-α) influences APOBEC3B levels by inhibiting *TFAP2A*. The rs12157810-C polymorphism in the *APOBEC3A* promoter reduces CCA and GBC risk (AORs of 0.80 and 0.75, respectively) and increases promoter activity in both cancer types. In vitro studies have shown that *APOBEC3A* overexpression induces apoptosis, unlike *APOBEC3B*, highlighting the impact of transcriptional regulation heterogeneity on cancer evolution (Table 4.1) [88]. Indeed, whole-exome sequencing studies of Chinese patients diagnosed with GBC have revealed that C>T/G>A changes constitute approximately 41% of the identified SNVs [16]. Furthermore, the trinucleotide signature was characterized predominantly by C>T mutations at TCN sites, particularly at TCA sites, which is the optimal motif for APOBEC3B [16, 89, 90]. Its presence in GBC underscores the significant role of APOBEC cytidine deaminase-mediated mutagenesis in the pathogenesis of GBC.

4.6 CONCLUSION

The GBC manifests a diverse spectrum of genetic alterations, primarily driven by mutations in key oncogenes such as *TP53*, *KRAS*, *ERBB2*, *CDKN2A*, *ARID1A*, *PIK3CA*, and *SMAD4*. These mutations play a fundamental role in carcinogenesis and significantly influence clinical outcomes.

While *TP53* mutations are prevalent across various populations, the mutation frequencies of *KRAS* and *ERBB2* exhibit geographic and ethnic variability. Additionally, the identification of SNPs associated with GBC risk in different populations highlights the heterogeneity of this pathology. This vast evidence obtained from studies conducted every year by different research groups around the world improves our understanding of the pathogenesis of GBC, opening the opportunity to develop new therapeutic strategies based mainly on the molecular profiles of patients, which vary significantly according to geographic location. These personalized approaches offer promising avenues to improve the clinical management of GBC patients, who currently lack effective therapeutic options.

REFERENCES

1. Roa JC, García P, Kapoor VK, Maithel SK, Javle M, Koshiol J. Publisher Correction: Gallbladder cancer. *Nat Rev Dis Primers*. 2022;8(1):75. Published 2022 Nov 18. doi:10.1038/s41572-022-00408-z.
2. Bray F, Laversanne M, Sung H, Ferlay J, Siegel RL, Soerjomataram I, Jemal A. Global cancer statistics 2022: GLOBOCAN estimates of incidence and mortality worldwide for 36 cancers in 185 countries. *CA Cancer J Clin*. 2024 May–Jun;74(3):229–263. doi:10.3322/caac.21834. Epub 2024 Apr 4. PMID: 38572751.
3. Kuipers H, de Bitter TJJ, de Boer MT, et al. Gallbladder cancer: Current insights in genetic alterations and their possible therapeutic implications. *Cancers (Basel)*. 2021;13(21):5257. Published 2021 Oct 20. doi:10.3390/cancers13215257.
4. Zhou Y, Yuan K, Yang Y, et al. Gallbladder cancer: Current and future treatment options. *Front Pharmacol*. 2023;14:1183619. Published 2023 May 11. doi:10.3389/fphar.2023.1183619.
5. Pavlidis ET, Galanis IN, Pavlidis TE. New trends in diagnosis and management of gallbladder carcinoma. *World J Gastrointest Oncol*. 2024;16(1):13–29. doi:10.4251/wjgo.v16.i1.13.
6. Chen HZ, Bonneville R, Roychowdhury S. Implementing precision cancer medicine in the genomic era. *Semin Cancer Biol*. 2019;55:16–27. doi:10.1016/j.semcancer.2018.05.009.
7. Chakravarthi BV, Nepal S, Varambally S. Genomic and epigenomic alterations in cancer. *Am J Pathol*. 2016;186(7):1724–1735. doi:10.1016/j.ajpath.2016.02.023.
8. Carrio-Cordo P, Acheson E, Huang Q, Baudis M. Geographic assessment of cancer genome profiling studies. *Database (Oxford)*. 2020;2020:baaa009. doi:10.1093/database/baaa009.
9. Mardis ER. The impact of next-generation sequencing on cancer genomics: From discovery to clinic. *Cold Spring Harb Perspect Med*. 2019;9(9):a036269. doi:10.1101/cshperspect.a036269.
10. Debyani C, Gao J, Phillips S, et al. OncoKB: A precision oncology knowledge base. *JCO Precis Oncol*. 2017;1:1–16. doi:10.1200/PO.17.00011.
11. Stadler ZK, Thom P, Robson ME, et al. Genome-wide association studies of cancer. *J Clin Oncol*. 2010;28(27):4255–4267. doi:10.1200/JCO.2009.25.7816.
12. Sato G, Shirai Y, Namba S, et al. Pancancer and cross-population genome-wide association studies dissect shared genetic backgrounds underlying carcinogenesis. *Nat Commun*. 2023;14:3671. doi: 10.1038/s41467-023-39136-7.
13. Song ST, Shi J, Wang XH, et al. Prevalence and risk factors for gallstone disease: A population-based cross-sectional study. *J Dig Dis*. 2020;21(4):237–245. doi:10.1111/1751-2980.12857.
14. Giraldo NA, Drill E, Satravada BA, et al. Comprehensive molecular characterization of gallbladder carcinoma and potential targets for intervention. *Clin Cancer Res*. 2022;28(24):5359–5367. doi:10.1158/1078-0432.CCR-22-1954.
15. Roessler S, Edeline J, Schirmacher P, Coulouarn C. Integrative genomics highlights opportunities for innovative therapies targeting the tumor microenvironment in gallbladder cancer. *J Hepatol*. 2021;74(5):1018–1020. doi:10.1016/j.jhep.2020.12.024.
16. Li M, Zhang Z, Li X, et al. Whole-exome and targeted gene sequencing of gallbladder carcinoma identifies recurrent mutations in the ErbB pathway. *Nat Genet*. 2014;46(8):872–876. doi:10.1038/ng.3030.
17. Roa I, Melo A, Roa J, et al. P53 gene mutation in gallbladder cancer. *Revista Medica de Chile*. 2000;128(3):251–258.
18. Mishra S, Kumari S, Srivastava P, Pandey A, Shukla S, Husain N. Genomic profiling of gallbladder carcinoma: Targetable mutations and pathways involved. *Pathol Res Pract*. 2022;232:153806. doi:10.1016/j.prp.2022.153806.

19. Yadav S, DE Sarkar N, Kumari N, Krishnani N, Kumar A, Mittal B. Targeted gene sequencing of gallbladder carcinoma identifies high-impact somatic and rare germline mutations. *Cancer Genomics Proteomics*. 2017;14(6):495–506. doi:10.21873/cgp.20059.
20. Nigam P, Misra U, Negi TS, Mittal B, Choudhuri G. Alterations of p53 gene in gallbladder cancer patients of North India. *Trop Gastroenterol*. 2010;31(2):96–100.
21. Narayan RR, Creasy JM, Goldman DA, et al. Regional differences in gallbladder cancer pathogenesis: Insights from a multi-institutional comparison of tumor mutations. *Cancer*. 2019;125(4):575–585. doi:10.1002/cncr.31850.
22. Nepal C, Zhu B, O'Rourke CJ, et al. Integrative molecular characterization of gallbladder cancer reveals microenvironment-associated subtypes. *J Hepatol*. 2021;74(5):1132–1144. doi:10.1016/j.jhep.2020.11.033.
23. Vidaurre T, Casavilca S, Montenegro P, et al. Tumor protein p53 and K-ras gene mutations in peruvian patients with gallbladder cancer. *Asian Pac J Cancer Prev*. 2019;20(1):289–294. doi:10.31557/APJCP.2019.20.1.289.
24. Moreno M, Pimentel F, Gazdar AF, Wistuba II, Miquel JF. TP53 abnormalities are frequent and early events in the sequential pathogenesis of gallbladder carcinoma. *Ann Hepatol*. 2005;4(3):192–199.
25. Koshiol J, Gao YT, Dean M, et al. Association of aflatoxin and gallbladder cancer. *Gastroenterology*. 2017;153(2):488–494.e1. doi:10.1053/j.gastro.2017.04.005.
26. Hainaut P, Pfeifer GP. Somatic TP53 mutations in the era of genome sequencing. *Cold Spring Harb Perspect Med*. 2016;6(11):a026179. doi:10.1101/cshperspect.a026179.
27. Lugo A, Peveri G, Gallus S. Should we consider gallbladder cancer a new smoking-related cancer? A comprehensive meta-analysis focused on dose–response relationships. *Int J Cancer*. 2020;146(12):3304–3311. doi:10.1002/ijc.32681.
28. Kim HJ, Lee HN, Jeong MS, Jang SB. Oncogenic KRAS: Signaling and drug resistance. *Cancers (Basel)*. 2021;13(22):5599. doi:10.3390/cancers13225599.
29. Parikh K, Banna G, Liu SV, et al. Drugging KRAS: Current perspectives and state-of-art review. *J Hematol Oncol*. 2022;15(1):152. doi:10.1186/s13045-022-01375-4.
30. Gilson P, Franczak C, Dubouis L, et al. Evaluation of KRAS, NRAS and BRAF hotspot mutations detection for patients with metastatic colorectal cancer using direct DNA pipetting in a fully automated platform and Next-Generation Sequencing for laboratory workflow optimization. *PLoS One*. 2019;14(7):e0219204. doi:10.1371/journal.pone.0219204.
31. Javle M, Rashid A, Churi C, et al. Molecular characterization of gallbladder cancer using somatic mutation profiling. *Hum Pathol*. 2014;45(4):701–708. doi:10.1016/j.humpath.2013.11.001.
32. Yokoyama M, Ohnishi H, Ohtsuka K, et al. KRAS mutation as a potential prognostic biomarker of biliary tract cancers. *Jpn Clin Med*. 2016;7:33–39. doi:10.4137/JCM.S40549.
33. Hanada K, Tsuchida A, Iwao T, et al. Gene mutations of K-ras in gallbladder mucosae and gallbladder carcinoma with an anomalous junction of the pancreaticobiliary duct. *Am J Gastroenterol*. 1999;94(6):1638–1642. doi:10.1111/j.1572-0241.1999.01155.x.
34. Butte JM, Matsuo K, Gönen M, et al. Gallbladder cancer: Differences in presentation, surgical treatment, and survival in patients treated at centers in three countries. *J Am Coll Surg*. 2011;212(1):50–61. doi:10.1016/j.jamcollsurg.2010.09.009.
35. Kayahara M, Nagakawa T, Nakagawara H, Kitagawa H, Ohta T. Prognostic factors for gallbladder cancer in Japan. *Ann Surg*. 2008;248(5):807–814. doi:10.1097/SLA.0b013e31818a1561.
36. Zhao R, Choi BY, Lee MH, Bode AM, Dong Z. Implications of genetic and epigenetic alterations of CDKN2A (p16(INK4a)) in cancer. *EBioMedicine*. 2016;8:30–39. doi:10.1016/j.ebiom.2016.04.017.
37. Yang P, Javle M, Pang F, et al. Somatic genetic aberrations in gallbladder cancer: Comparison between Chinese and US patients. *Hepatobiliary Surg Nutr*. 2019;8(6):604–614. doi:10.21037/hbsn.2019.04.11.
38. Lin J, Dong K, Bai Y, et al. Precision oncology for gallbladder cancer: Insights from genetic alterations and clinical practice. *Ann Transl Med*. 2019;7(18):467. doi:10.21037/atm.2019.08.67.
39. Abdel-Wahab R, Yap TA, Madison R, et al. Genomic profiling reveals high frequency of DNA repair genetic aberrations in gallbladder cancer. *Sci Rep*. 2020;10(1):22087. doi:10.1038/s41598-020-77939-6.
40. Roa I, de Toro G, Schalper K, de Aretxabala X, Churi C, Javle M. Overexpression of the HER2/neu gene: A new therapeutic possibility for patients with advanced gallbladder cancer. *Gastrointest Cancer Res*. 2014;7(2):42–48.
41. Mondaca S, Razavi P, Xu C, et al. Genomic characterization of ERBB2-driven biliary cancer and a case of response to ado-trastuzumab emtansine. *JCO Precis Oncol*. 2019;3:PO.19.00223. doi:10.1200/PO.19.00223.

42. Li M, Liu F, Zhang F, et al. Genomic ERBB2/ERBB3 mutations promote PD-L1-mediated immune escape in gallbladder cancer: A whole-exome sequencing analysis. *Gut.* 2019;68(6):1024–1033. doi:10.1136/gutjnl-2018-316039.
43. Yoshida H, Shimada K, Kosuge T, Hiraoka N. A significant subgroup of resectable gallbladder cancer patients has an HER2 positive status. *Virchows Arch.* 2016;468(4):431–439. doi:10.1007/s00428-015-1898-1.
44. Mondaca S, Walch H, Sepúlveda S, Schultz N, Muñoz G, Yaqubie A, Macanas P, Pareja C, Garcia P, Chatila W, Nervi B, Li B, Harding JJ, Viviani P, Roa JC, Abou-Alfa GK. Clinical and genomic characterization of ERBB2-altered gallbladder cancer: Exploring differences between an American and a Chilean Cohort. *JCO Glob Oncol.* 2024 Oct;10:e2400090. doi:10.1200/GO.24.00090.
45. Mullen J, Kato S, Sicklick JK, Kurzrock R. Targeting ARID1A mutations in cancer. *Cancer Treat Rev.* 2021;100:102287. doi:10.1016/j.ctrv.2021.102287.
46. Conci S, Ruzzenente A, Simbolo M, et al. Multigene mutational profiling of biliary tract cancer is related to the pattern of recurrence in surgically resected patients. *Updates Surg.* 2020;72(1):119–128. doi:10.1007/s13304-020-00718-5.
47. Sasaki M, Nitta T, Sato Y, Nakanuma Y. Loss of ARID1A expression presents a novel pathway of carcinogenesis in biliary carcinomas. *Am J Clin Pathol.* 2016;145(6):815–825. doi:10.1093/ajcp/aqw071.
48. Ross JS, Wang K, Javle MM, et al. Comprehensive genomic profiling of biliary tract cancers to reveal tumor-specific differences and frequency of clinically relevant genomic alterations. *Journal of Clinical Oncology.* 2015;33(15_suppl):4009. doi:10.1200/jco.2015.33.15_suppl.4009.
49. Wardell CP, Fujita M, Yamada T, et al. Genomic characterization of biliary tract cancers identifies driver genes and predisposing mutations. *J Hepatol.* 2018;68(5):959–969. doi:10.1016/j.jhep.2018.01.009.
50. Salvo M, González-Feliú E, Toro J, et al. Validation of an NGS panel designed for detection of actionable mutations in tumors common in Latin America. *J Pers Med.* 2021;11(9):899. doi:10.3390/jpm11090899.
51. Samuels Y, Diaz LA Jr, Schmidt-Kittler O, et al. Mutant PIK3CA promotes cell growth and invasion of human cancer cells. *Cancer Cell.* 2005;7(6):561–573. doi:10.1016/j.ccr.2005.05.014.
52. Deshpande V, Nduaguba A, Zimmerman SM, et al. Mutational profiling reveals PIK3CA mutations in gallbladder carcinoma. *BMC Cancer.* 2011;11:60. doi:10.1186/1471-2407-11-60.
53. Zhao S, Cao Y, Liu SB, et al. The E545K mutation of PIK3CA promotes gallbladder carcinoma progression through enhanced binding to EGFR. *J Exp Clin Cancer Res.* 2016;35(1):97. doi:10.1186/s13046-016-0370-7.
54. Roa I, Garcia H, Game A, de Toro G, de Aretxabala X, Javle M. Somatic mutations of PI3K in early and advanced gallbladder cancer: Additional options for an orphan cancer. *J Mol Diagn.* 2016;18(3):388–394. doi:10.1016/j.jmoldx.2015.12.003.
55. Wan R, Feng J, Tang L. Consequences of mutations and abnormal expression of SMAD4 in tumors and T cells. *Onco Targets Ther.* 2021;14:2531–2540. doi:10.2147/OTT.S297855.
56. Pandey A, Stawiski EW, Durinck S, et al. Integrated genomic analysis reveals mutated ELF3 as a potential gallbladder cancer vaccine candidate. *Nat Commun.* 2020;11(1):4225. doi:10.1038/s41467-020-17880-4.
57. de Reuver PR, van der Post RS. Clinicopathological and molecular insights into gallbladder cancer. *Cancers (Basel).* 2023;15(10):2728. doi:10.3390/cancers15102728.
58. Rose S. Adagrasib moving ahead in GI cancers. *Cancer Discov.* 2022;12(3):590–591. doi:10.1158/2159-8290.CD-NB2022-0003.
59. Jänne PA, Riely GJ, Gadgeel SM, et al. Adagrasib in non-small cell lung cancer harboring a KRASG12C mutation. *N Engl J Med.* 2022;387(2):120–131. doi:10.1056/NEJMoa2204619.
60. Hong DS, Fakih MG, Strickler JH, et al. KRASG12C Inhibition with Sotorasib in advanced solid tumors. *N Engl J Med.* 2020; 383(13):1207–1217. doi:10.1056/NEJMoa1917239.
61. Bekaii-Saab TS, Spira AI, Yaeger R, et al. KRYSTAL-1: Updated activity and safety of adagrasib (MRTX849) in patients (Pts) with unresectable or metastatic pancreatic cancer (PDAC) and other gastrointestinal (GI) tumors harboring a KRASG12C mutation. *J Clin Oncol.* 2022;40(4_suppl):519. doi:10.1200/jco.2022.40.4_suppl.519.
62. de Bitter TJJ, de Reuver PR, de Savornin Lohman EAJ, et al. Comprehensive clinicopathological and genomic profiling of gallbladder cancer reveals actionable targets in half of patients. *NPJ Precis Oncol.* 2022;6(1):83. doi:10.1038/s41698-022-00327-y.
63. Song X, Hu Y, Li Y, Shao R, Liu F, Liu Y. Overview of current targeted therapy in gallbladder cancer. *Signal Transduct Target Ther.* 2020;5(1):230. doi:10.1038/s41392-020-00324-2.
64. Chakravarty D, Gao J, Phillips SM, et al. OncoKB: A precision oncology knowledge base. *JCO Precis Oncol.* 2017;2017:PO.17.00011. doi:10.1200/PO.17.00011.
65. Javle M, Borad MJ, Azad NS, et al. Pertuzumab and trastuzumab for HER2-positive, metastatic biliary tract cancer (MyPathway): A multicenter, open-label, phase 2a, multiple basket study. *Lancet Oncol.* 2021;22(9):1290–1300. doi:10.1016/S1470-2045(21)00336-3.

66. Ohba A, Morizane C, Ueno M, et al. Multicenter phase II trial of trastuzumab deruxtecan for HER2-positive unresectable or recurrent biliary tract cancer: HERB trial. *Future Oncol.* 2022;18(19):2351–2360. doi:10.2217/fon-2022-0214.
67. Ostwal V, Mandavkar S, Bhargava P, et al. Trastuzumab plus gemcitabine-cisplatin for treatment-naïve human epidermal growth factor receptor 2-positive biliary tract adenocarcinoma: A multicenter, open-label, phase II study (TAB). *J Clin Oncol.* 2024;42(7):800–807. doi:10.1200/JCO.23.01193.
68. Harding JJ, Fan J, Oh DY, et al. Zanidatamab for HER2-amplified, unresectable, locally advanced or metastatic biliary tract cancer (HERIZON-BTC-01): A multicenter, single-arm, phase 2b study. *Lancet Oncol.* 2023;24(7):772–782. doi:10.1016/S1470-2045(23)00242-5.
69. Iyer P, Shrikhande SV, Ranjan M, et al. ERBB2 and KRAS alterations mediate response to EGFR inhibitors in early stage gallbladder cancer. *Int J Cancer.* 2019;144(8):2008–2019. doi:10.1002/ijc.31916.
70. Engelman JA, Chen L, Tan X, et al. Effective use of PI3K and MEK inhibitors to treat mutant Kras G12D and PIK3CA H1047R murine lung cancers. *Nat Med.* 2008;14(12):1351–1356. doi:10.1038/nm.1890.
71. Merla A, Liu KG, Rajdev L. Targeted therapy in biliary tract cancers. *Curr Treat Options Oncol.* 2015;16(10):48. doi:10.1007/s11864-015-0366-0.
72. Jain A, Javle M. Molecular profiling of biliary tract cancer: A target rich disease. *J Gastrointest Oncol.* 2016;7(5):797–803. doi:10.21037/jgo.2016.09.01.
73. Bizama C, García P, Espinoza JA, et al. Targeting specific molecular pathways holds promise for advanced gallbladder cancer therapy. *Cancer Treat Rev.* 2015;41(3):222–234. doi:10.1016/j.ctrv.2015.01.003.
74. Hu J, Xu Z, Ye Z, Li J, Hao Z, Wang Y. The association between single nucleotide polymorphisms and ovarian cancer risk: A systematic review and network meta-analysis. *Cancer Med.* 2023;12(1):541–556. doi:10.1002/cam4.4891.
75. Zollner L, Torres D, Briceno I, et al. Native American ancestry and breast cancer risk in Colombian and Mexican women: Ruling out potential confounding through ancestry-informative markers. *Breast Cancer Res.* 2023;25(1):111. Published 2023 Oct 2. doi:10.1186/s13058-023-01713-5.
76. Dos Santos W, Sobanski T, de Carvalho AC, et al. Mutation profiling of cancer drivers in Brazilian colorectal cancer. *Sci Rep.* 2019;9(1):13687. doi:10.1038/s41598-019-49611-1.
77. Vuthaluru S, Sharma P, Chowdhury S, Are C. Global epidemiological trends and variations in the burden of gallbladder cancer. *J Surg Oncol.* 2023;128(6):980–988. doi:10.1002/jso.27450.
78. Zollner L, Boekstegers F, Barahona Ponce C, et al. Gallbladder cancer risk and indigenous South American mapuche ancestry: Instrumental variable analysis using ancestry-informative markers. *Cancers (Basel).* 2023;15(16):4033. doi:10.3390/cancers15164033.
79. Lorenzo Bermejo J, Boekstegers F, González Silos R, et al. Subtypes of Native American ancestry and leading causes of death: Mapuche ancestry-specific associations with gallbladder cancer risk in Chile. *PLoS Genet.* 2017;13(5):e1006756. doi:10.1371/journal.pgen.1006756.
80. Cha PC, Zembutsu H, Takahashi A, Kubo M, Kamatani N, Nakamura Y. A genome-wide association study identifies SNP in DCC is associated with gallbladder cancer in the Japanese population. *J Hum Genet.* 2012;57(4):235–237. doi:10.1038/jhg.2012.9.
81. Mhatre S, Wang Z, Nagrani R, et al. Common genetic variation and risk of gallbladder cancer in India: A case–control genome-wide association study. *Lancet Oncol.* 2017;18(4):535–544. doi:10.1016/S1470-2045(17)30167-5.
82. Boekstegers F, Marcelain K, Barahona Ponce C, et al. ABCB1/4 gallbladder cancer risk variants identified in India also show strong effects in Chileans. *Cancer Epidemiol.* 2020;65:101643. doi:10.1016/j.canep.2019.101643.
83. Bustos BI, Pérez-Palma E, Buch S, et al. Variants in ABCG8 and TRAF3 genes confer risk for gallstone disease in admixed Latinos with Mapuche Native American ancestry. *Sci Rep.* 2019;9(1):772. doi:10.1038/s41598-018-35852-z.
84. Yadav S, Chandra A, Kumar A, Mittal B. Association of TERT-CLPTM1L and 8q24 common genetic variants with gallbladder cancer susceptibility and prognosis in North Indian population. *Biochem Genet.* 2018;56(4):267–282. doi:10.1007/s10528-018-9843-z.
85. Srivastava K, Srivastava A, Mittal B. Polymorphisms in ERCC2, MSH2, and OGG1 DNA repair genes and gallbladder cancer risk in a population of Northern India. *Cancer.* 2010;116(13):3160–3169. doi:10.1002/cncr.25063.
86. Sakai K, Loza E, Roig GV, et al. CYP1A1, GSTM1, GSTT1 and TP53 polymorphisms and risk of gallbladder cancer in bolivians. *Asian Pac J Cancer Prev.* 2016;17(2):781–784. doi:10.7314/apjcp.2016.17.2.781.
87. Nakamura H, Arai Y, Totoki Y, et al. Genomic spectra of biliary tract cancer. *Nat Genet.* 2015;47(9):1003–1010. doi:10.1038/ng.3375.

88. Liu W, Ji H, Zhao J, et al. Transcriptional repression and apoptosis influence the effect of APOBEC3A/3B functional polymorphisms on biliary tract cancer risk. *Int J Cancer*. 2022;150(11):1825–1837. doi:10.1002/ijc.33930.
89. Kuong KJ, Loeb LA. APOBEC3B mutagenesis in cancer. *Nat Genet*. 2013;45(9):964–965. doi:10.1038/ng.2736.
90. Petljak M, Dananberg A, Chu K, et al. Mechanisms of APOBEC3 mutagenesis in human cancer cells. *Nature*. 2022;607:799–807. doi:10.1038/s41586-022-04972-y.

5 Molecular Genetics Alteration and Its Prognostic Significance in Gallbladder Carcinogenesis

Raviranjan Kumar Gupta,[1] *Ravi Bhushan,*[2]
and Shyam Babu Prasad[1]*
[1]Cancer Genomics and Therapeutics Laboratory, Department of Zoology, School of Life Sciences, Mahatma Gandhi Central University Bihar (MGCUB), India.
[2]Centre for Genetic Disorders, Institute of Science, Banaras Hindu University, Uttar Pradesh, India
*Corresponding author: shyambabuprasad@mgcub.ac.in

5.1 INTRODUCTION

Gallbladder cancer (GBC) is a rare and highly fatal type of malignancy characterized by the uncontrolled growth of cells in the gallbladder [1]. This aggressive cancer typically presents with late-stage symptoms, making early diagnosis challenging and significantly impacting patient prognosis [2]. GBC is a multifactorial disease, including chronic inflammation, environmental influences, dietary changes, and genetic factors [1, 3] (Figure 5.1). Among these, the role of genetic alterations in gallbladder pathogenesis remains less well understood. Research has demonstrated that abnormal gene expression caused by a single mutation can initiate events that ultimately result in malignant transformation [4]. This continuous evolution of mutations is recognized as a fundamental mechanism in cancer biology, specifically within molecular genetics. GBC accounts for nearly 50% of all digestive tract cancers and ranks as the fifth most common digestive tract cancer worldwide [5]. In India, GBC is most prevalent among the northern population, particularly those living near the rural Gangetic basin, where the age-standardized incidence rate is notably higher. This regional variation underscores the geographic and ethnic factors contributing to the disease's prevalence in specific populations [6, 7]. As of 2022, the Indian mortality rate for GBC is 1.1 age-standardized deaths per 100,000 people, with an incidence rate of 1.5 age-standardized cases per 100,000 people [8]. According to GLOBOCAN data from the same year, the total incidence of GBC worldwide is 122,491, while the total death count is 89,055 [8].

Surgical resection is the only potential cure for gallbladder GBC, but it is viable for only a small percentage of patients diagnosed at an early stage [9]. The 5-year survival rate for GBC is approximately 30%, dropping to around 5–8% for those diagnosed with advanced-stage disease [1, 10]. This cancer poses a significant global health challenge, particularly in areas with higher incidence rates, highlighting the need for improved early detection and treatment strategies. Molecular genetics plays a crucial role in understanding GBC's development and progression [11, 12]. However, diagnosis is often delayed due to symptoms that mimic other gastrointestinal issues. Advances in molecular genetics may lead to the development of targeted biomarkers for earlier and more accurate detection, though implementing these tests in clinical practice remains limited [13, 14]. As surgical resection is the only potentially curative treatment, many patients face advanced-stage diagnoses. Research into molecular genetics is vital for identifying genetic mutations and pathways

DOI: 10.1201/9781003506201-5

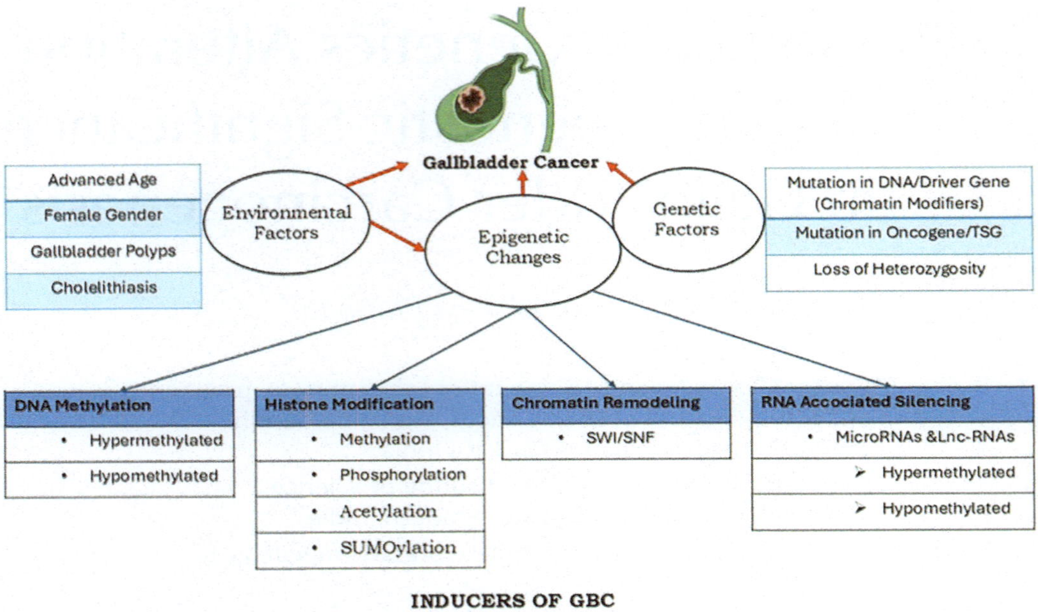

FIGURE 5.1 Inducers of GBC: Showing it's a multifactorial carcinoma.

for novel therapies [15]. Few effective systemic treatments are available for GBC, underscoring the necessity for ongoing research into precision medicine based on genetic profiling. Integrating these genetic insights into clinical protocols holds promise for improving outcomes and personalizing GBC management.

5.2 MOLECULAR MECHANISM OF GBC PROGRESSION

The molecular mechanisms underlying GBC involve a complex interaction of genetic mutations and epigenetic changes that drive the progression of this malignancy. Notable alterations include recurring mutations in oncogenes such as KRAS and PIK3CA and disruptions in key tumor suppressor genes like TP53 and CDKN2A. These genetic abnormalities, along with chromosomal changes, contribute to the aggressive nature of GBC and its resistance to standard therapies (Figure 5.2) [16].

5.2.1 Genetic Alterations

Mutations in the KRAS oncogene, specifically the KRAS (Kirsten rat sarcoma viral oncogene homolog) gene, are part of the mammalian ras gene family. Any single-nucleotide alteration can result in substituting an amino acid, ultimately leading to an activating mutation [17]. Furthermore, mutations in the NRAS oncogene, known as N-RAS (neuroblastoma RAS viral oncogene homolog), represent a proto-oncogene that encodes a membrane protein that is a transport mediator between the Golgi apparatus and the plasma membrane [15, 18]. Notably, mutations in this gene have been associated with various cancers, including rectal cancer and follicular thyroid cancer [19]. The PIK3CA gene, which encodes for the phosphatidylinositol 3-kinase (PI3K), plays a crucial role in cellular growth, survival, proliferation, motility, and morphology [20]. Importantly, KRAS and PIK3CA mutations frequently co-occur in human malignancies, collectively facilitating tumorigenesis through enhanced activation of the PI3K and MAPK signaling pathways. While KRAS

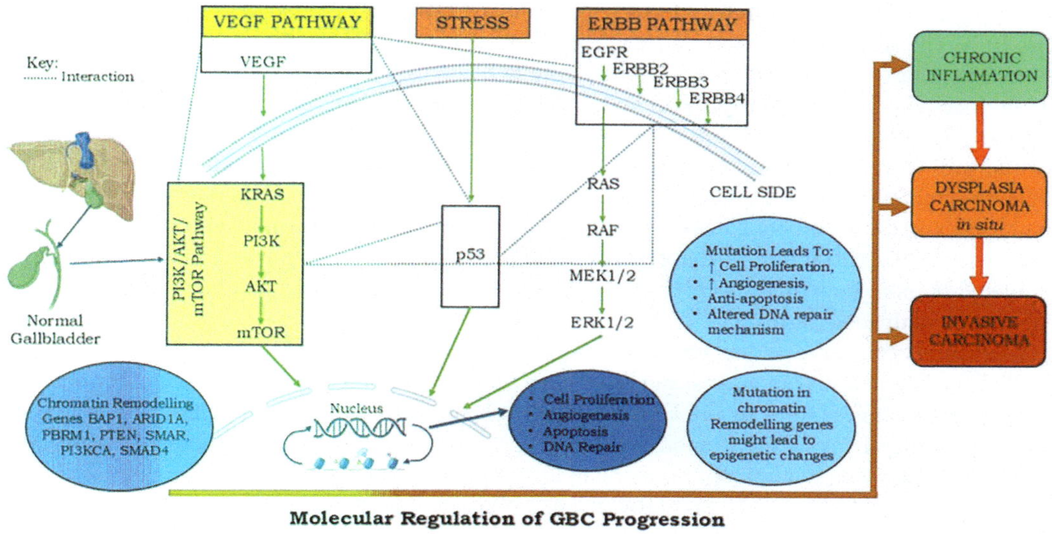

FIGURE 5.2 Molecular pathogenesis of GBC progression: A cascade of signaling which induces GB carcinomas.

mutations alone may initiate lung tumorigenesis, the presence of PIK3CA mutations significantly accelerates tumor formation by influencing cell cycle progression and survival factors [21].

A ligand-bound member of the EGFR family can form homo- or heterodimers to activate the downstream signaling pathways of PI3K–AKT–mTOR or RAS–RAF–MAPK, thereby evading apoptosis and promoting cellular proliferation [22, 23]. Notably, among all members of the EGFR family, HER2 lacks a ligand-binding domain and preferentially partners with other members for heterodimerization, even in the absence of a ligand [24]. The deregulation of the EGFR family receptor tyrosine kinase (RTK) signaling network endows tumor cells with capabilities that sustain malignant behavior and survival, which is frequently observed in GBC as well as breast cancer, lung cancer, pancreatic cancer, head and neck cancer, and colorectal cancer [25, 26]. Mutations in PIK3CA are known to enhance PI3K signaling, stimulate Akt activation, and promote growth factor-independent proliferation and metastasis [17]. Significantly, PIK3CA mutations can also confer resistance to MEK inhibitors in KRAS-mutant tumors by restoring cyclin D1 expression and facilitating cell cycle progression independently of KRAS and MEK/ERK signaling [27]. Research has shown that the somatic cell knock-in of both KRAS G12V and oncogenic PIK3CA mutations in human breast epithelial cells results in cooperative activation of the phosphoinositide 3-kinase (PI3K) and mitogen-activated protein kinase (MAPK) pathways in vitro, leading to tumor formation in immunocompromised mice [28, 29]. Xenografts derived from double-knock-in cells maintain single copies of the mutant KRAS and PIK3CA alleles, indicating that tumor formation does not necessitate an increased copy number of either oncogene; these findings have also been observed in human colorectal cancer specimens [29, 30].

Mechanistically, the cooperativity between mutant KRAS and PIK3CA is partially mediated by the binding of Ras to p110α, as inactivating point mutations within the Ras-binding domain of PIK3CA significantly reduce pathway signaling. Furthermore, the activation of the downstream effector p90RSK by Pdk1 is also augmented by the concurrent presence of mutant KRAS and PIK3CA. These findings provide new insights into the functional role of mutant KRAS in carcinogenesis [31, 32].

TP53 mutations are present in over 50% of GBC cases and are acknowledged as an early and pivotal event in the carcinogenic sequence. The overexpression of p53 has been documented in 56.25% of GBC instances and demonstrates a correlation with increasing tumor grade. Although inactivation of PTEN is infrequent (4.1% of advanced GBC), it is associated with a poor prognosis [33]. The nuclear expression of beta-catenin is evident in 75% of GBC cases and correlates with tumor invasiveness and grade. These findings accentuate the critical role of tumor suppressor gene inactivation, particularly TP53 and CDKN2A, in GBC progression, influencing cell cycle regulation and apoptosis [34, 35]. Overexpression of p53 is observed in 56.25% of GBC patients, while nuclear localization of β-catenin is noted in 75% of cases, a significantly increased prevalence compared to control tissues. P53 expression correlates positively with tumor grade, whereas nuclear expression of beta-catenin correlates with both tumor grade and depth of invasion, suggesting a contributory role of these genes in GBC tumor progression. The inactivation of TP53, mutated in more than 50% of GBC cases, emerges as one of the earliest and most significant carcinogenic pathways [36].

Enhanced cell turnover and oxidative stress contribute to early alterations in TP53, resulting in cell cycle deregulation, apoptosis, and replicative senescence. TP53 mutations are notably prevalent among both Chinese (69.4%) and U.S. (57.9%) GBC patients. Inactivation of CDKN2A may occur through mechanisms such as homozygous deletions, promoter methylation, or point mutations, with deletions being the predominant mechanism [37, 38]. Alterations in CDKN2A are detected in 21% of GBC cases and 26% of Chinese GBC patients. Promoter methylation of CDKN2A is associated with KRAS mutations but exhibits mutual exclusivity with EGFR mutations. Epigenetic silencing through methylation affects CDKN2A and CDKN2B genes in 36.7% and 31.4% of cases, respectively. These inactivation mechanisms correlate closely with the loss of mRNA and protein expression, underscoring their significance in the pathogenesis of GBC [39–41].

5.2.1.1 Chromosomal Instability and Genomic Alterations

Loss of heterozygosity (LOH) represents a prevalent genetic alteration within the cancer genome. This phenomenon can occur through events such as the heterozygous deletion of one of the two alleles, duplication of either a maternal or paternal chromosome or chromosomal region, accompanied by the concurrent loss of the other allele, thereby resulting in LOH [42]. Investigations aimed at detecting loss of heterozygosity in GBC have revealed frequent instances of heterozygous allelic loss across 18 distinct chromosomal regions. Specifically, cytogenetic locations associated with frequent loss of heterozygosity, namely regions 3p, 8p, 9p, and 22q, have been identified in GBC across diverse populations [43]; these regions have also been documented in various other malignancies such as retinoblastoma, melanoma, and squamous cell carcinoma of the larynx. Notably, gallbladder tumors exhibit numerous sites of allelic loss on the short arm of chromosome 3, which encompasses several known or putative tumor suppressor genes. As reported, a significant degree of microsatellite instability (MSI) has been observed in 10% of GBC cases [44]. Additionally, a distinctive pattern of allelic loss has been identified within the Japanese population, indicating that allelotype analysis of gallbladder carcinoma is intriguingly associated with an anomalous junction of the pancreaticobiliary duct [45]. Centrosome abnormalities, which can lead to multipolar spindles and aneuploidy, are frequently encountered in GBC, and their prevalence increases with advancing stages of the disease. DNA ploidy analysis reveals that 69% of GBCs are aneuploid, with a transition from tetraploid to triploid states as tumor progression occurs. Aneuploidy is correlated with higher tumor grade, more advanced disease stages, and reduced survival rates [46].

5.2.2 Epigenetic Modification

Epigenetics is currently understood and defined as the inheritance of gene expression patterns not determined by the nucleotide sequence. Alterations in the established epigenetic patterns generally lead to changes in gene expression, which can cause transcriptional repression. The classic epigenetic modifications include DNA methylation, histone modification (acetylation, methylation, phosphorylation, and so on), and chromatin remodeling [47].

5.2.2.1 DNA Methylation

DNA methylation serves a pivotal function in the processes of gene silencing and the progression of cancer. In healthy cellular environments, methylation patterns are meticulously regulated; however, in cancerous cells, there is a manifestation of anomalous methylation patterns, particularly the hypermethylation of CpG islands that are associated with tumor suppressor genes [48]. This hypermethylation, in conjunction with the activity of histone deacetylases, culminates in the transcriptional silencing of critical genes implicated in cell cycle regulation, signal transduction, and apoptosis. Specifically, in the context of GBC, DNA methylation impacts genes that govern cell adhesion, extracellular matrix degradation, and other vital pathways [49] (Figure 5.3). Aberrant methylation patterns present the potential to function as biomarkers for the early detection, diagnosis, and prognosis of various malignancies, including GBC. Notably, gene silencing may precede the onset of methylation in cancer, indicating a more intricate relationship between these phenomena. The hypermethylation observed in gene promoter regions represents a prevalent epigenetic mechanism for the inactivation of tumor suppressor genes. The methylation status of genes such as p73, MGMT, and DCL1 has been found to be significantly correlated with the survival outcomes of patients with GBC, whereas genes such as CDH13 and FHIT did not exhibit a statistically significant association with patient survival rates [12]. Promoter methylation in specific genes, including CDH1, CDKN2A-p16, REPRIMO (belonging to the tumor suppressor gene family), and UCHL1 (also referred to as PGP9.5), plays a crucial role in the pathogenesis of gallbladder carcinoma [12, 50]. Aberrant methylation may initiate at an early stage in tumor progression and contribute to numerous critical pathway abnormalities in cancer, encompassing loss of cell cycle control, altered transcription factor functionality, modifications in receptor activity, disruption of normal cell-cell and cell-substratum interactions, inactivation of signal transduction pathways, loss of apoptotic signals, and increased genetic instability [51]. The increasingly recognized active role of aberrant methylation in the transcriptional silencing of genes involves a synergistic interaction between methylation processes and histone deacetylase (HDAC) activity. This synergy may be facilitated directly by HDAC interactions with DNA methylating enzymes and via recruitment through complexes containing methyl-cytosine binding proteins [51–53]. From a translational perspective, the altered promoter methylation patterns present substantial

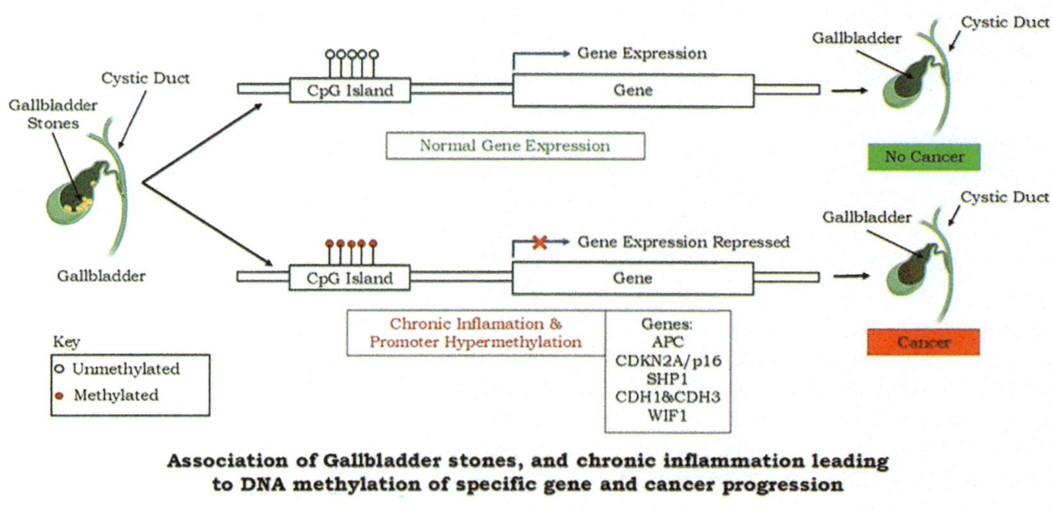

FIGURE 5.3 Association of gallbladder stones and chronic inflammation leading to methylation of GBC-specific genes and progression of cancer.

promise as DNA tumor markers, and their potentially reversible states create avenues for targeted therapeutic strategies aimed at gene reactivation. Notable genes such as TWIST1, HOXA1, SFRP1, PENK, GRIN2B, CDH13, NEUROG1, TMEFF2, TIMP3, MINT2, CCND2, RASSF1A, RUNX3, DLC1, CRABP1, GATA3, MT1G, SEZ6L, SOCS3, THBS1, and BCL2 have exhibited significant methylation in samples of extrahepatic cholangiocarcinoma (EHC) tissue, demonstrating higher levels compared to cholecystitis and cholangitis tissues [50, 54, 55]. Hypermethylation of CDKN2A stands as a primary mechanism inducing the loss of p16 expression. Analyzing p16 and p14 revealed a high methylation rate (between 40% and 60%) in cases of biliary duct carcinoma, GBC, and primary sclerosing cholangitis, whereas normal tissues and choledocholithiasis displayed either no or extremely low methylation rates for these two genes [40]. UCHL1 (also denoted as PGP9.5) is unique among the examined genes, as it has been found to be hypomethylated in the promoter region in cases of GBC, situated on chromosome 4p14, and was originally identified as a member of a gene family whose products hydrolyze small C-terminal adducts of ubiquitin to yield the ubiquitin monomer [56]. The gene product is a peptide that facilitates the removal of ubiquitin from proteins, thereby preventing their degradation by the proteasome. Proteins subjected to degradation through this mechanism play an active role in cell cycle control, including p53 and various cyclins. Ubiquitin, a protein composed of 76 amino acids, is present across all tested eukaryotic cells, and it is integral to the degradation of abnormal and short-lived proteins via ATP- and ubiquitin-dependent proteolytic pathways [57].

5.2.2.2 Histone Modifications

Histone modifications represent critical epigenetic regulators of gene expression and exert a considerable influence on tumorigenesis, particularly within the context of gastrointestinal cancers. Aberrant histone modifications, often designated as "histone onco-modifications," perturb normal cellular signaling pathways and contribute to the pathogenesis of cancer [58]. These modifications typically stem from the dysregulation of histone-modifying enzymes, which can precipitate significant epigenetic, transcriptional, and phenotypic alterations that facilitate tumor progression. Histone modifications are integral to chromatin architecture and play a pivotal role in the regulation of gene expression, thereby influencing cell fate determination and developmental processes [59, 60]. Disruptions in the patterns of these modifications can jeopardize genomic stability and modify gene expression profiles, thereby promoting the development of various diseases, including malignancies.

Histone modifications, which are classified as covalent post-translational modifications, predominantly occur at the N-terminal tails of histone proteins, namely H2A, H2B, H3, and H4. These modifications are catalyzed by enzymatic classes referred to as "writers" (including methyltransferases, acetyltransferases, etc.) and "erasers" (such as demethylases, deacetylases, etc.) [61]. Noteworthy histone modifications that have been extensively characterized in the context of cancer development encompass trimethylation at lysine 4 of histone H3 (H3K4me3) and trimethylation at lysine 36 of histone H3 (H3K36me3), both of which are associated with active transcription. Conversely, modifications such as H3K27me3, H3K9me2/3, and H4K20me3 are correlated with gene repression [62, 63]. Despite the existence of a comprehensive catalog of histone modifications, the functional implications of these changes remain insufficiently understood. Among the most prevalent alterations of histone tails are histone methylation, acetylation, and phosphorylation, alongside an expanding array of additional modifications, including ubiquitination, lactylation, propionylation, crotonylation, and formylation. Histone methylation, which entails the addition of methyl groups to lysine and arginine residues, is predominantly regulated by methyltransferases and demethylases; certain methylation marks, such as H3K4me3, serve to enhance transcription, whereas others, including H3K9me2/3, facilitate the formation of repressive chromatin [64]. Mutations in genes encoding histone-modifying enzymes have the potential to disrupt established methylation patterns, thereby contributing to the processes of oncogenesis and metastasis.

A thorough understanding of histone modifications is crucial for elucidating the mechanisms underlying tumorigenesis and for the development of targeted therapeutic approaches. The processes of histone acetylation and deacetylation involve the dynamic addition or removal of acetyl groups on lysine residues, which significantly impacts cellular functions such as DNA replication and transcription. Alterations in histone acetylation are associated with cancer; specifically, overactive histone deacetylases (HDACs) can promote tumorigenesis, while histone acetyltransferases (HATs), including p300 and CBP, may act as both oncogenes and tumor suppressors [65]. The approval of certain HDAC inhibitors for cancer treatment underscores the necessity for innovative strategies aimed at normalizing acetylation patterns in malignant cells. Furthermore, the regulation of DNA methylation during early developmental stages can be disrupted in cancer, resulting in abnormal methylation states. Notably, hypermethylation of CpG island promoters is typically linked to the silencing of tumor suppressor genes, although it may also be a secondary result of prior gene silencing. This indicates that gene silencing can precede and influence subsequent methylation changes, thereby contributing to chromatin inactivation [66]. Additionally, histone ubiquitination, especially of H2A and H2B, plays a significant role in gene regulation and interacts with other histone modifications, such as H3K27me3, through the Polycomb Repressive Complex 2 (PRC2). E3 ubiquitin ligases like CRL4B and deubiquitinases such as BAP1 regulate these histone marks and transcriptional activity, providing critical insights into the intricate mechanisms of gene regulation in cancer biology [59, 67].

5.2.3 MicroRNA Dysregulation

MicroRNAs (miRNAs), small non-coding RNA molecules approximately 18–25 nucleotides in length, are crucial in post-transcriptional gene regulation and have significant implications in the pathogenesis of GBC [68]. They modulate various biological processes, including cell proliferation, apoptosis, invasion, and migration. Dysregulation of miRNAs in GBC affects key signaling pathways such as Notch, JAK/STAT, AKT, and Hedgehog, where they can function as oncogenes or tumor suppressors [69] (Figure 5.4). A notable example is miR-21, one of the first aberrantly expressed miRNAs identified in GBC through expression profiling in BK5.erbB2 transgenic mice. Treatment with the histone deacetylase inhibitor PCI-24781 significantly downregulated miR-21 [70] and upregulated the tumor-suppressor miR-122, which is typically downregulated in GBC. Similarly, miR-20a, part of the oncogenic miR-17–92 cluster, has been linked to increased local invasion and distant metastasis in GBC, indicating its potential as a therapeutic target [71]. miR-155 is found to be elevated in GBC tissues compared to normal gallbladders, correlating with adverse tumor characteristics and poorer survival outcomes [72]. Furthermore, elevated levels of miR-182 have been associated with worse survival rates. The dysregulation of miRNAs in GBC thus presents promising opportunities for developing novel diagnostic and therapeutic strategies, underscoring their roles as potential biomarkers in GBC management [73].

Recent studies have highlighted the significant role of various microRNAs (miRNAs) in the progression and prognosis of GBC, categorizing them into oncogenic and tumor-suppressive groups based on their influence on cell invasion and tumor behavior. Oncogenic miRNAs, such as miR-182, exhibit elevated levels in metastatic tumors, enhancing cell migration and invasion; their inhibition has been shown to reduce pulmonary metastasis, with CADM1 identified as a target capable of mitigating these invasive effects [74]. Similarly, miR-218–5p is downregulated in GBC, where it inhibits cell invasion and migration by targeting the Bmi1 gene, further complicating its regulation through negative interaction with colon cancer-associated transcript-1 (CCAT1) [75]. Another oncogenic player, miR-130a, demonstrates tumor-suppressive activity but is downregulated in GBC, showing an inverse relationship with the poor prognostic factor HOTAIR; reducing HOTAIR levels leads to decreased invasion and proliferation, highlighting the significant interplay between these molecules [76].

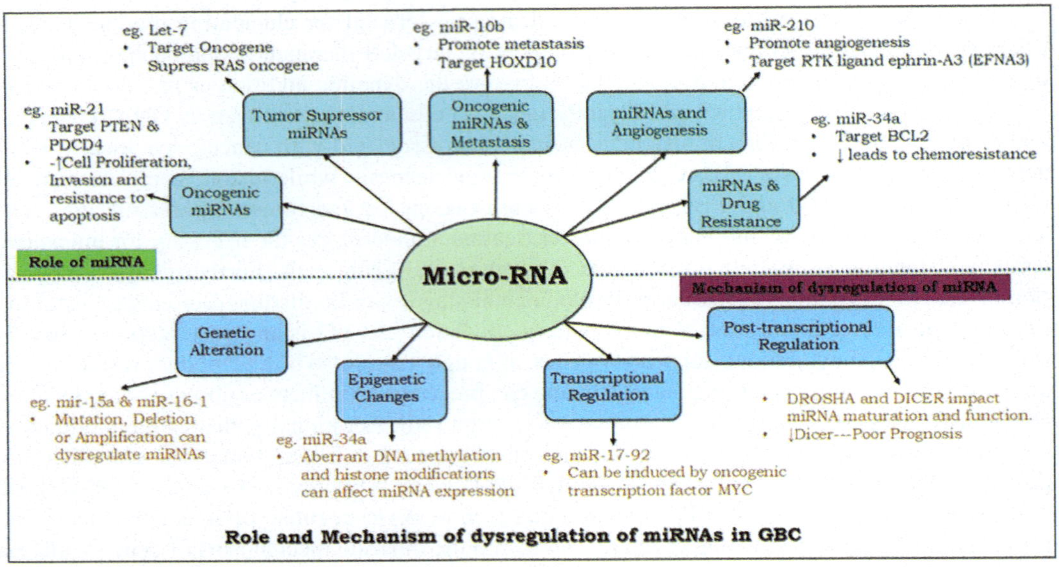

FIGURE 5.4 Role of miRNAs and their dysregulation by various biological cascade in GBC.

Tumor suppressor miRNAs such as miR-335, miR-34a, and miR-135a-5p have been linked to poor patient outcomes in GBC. miR-335 is notably downregulated, with its lower levels associated with poorer histological differentiation, advanced invasion, and worse survival outcomes, establishing it as an independent prognostic factor [77]. Similarly, low levels of miR-34a, which is directly regulated by p53, correlate with poor survival, while overexpression of miR-34a has shown potential to reduce telomere length and inhibit tumor growth [78]. Lastly, miR-135a-5p acts as a confirmed tumor suppressor, with evidence showing its downregulation correlating with unfavorable tumor histology and its overexpression capable of reducing cell proliferation through cell cycle arrest by targeting various genes [76]. Continued exploration of these miRNAs' regulatory networks is essential for further elucidating their roles in GBC and informing future treatment strategies.

5.3 GENETIC PREDISPOSITION IN GBC

5.3.1 Germline Mutations

Germline mutations are genetic alterations that occur in gametes—sperm or eggs—allowing these mutations to be inherited from parents to offspring. Present in every somatic cell from conception, these mutations can significantly elevate the risk of developing certain malignancies, including GBC [15]. Such mutations are foundational to familial cancer syndromes, where multiple family members may show increased susceptibility to cancer due to these inherited genetic changes [79]. The identification of specific germline mutations is critical for understanding the hereditary aspects of GBC, which aids in genetic counseling and can inform strategies for early detection and prevention in at-risk individuals. Deleterious germline mutations have been documented in genes such as ADAM metallopeptidase with thrombospondin type 1 motif 2 (ADAMTS2) [80] and fibroblast growth factor receptor 1 (FGFR1), with ADAMTS2 playing a key role in angiogenesis and tumor development [81]. Additionally, mutations in KMT2D—also known as MLL2, which encodes a lysine-specific methyltransferase—have been implicated in neoplastic cell proliferation [82]. Instances have also been reported where individuals carry double germline hits in the onco-protein coding gene MUC1 [83].

Recent research has highlighted two germline heterozygous likely deleterious mutations within the Fanconi anemia-related gene FANCG and the homologous recombination-associated gene

RAD50, which have been previously linked to the breast cancer phenotype due to their haploinsufficient effect on DNA repair mechanisms [84, 85]. Other mutations include alterations in the low-density lipoprotein receptor-related protein 1B (LRP1B) and two mutations found in the TET1 gene, associated with hypermethylation-dependent breast cancer carcinogenesis [86]. The findings also indicate deleterious mutations in genes with substantial clinical implications, such as ERCC2, FLT3, KMT2C, the phosphatidylinositol-4,5-bisphosphate 3-kinase catalytic subunit beta (PIK3CB), SYNE1, and transglutaminase 7 (TGM7) [81]. Moreover, a germline heterozygous mutation in MRE11A has been characterized alongside double hits in the PRKDC gene, which is crucial for non-homologous end-joining-mediated DNA repair [87]. Mutations in FN1, NOTCH4, KMT2C (related to histone methylation), and NLRP1 (involved in inflammasome suppression) highlight the complex genetic landscape associated with GBC [88].

5.3.2 Genetic Polymorphisms

Genetic polymorphism refers to the presence of two or more genetically distinct forms, or alleles, within a population [89]. This variation can manifest through differences in physical traits, behaviors, or susceptibility to diseases and is primarily a result of mutations in DNA. Polymorphisms may involve single nucleotide changes, known as single nucleotide polymorphisms (SNPs), or larger structural rearrangements of genetic material. The significance of genetic polymorphism lies in its contribution to biodiversity and the adaptive potential of populations, playing a crucial role in the processes of evolution and natural selection [90].

Recent studies have highlighted the potential association between specific genetic polymorphisms and the risk of developing various diseases, including GBC. For instance, the frequency of the C allele of the apoB rs693 polymorphism has been shown to be significantly higher in patients with GBC compared to both gallstone (GS) patients and healthy subjects. Further analyses indicate that the C/T and T/T genotypes are correlated with a reduced risk of GBC when compared to the C/C genotype, as evidenced by odds ratios (ORs) of 0.28 and 0.34, and 95% confidence intervals (CIs) of 0.17–0.46 and 0.13–0.89, respectively [91]. Moreover, polymorphisms in genes such as ERCC2, MSH2, XRCC1, and OGG1 have been associated with GBC susceptibility [92]. Other notable gene variants include those related to PTGS2 and TLR, which also show a link to GBC risk [92]. These findings underscore the importance of genetic variability in understanding the etiology of GBC and may guide future research into targeted preventative and therapeutic strategies.

5.4 FUTURE DIRECTIONS AND CLINICAL IMPLICATIONS

5.4.1 Precision Medicine

The advent of next-generation sequencing (NGS) technologies has markedly improved our comprehension of the molecular genetics associated with GBC. These sophisticated techniques facilitate the detection of specific genetic alterations that contribute to tumorigenesis, thereby paving the way for more personalized treatment strategies, commonly referred to as precision medicine. In the context of GBC, targeting particular mutations and molecular pathways has emerged as a promising therapeutic approach. For example, patients with KRAS mutations may benefit from customized therapies aimed at inhibiting the RAS-RAF-MEK-ERK signaling cascade, while those with alterations in the PIK3CA gene may show responsiveness to therapies that specifically target the PI3K pathway [93]. Moreover, comprehensive genomic analyses have the potential to unveil additional therapeutic targets, including disruptions in DNA repair mechanisms and immune checkpoint molecules. By elucidating the unique molecular profile of each patient's tumor, we can enhance treatment decision-making, leading to more effective and less toxic therapies. This personalized methodology signifies a transformative advancement in the management of GBC and highlights the critical role of molecular genetics in the development of innovative patient care strategies.

5.4.2 Early Detection and Prevention

The early detection of GBC presents significant challenges due to its frequently subtle presentation. However, advancements in molecular diagnostics are providing promising methods for identifying high-risk individuals and premalignant lesions at earlier stages. Germline genetic testing plays a crucial role in identifying individuals with inherited risks, such as those with Lynch syndrome or familial adenomatous polyposis. Moreover, the development of novel biomarkers, including circulating tumor DNA (ctDNA) and microRNAs, holds the potential for enhancing early GBC detection and monitoring disease progression [94]. Research indicates that elevated levels of specific microRNAs—such as miR-21, miR-221, miR-143, miR-335, miR-34a, miR-135a-5p, miR-20a, and miR-155—are correlated with an increased risk of GBC [71, 74, 76, 78]. Additionally, abnormal DNA methylation patterns, particularly in the promoter regions of tumor suppressor genes like p16INK4a and p53, have been associated with the development of GBC [40]. Furthermore, lifestyle modifications—such as adhering to a balanced diet, reducing alcohol consumption, and managing chronic inflammatory conditions like cholecystitis—may help mitigate the risk of developing GBC.

5.4.3 Novel Therapeutic Approaches and Future Direction

Emerging therapeutic approaches for GBC are capitalizing on advancements in molecular targeting and immunotherapy. Research is currently focused on targeted therapies that address specific mutations, such as KRAS and PIK3CA, to inhibit oncogenic pathways. Furthermore, immunotherapy strategies, including immune checkpoint inhibitors, aim to leverage the body's immune response to combat GBC.

Immunotherapy has proven to be a promising strategy for treating GBC. Immune checkpoint inhibitors, including PD-1 and CTLA-4 inhibitors, have demonstrated significant potential across various cancer types, although their effectiveness, specifically in GBC, has been limited [95]. To enhance the impact of immunotherapy, there is an ongoing exploration of combination therapies that integrate immune checkpoint inhibitors with chemotherapy or targeted therapies. Additionally, strategies for improving tumor immunogenicity, such as vaccination and adoptive cell therapy, are under investigation.

Combination therapies represent a forward-looking strategy to improve patient outcomes in GBC. By integrating targeted therapies, chemotherapy, and immunotherapy, clinicians can harness the synergistic effects of these treatments to address resistance mechanisms and enhance overall therapeutic efficacy. For instance, combining therapies like EGFR or HER2 inhibitors with chemotherapy has improved survival rates in specific patient populations [96]. Similarly, combining immunotherapy with chemotherapy or targeted therapy may enhance anti-tumor immune responses, resulting in more effective tumor control and better patient survival rates.

In addition, innovative drug delivery systems have significant potential to enhance the efficacy of GBC treatments while minimizing side effects. Nanocarriers, such as liposomes and polymeric nanoparticles, can be designed to specifically target tumor cells, thereby increasing drug concentration at the tumor site and reducing systemic toxicity [97]. These nanocarriers can also be loaded with multiple therapeutic agents, including chemotherapeutic drugs, targeted therapies, and immunomodulatory agents, to exploit synergistic effects [98]. By addressing the limitations of conventional drug delivery methods, these novel approaches offer promising avenues for improved outcomes in patients with GBC.

REFERENCES

1. Dwivedi, A.N., S. Jain, and R. Dixit, Gall bladder carcinoma: Aggressive malignancy with protean loco-regional and distant spread. *World J Clin Cases*, 2015. 3(3): p. 231–44.
2. Kanthan, R., et al., Gallbladder cancer in the 21st century. *J Oncol*, 2015. 2015: p. 967472.

3. Perez-Moreno, P., et al., Environmental and lifestyle risk factors in the carcinogenesis of gallbladder cancer. *J Pers Med*, 2022. 12(2).
4. Kumar, A., et al., Gallbladder cancer: Progress in the Indian subcontinent. *World J Clin Oncol*, 2024. 15(6): p. 695–716.
5. Rakic, M., et al., Gallbladder cancer. *Hepatobiliary Surg Nutr*, 2014. 3(5): p. 221–6.
6. Dutta, U., et al., Epidemiology of gallbladder cancer in India. *Chin Clin Oncol*, 2019. 8(4): p. 33.
7. Mishra, K., et al., Risk factors for gallbladder cancer development in northern India: A gallstones-matched, case-control study. *Indian J Med Res*, 2021. 154(5): p. 699–706.
8. Cancer, I.A.f.R.o., Globocan 2022, in *Globocan Cancer Dataset 2022*, GLOBOCAN, Editor. 2022.
9. Sun, J., et al., Current status and progress in laparoscopic surgery for gallbladder carcinoma. *World J Gastroenterol*, 2023. 29(16): p. 2369–79.
10. de Savornin Lohman, E., et al., Trends in treatment and survival of gallbladder cancer in the Netherlands; Identifying gaps and opportunities from a nation-wide Cohort. *Cancers (Basel)*, 2020. 12(4).
11. de Reuver, P.R., and R.S. van der Post, Clinicopathological and molecular insights into gallbladder cancer. *Cancers (Basel)*, 2023. 15(10).
12. Sharma, A., et al., Gallbladder cancer epidemiology, pathogenesis and molecular genetics: Recent update. *World J Gastroenterol*, 2017. 23(22): p. 3978–98.
13. Garcia, P., et al., Current and new biomarkers for early detection, prognostic stratification, and management of gallbladder cancer patients. *Cancers (Basel)*, 2020. 12(12).
14. Gupta, R.K., et al., In silico analysis unveiling potential biomarkers in gallbladder carcinogenesis. *Sci Rep*, 2024. 14(1): p. 14570.
15. Mishra, S., et al., Genomic profiling of gallbladder carcinoma: Targetable mutations and pathways involved. *Pathol Res Pract*, 2022. 232: p. 153806.
16. Song, X., et al., Overview of current targeted therapy in gallbladder cancer. *Signal Transduct Target Ther*, 2020. 5(1): p. 230.
17. Salem, M.E., et al., Landscape of KRAS(G12C), associated genomic alterations, and interrelation with immuno-oncology biomarkers in KRAS-mutated cancers. *JCO Precis Oncol*, 2022. 6: p. e2100245.
18. Whaby, M., et al., Inhibition and degradation of NRAS with a pan-NRAS monobody. *Oncogene*, 2024. 43(48): p. 3489–97.
19. Sharma, A., et al., Mutational frequency of KRAS, NRAS, IDH2, PIK3CA, and EGFR in North Indian gallbladder cancer patients. *Ecancermedicalscience*, 2017. 11: p. 757.
20. He, Y., et al., Targeting PI3K/Akt signal transduction for cancer therapy. *Signal Transduct Target Ther*, 2021. 6(1): p. 425.
21. Green, S., C.L. Trejo, and M. McMahon, PIK3CA(H1047R) accelerates and enhances KRAS(G12D)-driven lung tumorigenesis. *Cancer Res*, 2015. 75(24): p. 5378–91.
22. Bahar, M.E., H.J. Kim, and D.R. Kim, Targeting the RAS/RAF/MAPK pathway for cancer therapy: From mechanism to clinical studies. *Signal Transduct Target Ther*, 2023. 8(1): p. 455.
23. Wee, P., and Z. Wang, Epidermal growth factor receptor cell proliferation signaling pathways. *Cancers (Basel)*, 2017. 9(5).
24. Yu, X., et al., Targeting EGFR/HER2 heterodimerization with a novel anti-HER2 domain II/III antibody. *Mol Immunol*, 2017. 87: p. 300–7.
25. Iyer, P., et al., ERBB2 and KRAS alterations mediate response to EGFR inhibitors in early stage gallbladder cancer. *Int J Cancer*, 2019. 144(8): p. 2008–19.
26. Gazdar, A.F., and J.D. Minna, Deregulated EGFR signaling during lung cancer progression: Mutations, amplicons, and autocrine loops. *Cancer Prev Res (Phila)*, 2008. 1(3): p. 156–60.
27. Huang, L., et al., KRAS mutation: From undruggable to druggable in cancer. *Signal Transduct Target Ther*, 2021. 6(1): p. 386.
28. Gustin, J.P., et al., Knockin of mutant PIK3CA activates multiple oncogenic pathways. *Proc Natl Acad Sci USA*, 2009. 106(8): p. 2835–40.
29. Wang, G.M., et al., Single copies of mutant KRAS and mutant PIK3CA cooperate in immortalized human epithelial cells to induce tumor formation. *Cancer Res*, 2013. 73(11): p. 3248–61.
30. Liu, Y., et al., Patient-derived xenograft models in cancer therapy: Technologies and applications. *Signal Transduct Target Ther*, 2023. 8(1): p. 160.
31. Gagliardi, P.A., A. Puliafito, and L. Primo, PDK1: At the crossroad of cancer signaling pathways. *Semin Cancer Biol*, 2018. 48: p. 27–35.
32. Chandarlapaty, S., et al., AKT inhibition relieves feedback suppression of receptor tyrosine kinase expression and activity. *Cancer Cell*, 2011. 19(1): p. 58–71.

33. Kawakami, S., et al., Stepwise correlation of TP53 mutations from pancreaticobiliary maljunction to gallbladder carcinoma: A retrospective study. *BMC Cancer*, 2021. 21(1): p. 1245.
34. Ghosh, M., et al., p53 and beta-catenin expression in gallbladder tissues and correlation with tumor progression in gallbladder cancer. *Saudi J Gastroenterol*, 2013. 19(1): p. 34–9.
35. Trejo-Solis, C., et al., Crosstalk of the Wnt/beta-catenin signaling pathway in the induction of apoptosis on cancer cells. *Pharmaceuticals (Basel)*, 2021. 14(9).
36. Walvir, N.M., et al., p53 Expression in gallbladder lesions: A cross-sectional study from a Tertiary Care Centre in Northern India. *Journal of Clinical and Diagnostic Research*, 2024.
37. Liu, D., and Y. Xu, p53, oxidative stress, and aging. *Antioxid Redox Signal*, 2011. 15(6): p. 1669–78.
38. Kumari, R., and P. Jat, Mechanisms of cellular senescence: Cell cycle arrest and senescence associated secretory phenotype. *Front Cell Dev Biol*, 2021. 9: p. 645593.
39. Kohonen-Corish, M.R., et al., KRAS mutations and CDKN2A promoter methylation show an interactive adverse effect on survival and predict recurrence of rectal cancer. *Int J Cancer*, 2014. 134(12): p. 2820–8.
40. Shima, K., et al., Prognostic significance of CDKN2A (p16) promoter methylation and loss of expression in 902 colorectal cancers: Cohort study and literature review. *Int J Cancer*, 2011. 128(5): p. 1080–94.
41. Soltan, M.A., et al., Cyclin dependent kinase inhibitor 2A genetic and epigenetic alterations interfere with several immune components and predict poor clinical outcome. *Biomedicines*, 2023. 11(8).
42. Pedersen, B.S., et al., Copy neutral loss of heterozygosity is more frequent in older ovarian cancer patients. *Genes Chromosomes Cancer*, 2013. 52(9): p. 794–801.
43. Ye, H., et al., Genomic assessments of the frequent loss of heterozygosity region on 8p21.3-p22 in head and neck squamous cell carcinoma. *Cancer Genet Cytogenet*, 2007. 176(2): p. 100–6.
44. Moy, A.P., et al., Microsatellite instability in gallbladder carcinoma. *Virchows Arch*, 2015. 466(4): p. 393–402.
45. Nakayama, K., et al., Allelotype analysis of gallbladder carcinoma associated with anomalous junction of pancreaticobiliary duct. *Cancer Lett*, 2001. 166(2): p. 135–41.
46. Denu, R.A., et al., Centrosome amplification induces high grade features and is prognostic of worse outcomes in breast cancer. *BMC Cancer*, 2016. 16: p. 47.
47. Hamilton, J.P., Epigenetics: Principles and practice. *Dig Dis*, 2011. 29(2): p. 130–5.
48. Zhu, J., et al., DNA methylation profiles in cancer: Functions, therapy, and beyond. *Cancer Biol Med*, 2023. 21(2): p. 111–6.
49. Cai, C., et al., DNA methylation of RUNX3 promotes the progression of gallbladder cancer through repressing SLC7A11-mediated ferroptosis. *Cell Signal*, 2023. 108: p. 110710.
50. Letelier, P., et al., DNA promoter methylation as a diagnostic and therapeutic biomarker in gallbladder cancer. *Clin Epigenet*, 2012. 4(1): p. 11.
51. Baylin, S.B., et al., Aberrant patterns of DNA methylation, chromatin formation and gene expression in cancer. *Hum Mol Genet*, 2001. 10(7): p. 687–92.
52. Ghoshal, K., et al., Inhibitors of histone deacetylase and DNA methyltransferase synergistically activate the methylated metallothionein I promoter by activating the transcription factor MTF-1 and forming an open chromatin structure. *Mol Cell Biol*, 2002. 22(23): p. 8302–19.
53. Kazanets, A., et al., Epigenetic silencing of tumor suppressor genes: Paradigms, puzzles, and potential. *Biochim Biophys Acta*, 2016. 1865(2): p. 275–88.
54. Shin, S.H., et al., Bile-based detection of extrahepatic cholangiocarcinoma with quantitative DNA methylation markers and its high sensitivity. *J Mol Diagn*, 2012. 14(3): p. 256–63.
55. Tekcham, D.S., and P.K. Tiwari, Epigenetic regulation in gallbladder cancer: Promoter methylation profiling as emergent novel biomarkers. *Asia Pac J Clin Oncol*, 2016. 12(4): p. 332–48.
56. Lee, Y.M., et al., Hypomethylation of the protein gene product 9.5 promoter region in gallbladder cancer and its relationship with clinicopathological features. *Cancer Sci*, 2006. 97(11): p. 1205–10.
57. Chen, D., and Q.P. Dou, The ubiquitin-proteasome system as a prospective molecular target for cancer treatment and prevention. *Curr Protein Pept Sci*, 2010. 11(6): p. 459–70.
58. Georgieva, M., D. Staneva, and G. Miloshev, Epigenetic significance of chromatin organization during cellular aging and organismal lifespan. In *Epigenetics, the Environment, and Children's Health Across Lifespans*. 2016. p. 21–66. Springer International Publishing.
59. Yang, Y., M. Zhang, and Y. Wang, The roles of histone modifications in tumorigenesis and associated inhibitors in cancer therapy. *J Natl Cancer Cent*, 2022. 2(4): p. 277–90.
60. Gu, M., et al., Epigenetic regulation in cancer. *MedComm (2020)*, 2024. 5(2): p. e495.
61. Ohtomo, H., et al., The N-terminal tails of histones H2A and H2B adopt two distinct conformations in the nucleosome with contact and reduced contact to DNA. *J Mol Biol*, 2021. 433(15): p. 167110.

62. He, C., et al., High expression of trimethylated histone H3 lysine 4 is associated with poor prognosis in hepatocellular carcinoma. *Hum Pathol*, 2012. 43(9): p. 1425–35.
63. Naik, A., N. Dalpatraj, and N. Thakur, Global histone H3 lysine 4 trimethylation (H3K4me3) landscape changes in response to TGFbeta. *Epigenet Insights*, 2021. 14: p. 25168657211051755.
64. Black, J.C., C. Van Rechem, and J.R. Whetstine, Histone lysine methylation dynamics: Establishment, regulation, and biological impact. *Mol Cell*, 2012. 48(4): p. 491–507.
65. Xia, J.K., et al., Roles and regulation of histone acetylation in hepatocellular carcinoma. *Front Genet*, 2022. 13: p. 982222.
66. Esteller, M., CpG island hypermethylation and tumor suppressor genes: A booming present, a brighter future. *Oncogene*, 2002. 21(35): p. 5427–40.
67. Meas, R., and P. Mao, Histone ubiquitylation and its roles in transcription and DNA damage response. *DNA Repair (Amst)*, 2015. 36: p. 36–42.
68. Chandra, V., et al., MicroRNA aberrations: An emerging field for gallbladder cancer management. *World J Gastroenterol*, 2016. 22(5): p. 1787–99.
69. Doghish, A.S., et al., The potential role of miRNAs in the pathogenesis of gallbladder cancer—A focus on signaling pathways interplay. *Pathol Res Pract*, 2023. 248: p. 154682.
70. Li, Z., et al., MicroRNA expression and its implications for diagnosis and therapy of gallbladder cancer. *Oncotarget*, 2015. 6(16): p. 13914–21.
71. Chang, Y., et al., MiR-20a triggers metastasis of gallbladder carcinoma. *J Hepatol*, 2013. 59(3): p. 518–27.
72. Kono, H., et al., High expression of microRNA-155 is associated with the aggressive malignant behavior of gallbladder carcinoma. *Oncol Rep*, 2013. 30(1): p. 17–24.
73. Liu, X., et al., Elevated circulating miR-182 acts as a diagnostic biomarker for early colorectal cancer. *Cancer Manag Res*, 2018. 10: p. 857–65.
74. Sachdeva, M., et al., MicroRNA-182 drives metastasis of primary sarcomas by targeting multiple genes. *J Clin Invest*, 2014. 124(10): p. 4305–19.
75. Yang, G., et al., The role of microRNAs in gallbladder cancer. *Mol Clin Oncol*, 2016. 5(1): p. 7–13.
76. Pan, Y., et al., MicroRNA-130a inhibits cell proliferation, invasion and migration in human breast cancer by targeting the RAB5A. *Int J Clin Exp Pathol*, 2015. 8(1): p. 384–93.
77. Zu, Y., et al., Genetic variation in a miR-335 binding site in BIRC5 alters susceptibility to lung cancer in Chinese Han populations. *Biochem Biophys Res Commun*, 2013. 430(2): p. 529–34.
78. Jin, K., et al., miR-34 is associated with poor prognosis of patients with gallbladder cancer through regulating telomere length in tumor stem cells. *Tumour Biol*, 2014. 35(2): p. 1503–10.
79. Gomes, P., et al., Family adjustment to hereditary cancer syndromes: A systematic review. *Int J Environ Res Public Health*, 2022. 19(3).
80. Kesteloot, F., et al., ADAM metallopeptidase with thrombospondin type 1 motif 2 inactivation reduces the extent and stability of carbon tetrachloride-induced hepatic fibrosis in mice. *Hepatology*, 2007. 46(5): p. 1620–31.
81. Yadav, S., et al., Targeted gene sequencing of gallbladder carcinoma identifies high-impact somatic and rare germline mutations. *Cancer Genomics Proteomics*, 2017. 14(6): p. 495–506.
82. Guo, C., et al., KMT2D maintains neoplastic cell proliferation and global histone H3 lysine 4 monomethylation. *Oncotarget*, 2013. 4(11): p. 2144–53.
83. Ghosh, M., et al., MUC 1 core protein as a marker of gallbladder malignancy. *Eur J Surg Oncol*, 2005. 31(8): p. 891–6.
84. Verhagen, C.V.M., et al., Fanconi anemia and homologous recombination gene variants are associated with functional DNA repair defects in vitro and poor outcome in patients with advanced head and neck squamous cell carcinoma. *Oncotarget*, 2018. 9(26): p. 18198–213.
85. Huang, X., et al., Advanced ovarian clear cell carcinoma with RAD50 mutation treated by PARP inhibitor pamiparib combined with anti-angiogenesis therapy: A case report. *Anticancer Drugs*, 2023. 34(2): p. 290–3.
86. Brown, L.C., et al., LRP1B mutations are associated with favorable outcomes to immune checkpoint inhibitors across multiple cancer types. *J Immunother Cancer*, 2021. 9(3).
87. Rahman, S., M. Beikzadeh, and M.P. Latham, Biochemical and structural characterization of analogs of MRE11 breast cancer-associated mutant F237C. *Sci Rep*, 2021. 11(1): p. 7089.
88. Wang, F., et al., Mutations in the notch signalling pathway are associated with enhanced anti-tumour immunity in colorectal cancer. *J Cell Mol Med*, 2020. 24(20): p. 12176–87.
89. Loktionov, A., Common gene polymorphisms, cancer progression and prognosis. *Cancer Lett*, 2004. 208(1): p. 1–33.

90. Karki, R., et al., Defining "mutation" and "polymorphism" in the era of personal genomics. *BMC Med Genom*, 2015. 8: p. 37.
91. Baez, S., et al., Genetic variants involved in gallstone formation and capsaicin metabolism, and the risk of gallbladder cancer in Chilean women. *World J Gastroenterol*, 2010. 16(3): p. 372–8.
92. Srivastava, K., et al., Gallbladder cancer predisposition: A multigenic approach to DNA-repair, apoptotic and inflammatory pathway genes. *PLoS One*, 2011. 6(1): p. e16449.
93. McCubrey, J.A., et al., Mutations and deregulation of Ras/Raf/MEK/ERK and PI3K/PTEN/Akt/mTOR cascades which alter therapy response. *Oncotarget*, 2012. 3(9): p. 954–87.
94. Cheng, F., L. Su, and C. Qian, Circulating tumor DNA: A promising biomarker in the liquid biopsy of cancer. *Oncotarget*, 2016. 7(30): p. 48832–41.
95. Wojtukiewicz, M.Z., et al., Inhibitors of immune checkpoints-PD-1, PD-L1, CTLA-4-new opportunities for cancer patients and a new challenge for internists and general practitioners. *Cancer Metastasis Rev*, 2021. 40(3): p. 949–82.
96. Rebuzzi, S.E., et al., Combination of EGFR-TKIs and chemotherapy in advanced EGFR mutated NSCLC: Review of the literature and future perspectives. *Crit Rev Oncol Hematol*, 2020. 146: p. 102820.
97. Yu, B., et al., Receptor-targeted nanocarriers for therapeutic delivery to cancer. *Mol Membr Biol*, 2010. 27(7): p. 286–98.
98. Sun, L., et al., Smart nanoparticles for cancer therapy. *Signal Transduct Target Ther*, 2023. 8(1): p. 418.

6 Role of Metabolomics in Gallbladder Carcinoma

Vibha Singh,[1] *Adiba Nizam,*[2] *and Anurag Mishra*[1]*
[1]Department of Surgery, Maulana Azad Medical College, New Delhi, India
[2]Department of Surgery, Lady Hardinge Medical College, New Delhi, India
*Corresponding author: anurag.alok@gmail.com

6.1 INTRODUCTION

Gallbladder cancer (GBC) is the most common malignancy of the biliary tract and ranks as the sixth most prevalent gastrointestinal cancer globally [1]. Accounting for 1.2% of all cancers and 1.7% of all cancer-related deaths, GBC is a highly aggressive disease with poor outcomes, especially when it is diagnosed at an advanced stage [2]. Owing to the nonspecific nature of early symptoms, many patients present with late-stage disease, contributing to dismal survival rates. While surgery remains the only curative option, it is generally limited to patients with early-stage GBC. Conventional treatments such as chemotherapy and radiation therapy have shown limited success in improving survival and quality of life. For example, in cases where cancer is confined to the gallbladder mucosa, the 5-year survival rate is 32%, whereas for advanced disease, the 1-year survival rate decreases to 10% [1]. The development of novel diagnostics and therapeutics is critical for improving outcomes in patients with GBC.

One emerging field with significant potential in the management of GBC is metabolomics. Metabolomics involves the comprehensive study of metabolites—small molecules produced in biological systems—and offers a promising avenue for early diagnosis, therapeutic targeting, and personalized treatment. Compared with other "omics" fields, such as genomics, transcriptomics, and proteomics, metabolomics is less labor-intensive and provides direct insight into an organism's current physiological state. This chapter explores the role of metabolomics in GBC, from its potential in biomarker discovery to understanding the pathogenesis of the disease and improving therapeutic strategies.

6.1.1 What Is Metabolomics?

Metabolomics is the study of metabolites—small molecules involved in cellular processes—within biological samples such as blood, urine, or tissue. It is a complementary discipline to genomics, transcriptomics, and proteomics, which analyze genes, RNA, and proteins, respectively. Unlike these other methods, metabolomics provides a direct snapshot of an organism's phenotype, revealing the cumulative effects of genetic expression, environmental factors, and cellular processes at a given moment. Metabolomics provides phenotypic information, whereas other omics techniques provide information about genetic and functional potential. Our genomes, transcriptomes, proteomes, and metabolomes are impacted by one another and the environment (Figure 6.1) [3].

6.1.2 What Are the Metabolites?

Metabolites are small molecules less than 1.5 kDa in size [4]. They are intermediates or end products of enzymatic biochemical reactions in an organism and encompass a diverse range of functions,

DOI: 10.1201/9781003506201-6

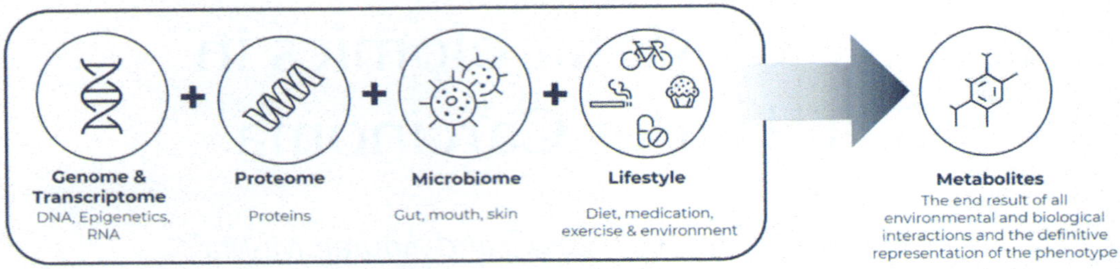

FIGURE 6.1 Complex interrelationships among genes, the environment, and functional capacity.

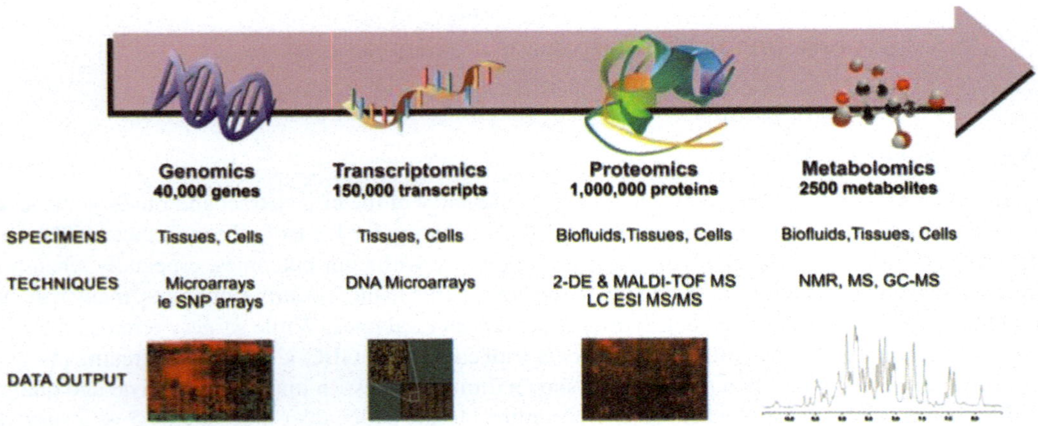

FIGURE 6.2 The omic sciences: side-by-side comparison. SNP arrays, single-nucleotide polymorphism array; 2-DE and MALDI TOF, 2-D electrophoresis and matrix-assisted laser desorption/ionization time-of-flight mass spectrometer; LC ESI MS/MS, liquid chromatography–electrospray ionization–tandem mass spectrometry; NMR, nuclear magnetic resonance; MS, mass spectrometry; GC–MS, gas chromatography/mass spectrometry [10].

including cell growth, inhibition, and stimulation. Metabolites may include amino acids, alcohols, vitamins, polyols, organic acids, and many other types of molecules and are often the building blocks for larger compounds [5]. The number of metabolites in any given biological system is estimated to be in the range of 2000–3000 [6]. In contrast, the number of transcripts and proteins ranges between 40,000 and 100,000 products, leading to reduced complexity and simplified analysis with a greater probability of detecting meaningful data. Identification of metabolites, their interactions, changes in the disease state, and responses to treatments can aid future discoveries in novel biomarkers, prognostication, and the development of newer therapeutic modalities [7–9]. Figure 6.2 summarizes the various omics sciences, showing how SNP arrays enable genomic analysis, while 2-DE, MALDI-TOF and LC-ESI-MS/MS support proteomic profiling, and NMR and GC-MS are used for metabolomic assessment.

Metabolites are the downstream products of genes, transcripts, and enzymes; type and levels represent sensitive measurements of the effects of disease, drugs, toxins, and the environment, making metabolomics the true reflection of the phenotype at any given time. Any changes in or interactions between gene expression, protein expression, and the environment can be directly observed through the metabolome, making metabolomics the most complex and informative of all the omics techniques. The application of metabolomics covers diverse research areas, including but not limited to biomedical research, pharmaceutical drug discovery, agricultural research, and toxicity studies.

6.1.3 What Is the Metabolome?

The metabolome represents the complete set of metabolites found in an organism at any given time. This includes both native metabolites and those produced by the organism's microbiome, further broadening the scope of metabolic analysis. By profiling the metabolome, researchers can identify changes in metabolism that correlate with disease progression, environmental exposures, or therapeutic interventions.

6.1.4 Advantages of Metabolomics

- *Reflects the current physiological state*: Metabolites provide real-time information about an organism's health.
- For simplified analysis, with fewer target molecules than genes or proteins, metabolomics allows for streamlined analysis with a greater chance of revealing meaningful data.
- *Conserved among species*: Metabolites are highly conserved across organisms, making metabolomic techniques applicable across species.

6.1.5 Types of Metabolomic Studies

Untargeted Metabolomics—These studies involve the simultaneous measurement of many metabolites in a biological sample. This strategy, known as the top-down approach, avoids the need for a prior hypothesis on a particular set of metabolites and analyzes the global metabolic profile [11]. This approach measures a wide range of metabolites in a sample without a specific focus, allowing for the discovery of novel metabolic alterations.

Targeted metabolomics focuses on measuring specific metabolites, often on the basis of previous knowledge of disease mechanisms.

6.1.6 Analytical Methods Used in Metabolomics

The metabolome contains a large number of metabolites with widely differing physicochemical properties, such as amino acids, amines, nucleotides, sugars, and lipids. At present, a single method for simultaneously analyzing all metabolites via just one instrument does not exist. Two main approaches for generating metabolomic data are nuclear magnetic resonance (NMR) and mass spectrometry (MS).

6.1.6.1 NMR Spectroscopy

NMR is a fast and highly reproducible spectroscopic technique that is based on the energy absorption and re-emission of the atomic nuclei due to variations in the external magnetic field [12]. Depending on the atomic nuclei being targeted by the applied magnetic field, different types of metabolomic data are generated. However, in the analysis of samples of biological origin, hydrogen is the most commonly targeted nucleus (^{1}H-NMR) because of its natural abundance in biological samples. Although less common, other atoms, such as carbon (^{13}C-NMR) and phosphorus (^{31}P-NMR), are also targeted by NMR, providing additional information on specific metabolite types [13]. The spectral peak areas generated by each molecule are used as an indirect measure of the quantity of the metabolite in the sample, whereas the pattern of spectral peaks informing the physical properties of the molecule is used to identify the type of metabolite. The spectral data obtained with NMR techniques can be referenced to one or two frequency axes. One-dimensional NMR (1D-NMR) spectra are based on a single frequency axis, where the peaks of each molecule are placed within its resonant frequencies. 1D-NMR is the most commonly used method in high-throughput metabolomics studies. Conversely, two-dimensional NMR (2D-NMR) spectra are based on two frequency axes, and their use is often restricted to the characterization of those compounds that cannot be identified with 1D-NMR spectra. The second dimension in 2D-NMR allows the separation of otherwise overlapping spectral peaks and, therefore, provides additional and important information on the chemical properties of the metabolite [14].

NMR is emerging as an important diagnostic tool for many different diseases, as it is noninvasive, highly reproducible, and rapid, and single experiments provide much information on all the constituents/molecules present in a sample without any ambiguity [15–17]. NMR-based metabolic profiling followed by pattern recognition statistical techniques provides comprehensive metabolic information on various components in biofluids, reflecting the levels of endogenous metabolites/biomarkers involved in key cellular pathways, which indicate physiological and pathophysiological status and are further used in diagnosis [18].

High-resolution magic angle spinning (HR-MAS) NMR spectroscopy is a further advancement in this technique and provides fast, easy, and direct analysis of intact tissue cells, foods, pastes, soils, fruits, plants, tissues, and so on [19, 20].

6.1.6.2 Mass Spectrometry

Mass spectrometry is an analytical technique that acquires spectral data in the form of a mass-to-charge ratio (m/z) and a relative intensity of the measured compounds. For the spectrometer to generate the peak signals for each metabolite, the biological sample first needs to be ionized. The resulting ionized compounds from each molecule will then generate different peak patterns that define the fingerprint of the original molecule. A wide range of instrumental and technical variants are currently available for MS. These variants are characterized mainly by different ionization and mass selection methods [21]. In metabolomics, MS is generally preceded by a separation step. This step reduces the high complexity of the biological sample and allows the MS analysis of different sets of molecules at different times. Liquid and gas chromatography columns (LC and GC, respectively) are the most commonly used separation techniques [22]. This chromatographic separation technique is based on the interaction of the different metabolites in the sample with the adsorbent materials inside the chromatographic column. In this way, metabolites with different chemical properties will require different amounts of time to pass through the column. The time that each metabolite requires, called the retention time, is used together with the m/z MS values to generate the two axes of the LC–MS and GC–MS spectral data.

6.2 STEPS IN METABOLOMICS ANALYSIS

The typical pipeline for untargeted metabolomics involves several steps, starting with sample preparation and data acquisition through NMR or MS. Once the raw spectral data are generated, statistical methods are applied to identify significant metabolic features. These features are then correlated with phenotypic data, such as disease status or treatment outcomes, to draw meaningful biological conclusions.

This methodological pipeline starts with the processing of the spectral data to generate the sample metabolic information (i.e., metabolic features), as shown in Figure 6.3 [23]. The different methods available to process spectral data are revised in the "Spectral Processing" section. Together with metabolite identification methods, spectral processing methods are highly dependent on the analytical technique used (e.g., NMR, LC–MS, or GC–MS). Once the complete set of metabolomic features has been generated, univariate and multivariate data analysis methods can be applied to investigate (a) the general structure of the metabolomics data in the dataset and (b) how the different metabolic features are related to the phenotypic data associated with the samples. These groups of techniques are commonly known as chemometric methods.

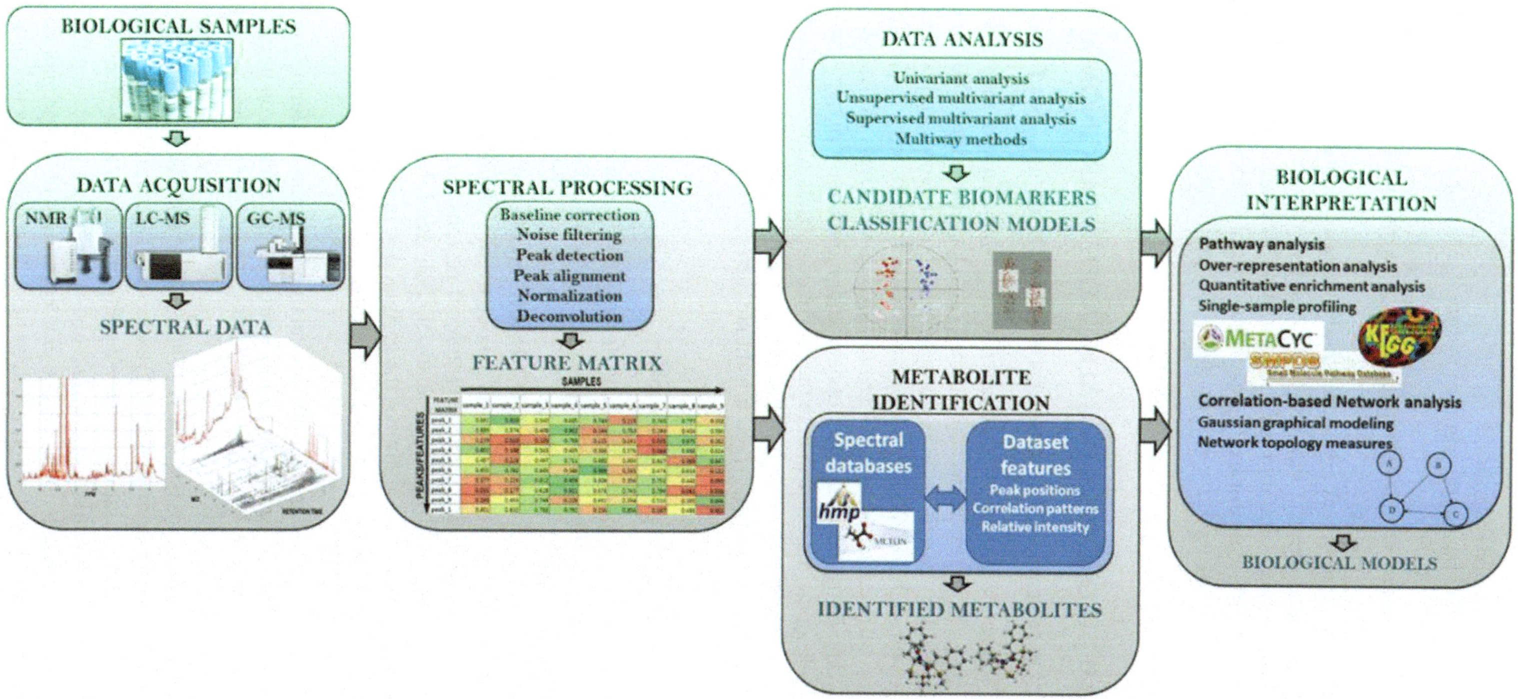

FIGURE 6.3 Different steps of the metabolomic analysis pipeline in untargeted metabolomic studies.

6.3 METABOLOMICS AS A BIOMARKER IN ONCOLOGY

- Analysis of the cancer metabolome
- Cancer diagnosis—The best application of metabolomics thus far in cancer diagnostics is in breast cancer
- Assessment of the response to traditional therapy as a predictive measure of efficacy and as a pharmacodynamic marker
- Novel therapeutics: Biomarkers are increasingly used to identify, validate, and optimize therapeutic targets and agents; determine and confirm the mechanism of drug action as a pharmacodynamic endpoint; and predict or monitor responsiveness to treatment, toxicity, and resistance [24]. Current examples of the use of metabolomics in developmental therapeutics involve tyrosine kinase inhibitors, proapoptotic agents, and heat shock protein inhibitors [25–31].

6.4 ROLE OF METABOLOMICS IN GALLBLADDER CARCINOMA

6.4.1 Biomarker Discovery: Metabolomics for Early Detection and Prognosis

Identifying unique metabolic alterations associated with gallbladder carcinoma is critical for early diagnosis and prognosis. With respect to carcinoma gallbladder, one study (Xu X et al.) [32] demonstrated a clear separation between the disease groups (biliary tract cancer and benign biliary tract diseases) in bile samples through a liquid chromatography/mass spectrometry (LC/MS)-based approach. Patients with biliary tract cancer presented significantly lower levels of lysophosphatidylcholine, phenylalanine, 2-octanoylcarnitine, and tryptophan, and significantly higher levels of taurine- and glycine-conjugated bile acids in the bile than in the bile from patients with benign biliary tract disease. These metabolites may have potential as novel biomarkers for the early detection of biliary tract cancer, and further studies need to be performed to validate their individual clinical applicability.

6.5 UNDERSTANDING PATHOGENESIS

Gallstones are a major risk factor for gallbladder disease [33]. The risk of gallbladder carcinoma is 4–5 times greater in patients with gallstones [34, 35]. Several studies have related gallstones with gallbladder carcinoma [36–40].

Analysis of gallbladder stones via solid-state NMR from different geographical regions revealed correlations between environmental factors and the occurrence of gallbladder stones. A comparison of gallstones from patients from different geographical regions, such as north India, south India, and UAE, revealed different constituents and levels of cholesterol, bilirubin, and calcium carbonate [41]. Chronic cholecystitis, an inflammation of the gallbladder, may be closely associated with GBC. Therefore, serum profiling of chronic cholecystitis may provide avenues for the early detection, prevention, and treatment of GBC.

Serum profiling of gallbladder carcinoma patients and patients suffering from other types of inflammatory disorders (chronic cholecystitis and xanthogranulomatous cholecystitis [XGC] in the gallbladder via one-dimensional ^{1}H NMR (NOESY–1D nuclear Overhauser enhancement spectroscopy and CPMG–Carr–Purcell–Meiboom–Gill pulse sequences) as well as two-dimensional ^{1}H–^{1}H COSY (correlated spectroscopy), ^{1}H–^{1}H TOCSY (total correlation spectroscopy), and ^{1}H–^{13}C HSQC (heteronuclear single quantum coherence) experiments revealed important alterations in certain metabolites [42, 43]. Concentrations of low-density lipids, very-low-density lipids, branched-chain amino acids, alanine, glutamine, creatinine, and tyrosine are lower in gallbladder carcinoma patients than in healthy individuals, whereas the concentrations of 1,2-propanediol, pyruvate, glutamate, and formate are higher [43]. Such changes have also been detected in gallbladder tissues via HR-MAS NMR. In particular, differences in the levels of triacylglycerol (TAG) and other small metabolites were detected among chronic cholecystitis, xanthogranulomatous cholecystitis, and gallbladder carcinoma patients [44].

Lipid profiling of gallbladder disease patients via NMR of tissue extracts revealed alterations in lipid metabolism between benign and malignant gallbladder diseases [45]. Solid-state NMR studies of gallbladder stones via ^{13}C CPMAS (Phoenix MAS Speed Controller) experiments revealed that different gallbladder diseases involve different compositions and microstructures of gallstones [46].

Using ^{1}H NMR, minerals such as calcium and magnesium were also found to be present in greater proportions in the extracts of gallstones from gallbladder carcinoma patients than in those from noncancerous gallbladder inflammation patients [34]. 1D^{1}H and 2D^{1}H–^{13}C HSQC NMR and metabolite profiling of bile fluid revealed that the levels of cholic acid and phospholipids could be used to differentiate between different types of gallbladder disease [47]. ^{31}P NMR of bile provides information about the altered levels of phosphorus-containing metabolites such as inorganic phosphate, lysophosphatidylcholine, and phosphatidylcholine in cholelithiasis patients, liver transplant recipients, and liver donors [48].

The classification of gallstones on the basis of cholesterol content revealed that all the stones associated with gallbladder carcinoma are cholesterol non-predominant mixed stones (<50% cholesterol), whereas the majority of stones associated with benign disease are either cholesterol or cholesterol predominant [40].

Recent methodological developments have enabled the detection and quantitative analyses of several bile metabolites, including six individual bile acids, which were hitherto not accessible by NMR spectroscopy [15, 16, 49, 50]. In fact, because a large amount of bile fluid exists in the gallbladder (40–50 mL in humans), attempts to determine bile metabolites noninvasively via in vivo localized magnetic resonance spectroscopy have yielded highly encouraging results [51, 52].

^{1}H NMR-based metabolic profiling with a multivariate statistical analysis approach has been utilized for the investigation of serum from controls (healthy persons) and those associated with gallbladder diseases (benign: chronic cholecystitis and xanthogranulomatous cholecystitis; malignant: gallbladder carcinoma). The metabolic profiles of the sera from the control and diseased groups differed in the relative concentrations of various lower-molecular-weight metabolites.

The lactate concentration varied significantly between the control and diseased groups; it was highest for the gallbladder carcinoma samples among the diseased groups. The glucose concentration was lower in all the diseased groups than in the control group, but there were no significant differences among the three diseased groups. As reported in various studies, lactate levels are usually higher under malignant conditions because of the increased rates of glycolysis in malignant cells [53]. Increased glycolysis levels are also indicative of hypoglycemic conditions, as the carbon skeleton of lactate comes from glucose. These variations in glucose and lactate levels are indicative of aerobic glycolysis, or the Warburg effect [54]. Increased levels of pyruvate, along with increased lactate, are considered to be the result of retrograde regulation due to mitochondrial dysfunction, a common feature of carcinogenic cells [55].

As reported for other malignancies and cell line studies, branched-chain amino acids are reduced due to increased protein synthesis, which frequently utilizes these amino acids [56]. Glutamate was significantly higher in gallbladder carcinoma samples than in normal samples, whereas glutamine was depleted significantly. The increased activity of the enzyme glutaminase is associated with depleted synthesis and very rapid utilization rates [57]. These findings are consistent with other reports on different cancers showing decreased glutamine levels because glutamine serves not only as a major source of respiratory fuel contributing to energy regeneration in carcinogenic cells but also as a provider of important precursor molecules of the nucleic acid synthesis pathway.

^{1}H NMR analysis of lipid extracts from gallbladder tissue revealed elevated levels of phospholipids, cholesterol esters, and plasmalogens and depleted triacylglycerols. Compared with xanthogranulomatous cholecystitis and chronic cholecystitis, gallbladder carcinoma is associated with increased amounts of cholesterol esters. The accumulation of cholesterol esters was also greater in xanthogranulomatous cholecystitis than in chronic cholecystitis [45]. Cholesterol is an important constituent of the cell membrane and maintains membrane permeability and fluidity. In addition, cholesterol is a precursor molecule for the biosynthesis of bile acids and steroid hormones [58]. Lipoproteins are mainly transported as plasma lipoproteins in an esterified form through the blood to the tissue. The possible role of cholesterol esters in various cancer types has been well documented [59–62]. The accumulation of cholesterol esters in rapidly proliferating cancer cells may be a result of the high uptake of cholesterol via receptor-mediated endocytosis by proliferating tumor cells and has been shown to increase low-density lipoprotein receptor activity and the number of low-density lipoprotein receptors on the surface of each cell [63, 64].

Triacylglycerols (TAGs) are drastically depleted in gallbladder carcinoma. This may be caused by an alteration in TAG metabolism. TAG mainly serves as an energy store, being present as a separate phase of microscopic oily droplets in the aqueous cytosol and serving as a store of metabolic fuel. In eukaryotic cells, TAG is stored in adipocytes, which contain lipase enzymes that catalyze the hydrolysis of stored TAG, releasing fatty acids for export to sites at which they are required as fuels. Lipoprotein lipase activity was found to be decreased in cancer cells. This decreased lipoprotein lipase activity may be responsible for the depleted concentration of TAG in cancerous tissue. ^{1}H NMR analysis was also used to explore alterations in the degree of saturated and unsaturated fatty acids. An increased amount of polyunsaturated fatty acids was observed in gallbladder carcinoma patients relative to chronic cholecystitis patients and xanthogranulomatous cholecystitis patients. The increased ratios of diallylic methylene/olefinic CH and diallylic methylene/monallylic methylene and the decreased ratios of saturated CH_2/terminal CH_3 and saturated CH_2/diallylic methylene point to the accumulation of high concentrations of polyunsaturated fatty acids in gallbladder carcinoma.

6.6 THERAPEUTIC INSIGHTS

Metabolomic profiling of GBC tissues and serum has revealed significant changes in energy metabolism, lipid synthesis, and amino acid utilization in cancerous cells. These insights could pave the way for targeted therapies aimed at disrupting the metabolic pathways that sustain tumor growth.

6.7 CONCLUSION

This chapter thoroughly explores how metabolomics can revolutionize the diagnosis, treatment, and understanding of gallbladder cancer. With the use of analytical methods such as NMR and MS, we are able to detect subtle metabolic changes; hence, metabolomics holds potential for early detection, prognosis, and personalized treatments. Techniques such as NMR and MS are central to this approach, and they reveal distinct metabolic patterns that distinguish cancerous conditions from benign gallbladder diseases. By understanding metabolic dysregulation in GBC, such as the Warburg effect, lipid alterations, and amino acid metabolism, metabolomics may also facilitate the development of new therapies that target specific metabolic pathways.

REFERENCES

1. Mishra A., Nizam A., Singh R., MicroRNA and their role in carcinoma gallbladder. In: Kumar Shukla V., Pandey M., Dixit R. (eds.), *Gallbladder Cancer*. Singapore: Springer, 2023.
2. Rawla P., Sunkara T., Thandra K.C., Barsouk A. Epidemiology of gallbladder cancer. *Clin Exp Hepatol.* 2019 May; 5(2): 93–102.
3. Chapter 1—Metabolomics, metabolites, and the metabolome. In: *Your Guide to Metabolomics*. Ebook. www.metabolon.com
4. Wishart D.S., Tzur D., Knox C., Eisner R., Guo A.C., Young N., et al. HMDB: The human metabolome database. *Nucl Acids Res*. 2007; 35: D521–6.

5. Ashrafian H., Sounderajah V., Glen R., Ebbels T., Blaise B.J., Kalra D., et al. Metabolomics: The stethoscope for the twenty-first century. *Med Princ Pract.* 2021; 30(4): 301–10.
6. Di Leo A., Claudine W., Colangiuli D., Bessi S., Pestrin M., Biganzoli. New strategies to identify molecular markers predicting Chetnik therapy activity and toxicity in breast cancer. *ANN Oncol.* 2007; 18: xii8–14.
7. Ryals J. Metabolomics: An important emerging science. *Journal.* 2004; 92(4): 549–55.
8. Nordstrom A., Lewensohn R. Metabolomics: Moving to the clinic. *J Neuroimmune Pharmacol.* 2009; 5: 4–17.
9. Van Ravenzwaay B., Cunha G.C., Leibold E., Looser R., Mellert W., Prokoudine., et al. The use of metabolomics for the discovery of new biomarkers of effect. *Toxicol Lett.* 2007; 172: 21–8.
10. Davis V.W., Bathe O.F., Schiller D.E., Slupsky C.M., Sawyer M.B. Metabolomics and surgical oncology: Potential role for small molecule biomarkers. *J Surg Oncol.* 2011; 103: 451–9.
11. Alonso A., Julià A., Beltran A., Vinaixa M., Díaz M., Ibañez L., et al. AStream: An R package for annotating LC/MS metabolomic data. *Bioinformatics.* 2011; 27: 1339–40.
12. Bothwell J.H., Griffin J.L. An introduction to biological nuclear magnetic resonance spectroscopy. *Biol Rev Camb Philos Soc.* 2011; 86: 493–510.
13. Reo N.V. NMR-based metabolomics. *Drug Chem Toxicol.* 2002; 25: 375–82.
14. Ward J.L., Baker J.M., Beale M.H. Recent applications of NMR spectroscopy in plant metabolomics. *FEBS J.* 2007; 274: 1126–31.
15. Nagana Gowda G.A., Somashekar B.S., Ijare O.B., Sharma A., Kapoor V.K., Khetrapal C.L. Single-step analysis of individual conjugated bile acids in human bile using 1H NMR spectroscopy. *Lipids.* 2006; 41: 591–603.
16. Ijare O.B., Somashekar B.S., Jadegoud Y., Nagana Gowda G.A. 1H and 13C NMR characterization and stereochemical assignments of Bile acids in aqueous media. *Lipids.* 2005; 40: 1031–41.
17. Srivastava N.K., Pradhan S., Mittal B., Kumar R., Pandey C.M., Nagana Gowda G.A. Novel corrective equations for complete estimation of human tissue lipids after their partial destruction by perchloric acid pre-treatment: high-resolution 1H-NMR-based study. *NMR Biomed.* 2007; 21: 89–100.
18. Lindon J.C., Nicholson J.K. Spectroscopic and statistical techniques for information recovery in metabonomics and metabolomics. *Annu Rev Anal Chem.* 2008; 1(1): 45–69.
19. Beckonert O., Coen M., Keun H.C., Wang Y., Ebbels T.M., Holmes E., et al. High-resolution magic-angle-spinning NMR spectroscopy for metabolic profiling of intact tissues. *Nat Protoc.* 2010; 5(6): 1019–32.
20. Bharti S.K., Bhatia A., Tewari S.K., Sidhu O.P., Roy R. Application of HR-MAS NMR spectroscopy for studying chemotype variations of Withania somnifera (L.) Dunal. *Magnetic Resonance Chem.* 2011; 49(10): 659–67.
21. El-Aneed A., Cohen A., Banoub J. Mass spectrometry, review of the basics: Electrospray, MALDI, and commonly used mass analyzers. *Appl Spectrosc Rev.* 2009; 44: 210–30.
22. Theodoridis G., Gika H.G., Wilson I.D. Mass spectrometry-based holistic analytical approaches for metabolite profiling in systems biology studies. *Mass Spectrom Rev.* 2011; 30: 884–906.
23. Alonso A., Marsal S., Julia A. Analytical methods in untargeted metabolomics: State of the art in 2015. *Front Biotechnol.* 2015; 3: 23.
24. Park J.W., Kerbel R.S., Kelloff G.J., Barrett J.C., Chabner B.A., Parkinson D.R., et al. Rationale for biomarkers and surrogate end points in mechanism-driven oncology drug development. *Clin Cancer Res.* 2004; 10: 3885–96.
25. Gottschalk S., Anderson N., Hainz C., Eckhardt S.G., Serkova N.J. Imatinib (STI571)-mediated changes in glucose metabolism in human leukemia BCR-ABL-positive cells. *Clin Cancer Res.* 2004; 10: 6661–8.
26. Serkova N., Boros L.G. Detection of resistance to imatinib by metabolic profiling: Clinical and drug development implications. *Am J Pharmacogenom.* 2005; 5: 293–302.
27. Hasmann M., Schemainda I. FK866, a highly specific noncompetitive inhibitor of nicotinamide phosphori-bosyltransferase, represents a novel mechanism for induction of tumor cell apoptosis. *Cancer Res.* 2003; 63: 7436–42.
28. Muruganandham M., Alfieri A.A., Matei C., Chen Y., Sukenick G., Schemainda I., et al. Metabolic signatures associated with a NAD synthesis inhibitor-induced tumor apoptosis identified by 1H-decoupled-31P magnetic resonance spectroscopy. *Clin Cancer Res.* 2005; 11: 3503–13.
29. Lyng H., Sitter B., Bathen T.F., Jensen L.R., Sundfor K., Kristensen G.B., et al. Metabolic mapping by use of high-resolution magic angle spinning 1H MR spectroscopy for assessment of apoptosis in cervical carcinomas. *BMC Cancer.* 2007; 7: 11.
30. Blankenberg F.G., Katsikis P.D., Storrs R.W., Beaulieu C., Spielman D., Chen J.Y., et al. Quantitative analysis of apoptotic cell death using proton nuclear magnetic resonance spectroscopy. *Blood.* 1997; 89: 3778–86.

31. Chung Y.L., Troy H., Banerji U., Jackson L.E., Walton M.I., Stubbs M., et al. Magnetic resonance spectroscopic pharmacodynamic markers of the heat shock protein 90 inhibitor 17-allylamino,17-demethoxygeldanamycin (17AAG) in human colon cancer models. *J Natl Cancer Inst*. 2003; 95: 1624–33.
32. Xu X., Cheng S., Ding C., Lv Z., Chen D., Wu J., et al. Identification of bile biomarkers of biliary tract cancer through a liquid chromatography/mass spectrometry-based metabolomic method. *Mol Med Rep*. 2015; 11: 2191–8.
33. Ranjan R., Sinha N. NMR-based metabolomics in gallbladder cancer research. *Methods Mol Biol*. 2019; 2037: 231–41.
34. Srivastava M., Sharma A., Kapoor V.K., Nagana Gowda G.A. Stones from cancerous and benign gallbladders are different: A proton nuclear magnetic resonance spectroscopy study. *Hepatol Res*. 2008; 38(10): 997–1005.
35. Lowenfels A.B., Maisonneuve P., Boyle P., Zatonski W.A. Epidemiology of gallbladder cancer. *Hepatogastroenterology*. 1999; 46: 1529–32.
36. Diehl A.K. Gallstone size and the risk of gallbladder cancer. *JAMA*. 1983; 250: 2323–6.
37. Lowenfels A.B., Walker A.M., Althaus D.P., Townsend G., Domellof L. Gallstone growth, size, and risk of gallbladder cancer: An interracial study. *Int J Epidemiol*. 1989; 18: 50–4.
38. Moerman C.J., Lagerwaard F.J., Bueno de Mesquita H.B., van Dalen A., van Leeuwen M.S., Schrover PA. Gallstone size and the risk of gallbladder cancer. *Scand J Gastroenterol*. 1993; 28: 482–6.
39. Strom B.L., Soloway R.D., Rios-Dalenz J., Rodriguez-Martinez H.A., West S.L., Kinman J.L., et al. Biochemical epidemiology of gallbladder cancer. *Hepatology*. 1996; 23: 1402–11.
40. Csendes A., Becerra M., Rojas J., Medina E. Number and size of stones in patients with asymptomatic and symptomatic gallstones and gallbladder carcinoma: A prospective study of 592 cases. *J Gastrointest Surg*. 2002; 6: 272–3.
41. Sharma R.K., Sonkar K., Sinha N., Rebala P., Albani A.E., Behari A., et al. Gallstones: A worldwide multifaceted disease and its correlations with gallbladder carcinoma. *PLoS One*. 2016; 11(11): e0166351.
42. Sharma R.K., Mishra K., Farooqui A., Behari A., Kapoor V.K., Sinha N. 1H nuclear magnetic resonance (NMR)-based serum metabolomics of human gallbladder inflammation. *Inflamm Res*. 2017; 66(1): 97–105.
43. Sonkar K., Behari A., Kapoor V., Sinha N. 1H NMR metabolic profiling of human serum associated with benign and malignant gallstone diseases. *Metabolomics*. 2013; 9(2): 515–28.
44. Bharti S.K., Behari A., Kapoor V.K., Kumari N., Krishnani N., Roy R. Magic angle spinning NMR spectroscopic metabolic profiling of gall bladder tissues for differentiating malignant from benign disease. *Metabolomics*. 2013; 9(1): 101–18.
45. Jayalakshmi K., Sonkar K., Behari A., Kapoor V.K., Sinha N. Lipid profiling of cancerous and benign gallbladder tissues by 1H NMR spectroscopy. *NMR Biomed*. 2011; 24(4): 335–42.
46. Jayalakshmi K., Sonkar K., Behari A., Kapoor V.K., Sinha N. Solid-state 13C NMR analysis of human gallstones from cancer and benign gallbladder diseases. *Solid State Nucl Magn Reson*. 2009; 36(1): 60–5.
47. Ko H., Choi I., Chang K., Jeong G., Gong G., Seo H., et al. Amphiphilic metabolites in gallbladder bile: Potential biomarkers for gallbladder diseases. *Appl Spectrosc Rev*. 2016; 51(7–9): 706–17.
48. Toczyłowska B., Piotrowski M., Chalimoniuk M. 31P high resolution NMR spectroscopy in analysis of phosphate-containing compounds of bile. *Biull Eksp Biol Med*. 2011; 31 (1): 63–71.
49. Ijare O.B., Somashekar B.S., Nagana Gowda G.A., Sharma A., Kapoor V.K., Khetrapal C.L. Quantification of glycine and taurine conjugated bile acids in human bile using 1H NMR spectroscopy. *Magn Reson Med*. 2005; 53: 1441–6.
50. Nagana Gowda G.A., Somashekar B.S., Ijare O.B., Sharma A., Kapoor V.K., Khetrapal C.L. One step analysis of major bile metabolites in human bile using 1H NMR spectroscopy. *Lipids*. 2006; 41: 577–89.
51. Prescot A.P., Collins D.J., Leach M.O., Dzik-Jurasz A.S. Human gallbladder bile: Noninvasive investigation in vivo with single-voxel 1H MR spectroscopy. *Radiology*. 2003; 229: 587–92.
52. Künnecke B., Bruns A., von Kienlin M. Noninvasive analysis of gallbladder bile composition in cynomolgus monkeys using in vivo 1H magnetic resonance spectroscopy. *Biochim Biophys Acta*. 2007; 1771: 544–9.
53. Semenza G.L. Tumor metabolism: Cancer cells give and take lactate. *J Clin Investig*. 2008; 118(12): 3835–7.
54. Vander Heiden M.G., Cantley L.C., Thompson C.B. Understanding the Warburg effect: The metabolic requirements of cell proliferation. *Science*. 2009; 324(5930): 1029–33.
55. Erol A. Retrograde regulation due to mitochondrial dysfunction may be an important mechanism for carcinogenesis. *Med Hypotheses*. 2005; 65(3): 525–9.

56. Baracos V.E., Mackenzie M.L. Investigations of branched-chain amino acids and their metabolites in animal models of cancer. *J Nutr.* 2006; 136(1): 237S–242S.
57. Curthoys N.P., Watford M. Regulation of glutaminase activity and glutamine metabolism. *Annu Rev Nutr.* 1995; 15(1): 133–59.
58. Nelson D.L., Cox M.M. Chapter title. In: *Lehninger Principles of Biochemistry*, Ryan M. (ed.). New York: Worth Publisher, 2008, 805–50.
59. Tosi M.R., Fini G., Tinti A., Reggiani A., Tugnoli V. Molecular characterization of human healthy and neoplastic cerebral and renal tissues by in vitro (1)H NMR spectroscopy (review). *Int J Mol Med.* 2002; 9: 299–310.
60. Singer S., Sivaraja M., Souza K., Millis K., Corson J.M. 1H-NMR detectable fatty acyl chain unsaturation in excised leiomyosarcoma correlates with grade and mitotic activity. *J Clin Invest.* 1996; 98: 244–50.
61. Tosi M.R., Ricci R., Bottura G., Tugnoli V. In vivo and in vitro nuclear magnetic resonance spectroscopy investigation of an intracranial mass. *Oncol Rep.* 2001; 8: 1337–9.
62. Pollesello P., Eriksson O., Hockerstedt K. Analysis of total lipid extracts from human liver by 13C and 1H nuclear magnetic resonance spectroscopy. *Anal Biochem.* 1996; 236: 41–8.
63. Rudling M.J., Angelin B., Peterson C.O., Collins V.P. Low density lipoprotein receptor activity in human intracranial tumors and its relation to the cholesterol requirement. *Cancer Res.* 1990; 50: 483–7.
64. Murakami M., Ushio Y., Mihara Y., Kuratsu J., Horiuchi S., Morino Y. Cholesterol uptake by human glioma cells via receptor-mediated endocytosis of low-density lipoprotein. *J Neurosurg.* 1990; 73: 760–7.

7 Metabolomics Approach to Understanding Gallbladder Cancer Progression Due to Gallstone Disease

Jiya Mishra,[1,2] *Anu Behari,*[3] *and Neeraj Sinha*[1,2*]
[1]Centre of Biomedical Research, Sanjay Gandhi Postgraduate Institute of Medical Sciences Campus, Uttar Pradesh, India
[2]Academy of Scientific and Innovative Research, Uttar Pradesh, India
[3]Department of Surgical Gastroenterology, Sanjay Gandhi Postgraduate Institute of Medical Sciences, Uttar Pradesh, India
*Corresponding author: neerajcbmr@gmail.com, neeraj.sinha@cbmr.res.in

7.1 INTRODUCTION

The analysis of biofluids for illness detection, which dates back to the medieval use of urine charts on the basis of color, smell, and taste, has evolved with technological advancements into the powerful field known as metabolomics (1). Oliver et al. were the first to introduce the term "metabolomics," defining it as a thorough and quantitative examination of all metabolites present within the biological system under investigation (2). A thorough assessment of the dynamic, multiparameter metabolic reactions of biological systems to pathophysiological triggers or genetic mutations, as outlined by Nicholson and colleagues, is occasionally synonymous with the term metabolomics (3). By concentrating on the comprehensive exploration, classification, and quantitative analysis of intrinsic small molecules (metabolites) within the human metabolome, metabolomics offers valuable insights into the interconnected metabolism of organisms. This contributes significantly to our understanding of the pathophysiological mechanisms linked to various illnesses and diseases (1). Metabolites, which are small molecules with molecular weights under 1500 Da, include steroids, nucleotides, vitamins, carbohydrates, organic acids, amino acids, and peptides. They play crucial roles in the biochemical processes of systems biology (4). Metabolites, also known as biological traits, are identified as indicators of changes in processes, including those underlying the pathogenic causes of diseases. They are frequently utilized as biomarkers (5).

In recent years, various metabolites have been linked to many diseases in terms of their ability to predict outcomes and mortality, as well as those indicative of recovery. This chapter delves into the advancements in the use of metabolomics in critical care, particularly emphasizing its importance in gallbladder carcinoma and its role in predicting patient outcomes.

7.2 BIOANALYTICAL METABOLOMICS TECHNIQUES

Mass spectrometry (MS) and nuclear magnetic resonance spectroscopy (NMR), when combined with multivariate analysis, are widely used bioanalytical techniques in metabolomics for the

DOI: 10.1201/9781003506201-7

TABLE 7.1
Features and differences between MS and NMR

Features	Mass spectrometry	Nuclear magnetic resonance spectrometry
Principle	By determining the mass-to-charge ratio (m/z) of the ions, mass spectrometry identifies the molecules within the mixture.	Nuclear magnetic resonance (NMR) is a physical phenomenon that can be used to identify metabolites when a nucleus with a magnetic moment (^{1}H, ^{13}C) is exposed to a strong magnetic field and radiofrequency pulses.
Component	Ion source, mass analyzer, ion detector	Magnet, Radiofrequency emitter, Receiver, Console, Computer with integrated software
Sample	• Liquid • Gas	• Solid • Liquid • Intact tissue
Analyzer	• Quadrupole • Time of flight (TOF) • Fourier transform ion cyclotron resonance (FTICR) • Orbitrap	• ^{1}H • ^{13}C • ^{15}N • ^{31}P
Advantage	• High sensitivity • Many metabolites can be discovered and identified • Offers a specialized research platform • Can be coupled with LC and GC	• Non-destructible • Highly reproducible • Inherently quantitative • Minimal or no sample preparation • Quick measurement
Disadvantage	• Less reproducible • Tedious sample preparation • Destructible • Similarity in isomers makes the identification of metabolites difficult.	• Not as sensitive as MS • Overlapping peaks

identification and quantification of metabolites in a sample. The primary differences between MS and NMR are outlined in Table 7.1.

In metabolomics-based research, two profiling techniques are employed:

1. *Targeted Profiling:* This method focuses on measuring a specific panel of relevant metabolites.
2. *Untargeted Profiling*: This approach does not depend on prior knowledge of metabolites and provides a more comprehensive analysis by measuring as many metabolites as possible.

7.3 CLINICAL UTILITY OF METABOLOMICS

With continually evolving techniques, instruments, and methodologies, metabolomics is an emerging field that enhances our understanding of systems biology at the molecular level, aiding in the investigation of severe disorders. The overall workflow of metabolomics is summarized (Figure 7.1). Globally, numerous organisations employ metabolomics to study a wide array of illnesses, including diabetes, heart disease, renal disease, liver failure, neurological disorders, chronic obstructive pulmonary disease (COPD) (6), and various cancers. Metabolic fingerprinting has proven to be an effective biomarker technique for several neurological conditions, such as spinal cord injury, stroke, and sports-related concussions.

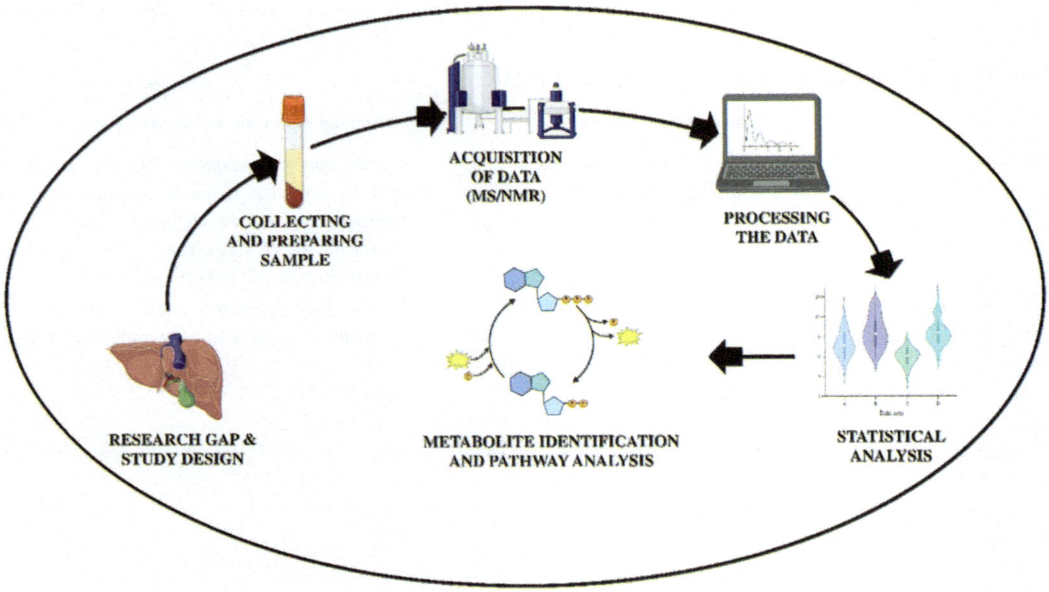

FIGURE 7.1 Workflow of metabolomics.

Metabolomics has also been utilized to understand and diagnose various critical care conditions, including acute respiratory distress syndrome (ARDS), severe acute pancreatitis (SAP), and sepsis with septic shock. For example, in SAP, the metabolic profiles of EVs released by the pancreas were analyzed, revealing distinct differences among healthy controls, SAP patients, and those with mild acute pancreatitis (MAP) (7). This technique shows promise for improving early detection and distinguishing the severity of acute pancreatitis. Similarly, in ARDS, metabolomic profiles are linked to the underlying molecular processes, aiding in differentiating ARDS patients from controls and identifying potential therapeutic targets. In cases of sepsis and septic shock, various studies have utilized metabolomics to predict mortality, distinguish between sepsis and systemic inflammatory response syndrome, and identify metabolic markers for early detection (1). The focus on lipid mediators and specific metabolites in sepsis patients has further enhanced the understanding of clinical outcomes and survival rates. Similarly, metabolomics has provided insight into various conditions related to hepatobiliary diseases. In particular, metabolomics has emerged as a powerful tool for studying the progression from gallbladder stones to gallbladder cancer. By analyzing the metabolic profiles associated with gallbladder stones, researchers have been able to identify specific biomarkers that may indicate a greater risk of developing gallbladder cancer (8).

7.4 GALLBLADDER STONE

Gallstones are solid particles that form in the gallbladder, a small organ located under the liver. The gallbladder stores bile, a digestive fluid produced by the liver that helps break down fats in the small intestine. Gallstone disease is prevalent in the Western world, affecting an estimated 10% to 15% of white adults and resulting in significant morbidity, mortality, and substantial healthcare costs. Among those with gallstones, approximately 70% have cholesterol gallbladder stones (with cholesterol content exceeding 50%), whereas the remaining 30% have black pigment gallbladder stones. In East Asia, brown pigment stones are commonly found in the bile ducts and can cause potentially severe cholangitis (9). The increase in cholesterol gallstones in East Asia is thought to be associated with the increasing adoption of a Western diet.

Gallstones can vary in size and number and can be composed of different materials. Gallbladder stones are classified into eight types with more than ten subtypes, including cholesterol stones, pigment stones, calcium carbonate stones, phosphate stones, calcium stearate stones, protein stones, cystine stones, and mixed stones. Mixed stones contain two or more types of materials, with each component present in similar amounts (10).

Chronic inflammation of the gallbladder, caused by the presence of gallstones (in more than 90% of cases) or infections, is known as chronic cholecystitis (CC), a benign condition. Over time, this persistent inflammation thickens the gallbladder wall, leading to a condition called xanthogranulomatous cholecystitis (8). The progression from gallbladder stones (cholelithiasis) to gallbladder cancer (GBC) is associated with different types of gallstones at various stages. Cholesterol stones, which are typically present in the early stages of gallbladder disease, form due to cholesterol supersaturation in bile and can lead to chronic inflammation. This inflammation is a significant risk factor for gallbladder cancer, as it induces changes in the gallbladder epithelium that may result in malignant transformation (11). Pigment stones, which are often associated with chronic gallbladder inflammation due to chronic infections or hemolytic disorders, also play a role in cancer risk. Their presence contributes to epithelial changes and dysplasia, further increasing the likelihood of malignancy. Thus, both cholesterol and pigment stones are crucial in understanding the progression of gallbladder cancer (12). Mixed stones, composed of cholesterol, calcium, bilirubin, and other elements, are found in advanced gallstone disease and are linked to severe chronic inflammation, fibrosis, and a high risk of gallbladder carcinoma (13). Similarly, a porcelain gallbladder, characterized by extensive wall calcification and chronic cholecystitis, also represents a late-stage condition. Both mixed stones and a porcelain gallbladder significantly increase the risk of gallbladder cancer, particularly when accompanied by large stones (14).

7.5 PATHOGENESIS OF GALLBLADDER STONES

Gallstones typically form in the gallbladder rather than in the bile ducts. In patients with cholesterol gallstones, increased levels of the hydrophobic bile salt deoxycholate in gallbladder bile are linked to rapid crystallization. In cholesterol gallstone patients, impaired gallbladder emptying after meals can prolong bile retention, providing more time for cholesterol crystals to nucleate and grow into stones (15). Inflammation of the gallbladder wall, caused by exposure to detergent bile salts, unesterified cholesterol, and bacteria, may play a crucial role in gallstone formation (16).

Gallbladder motility refers to the ability of the gallbladder to contract and release bile into the small intestine. This process is crucial for the digestion and absorption of fats. Proper gallbladder motility ensures efficient storage, concentration, and release of bile, which plays a significant role in digestive health. Gallbladder motility is frequently impaired in high-risk scenarios for gallstone formation, including pregnancy, obesity, diabetes, gastric surgery, octreotide treatment, very low-calorie diets, and total parenteral nutrition.

The transition from gallstones to gallbladder cancer involves a complex interplay of chronic inflammation, genetic mutations, and cellular changes. While not all individuals with gallstones develop gallbladder cancer, the presence of gallstones significantly increases the risk (Figure 7.2).

7.6 GALLBLADDER CANCER

Gallbladder cancer (GBC) is a malignant neoplasm originating from gallbladder tissues that store bile, a digestive fluid produced by the liver. This type of cancer is relatively rare but can be aggressive and difficult to diagnose early because of the often asymptomatic nature of its early stages.

7.7 PATHOGENESIS OF GALLBLADDER CARCINOMA

Gallbladder carcinoma typically arises through a series of stages over approximately 15 years, beginning with exposure to carcinogens (17). The progression of gallbladder carcinoma (GBC) is

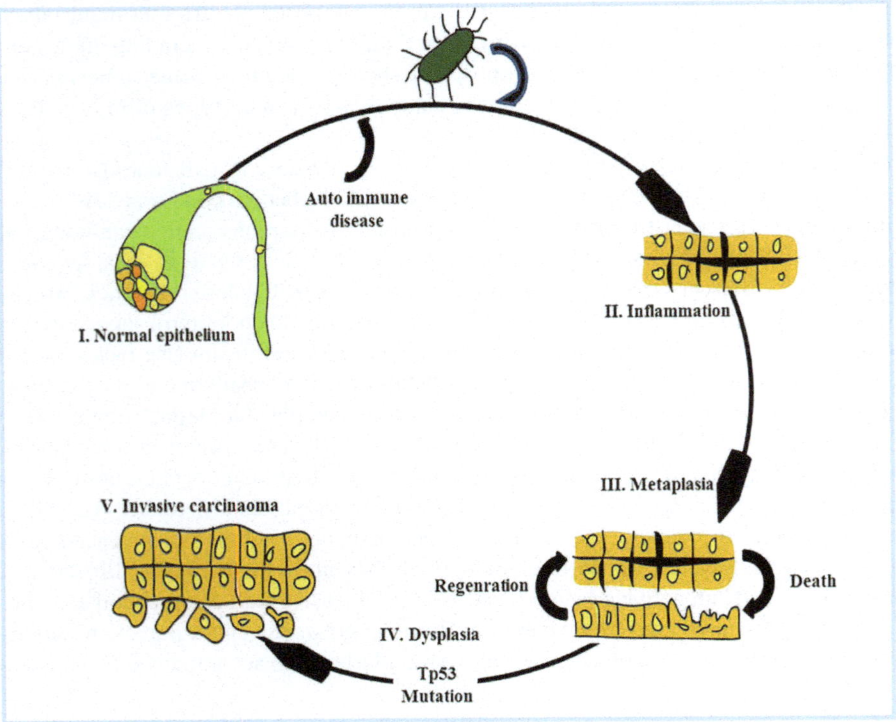

FIGURE 7.2 Transition from gallstone to GBC including different stages.

driven by factors such as chronic inflammation from gallstones or microbial infections, resulting in genomic changes that contribute to cancer development. Genetic alterations play crucial roles in the aggressiveness and prognosis of GBC, prompting ongoing research to identify suitable biomarkers for early diagnosis and targeted therapies (18).

This exposure transforms the normal gallbladder epithelium into metaplasia, dysplasia, carcinoma in situ (CIS), and ultimately invasive carcinoma. The multistage process often begins with the formation of gallstones, which can lead to chronic cholecystitis, thereby increasing the risk of gallbladder cancer. Over 90% of gallbladder carcinoma patients exhibit dysplasia and CIS, highlighting the disease's progressive nature (Figure 7.2) (17).

Gallbladder cancer is predominantly adenocarcinoma (80–95%), with subtypes such as papillary, tubular, mucinous, or signet cell types. Less common variants include squamous cell carcinoma, undifferentiated/anaplastic carcinoma, and adenosquamous carcinoma. The development of gallbladder cancer is linked to various genetic and environmental factors, including chronic gallbladder infections, exposure to certain chemicals, heavy metals, and dietary influences. Risk factors also include female sex, specific geographic regions (primarily in developing countries), and elements such as female hormones, cholesterol cycling, and *Salmonella* infections, although the precise role of female hormones in gallbladder cancer development remains under investigation (17).

7.8 METABOLOMICS APPROACH TOWARD GALLSTONE DIAGNOSIS

Metabolomics is essential for detecting quantitative and qualitative differences in serum metabolic profiles between control samples and samples from various gallbladder conditions, such as chronic cholecystitis, xanthogranulomatous cholecystitis, and gallbladder carcinoma.

The metabolomics approach was used in different studies, including different samples. This study offers valuable insights into the metabolic changes linked to both benign and malignant gallstone diseases, presenting a promising path for understanding their pathophysiology and identifying potential biomarkers. Previous research has shown that NMR spectroscopy-based metabolomics is an exceptionally valuable method for diagnosing diseases, allowing for the early detection of conditions that might otherwise remain undetected. In recent years, metabolites have been shown to play significant roles in gallbladder diseases, assisting in their diagnosis and understanding their pathophysiology (Table 7.2).

TABLE 7.2
Recent NMR-based Studies on Gallbladder Disease

Paper	Sample	Analytical tool	Key metabolites
Lipid profiling of cancerous and benign gallbladder tissues by ^{1}H NMR spectroscopy. DOI:10.1002/nbm.1594	Tissue	^{1}H NMR spectroscopy	Total Cholesterol (Chol), Choline-Containing Phospholipids (PL), Cholesterol Ester, Triacylglycerides (TAGs), and Plasmalogen.
^{1}H NMR metabolic profiling of human serum associated with benign and malignant gallstone diseases. DOI: 10.1007/s11306-012-0468-7	Serum	^{1}H NMR spectroscopy	Increased levels of lactate and pyruvate, whereas decreased levels of glucose, some amino acids, and low-density lipoprotein/very-low-density lipoprotein (LDL/VLDL) were observed.
^{1}H NMR-based serum meta-bolomics of human gallbladder inflammation. DOI: 10.1007/s00011–016–0998-y	Serum	^{1}H NMR spectroscopy	Depleted metabolites: glutamine, LDL, VLDL, alanine, branched-chain amino acids, histidine, and tyrosine. Elevated metabolites: formate, lactate, 1,2-propanediol.
Gallstones: A worldwide multifaceted disease and its correlations with gallbladder carcinoma. DOI: 10.1371/journal. pone.0166351	Gallstones	Solid State NMR spectroscopy	Cholesterol was predominantly found in gallstones from North India, while bilirubin was a major constituent in gallstones from South India. Gallstones from the UAE contained both cholesterol and bilirubin as their main components.
Amphiphilic metabolites in gallbladder bile: Potential biomarkers for gallbladder diseases. DOI: 10.1080/05704928.2016.1167707	Bile	1H NMR spectroscopy.	Cholic acids and phospholipids were identified as significant metabolites involved in differentiating gallbladder diseases.
Magic angle spinning NMR spectro-scopic metabolic profiling of gall bladder tissues for differentiating malignant from benign disease DOI 10.1007/s11306-012-0431-7	Tissue	^{1}H NMR spectroscopy	Malignant gallbladder tissues showed an increase in choline-containing compounds, amino acids, taurine, nucleotides, and lactate compared to benign tissues. Elevated levels of beta-hydroxybutyrate, lactate, alanine, lysine, creatine, choline compounds, taurine, glycine, and glucose were observed in malignant tissues.
NMR spectroscopy for discovery and quantitation of biomarkers of disease in human bile. DOI: 10.4155/bio.11.152	Bile	^{1}H NMR spectroscopy	Cholesterol, urea, total glycine-conjugated bile acids, and total taurine-conjugated bile acids.

Recent metabolomics studies using ^{1}H NMR spectroscopy have provided insights into gallbladder diseases. Lipid profiling of gallbladder tissues has identified key metabolites such as total cholesterol, choline-containing phospholipids, and cholesterol esters in both cancerous and benign tissues (19). Serum profiling associated with gallstone diseases revealed increased levels of lactate and pyruvate, with decreased glucose and LDL/VLDL levels. Studies on gallbladder inflammation have shown depleted glutamine and LDL/VLDL levels, alongside elevated formate and lactate levels (20). Gallstone analysis highlighted regional differences, with cholesterol and bilirubin as major components (21). Bile analysis identified cholic acids and phospholipids as significant biomarkers, whereas malignant gallbladder tissues presented elevated levels of choline compounds, amino acids, and lactate (22). These findings underscore the role of metabolomics in understanding and differentiating gallbladder diseases.

7.9 METABOLOMICS ANALYTICAL PIPELINE

Sample collection involves the collection of blood, tissue, bile, stone, and other samples from patients diagnosed with gallbladder cancer (GBC) and gallstones at the hospital (Figure 7.3).

Additionally, samples are collected from patients who have undergone surgical procedures for gallstones and GB (23, 24).

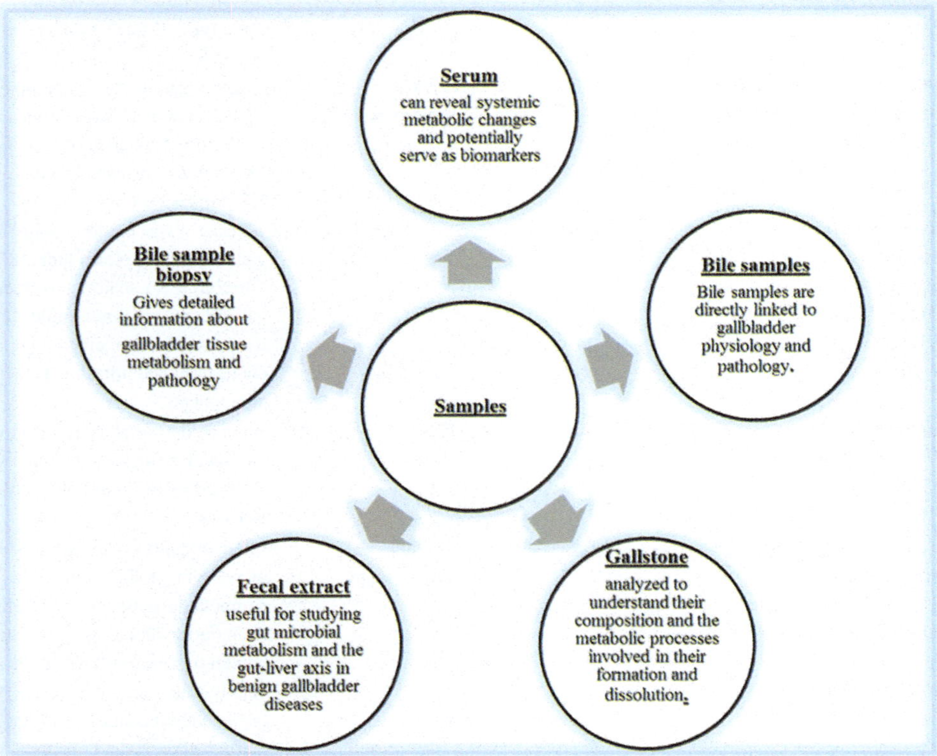

FIGURE 7.3 Different samples used for the metabolomics study of GBC and gallstone.

7.9.1 Different Types of Samples and Their Preparation for NMR Spectroscopy

7.9.1.1 Blood Samples

Preparation:

- Blood samples were collected via standard venipuncture techniques.
- The blood was centrifuged at 3,000 rpm for 10 minutes to separate the plasma or serum.
- The plasma or serum was transferred to clean vials and frozen at -80°C until further analysis.
- The samples were shaken on ice, and D_2O (deuterium oxide) was added to each sample for the NMR experiments.

7.9.1.2 Tissue Samples

Preparation:

- Tissue samples are collected during surgical procedures or biopsies.
- The tissues were immediately snap-frozen in liquid nitrogen and stored at -80°C.
- The tissue samples were homogenized in a suitable buffer (e.g., phosphate buffer) via a tissue homogenizer.
- The homogenate was centrifuged at 13,000 rpm for 10 minutes at 4°C.
- The supernatant was transferred to a clean vial, and D_2O was added before NMR analysis.

7.9.1.3 Bile Samples

Preparation:

- Bile samples are collected during surgical procedures or via drainage.
- The bile was centrifuged at 3,000 rpm for 10 minutes to remove any particulate matter.
- The clear supernatant was transferred to clean vials and frozen at -80°C.
- The samples were placed on ice and mixed with D_2O before NMR measurement.

7.9.1.4 Stone Samples

Preparation:

- Stone samples are collected during surgical procedures or through endoscopic retrieval.
- The stones were cleaned with distilled water to remove any surface contaminants.
- The stones were dried at room temperature or in a low-temperature oven.
- The dried stones were crushed into a fine powder via a mortar and pestle.
- The powder was suspended in a suitable solvent (e.g., D_2O) for NMR spectroscopy.

General Preparation Steps for NMR Spectroscopy

1. *Sample Thawing:* All frozen samples (plasma, serum, tissue homogenate, and bile) were thawed on ice to maintain sample integrity.
2. *Sample Mixing:* The samples were mixed thoroughly to ensure homogeneity.
3. *Buffer Addition:* An appropriate buffer (e.g., phosphate buffer) was added to the sample to stabilize the pH and improve the spectral quality.
4. *D_2O Addition:* D_2O was added to the sample to provide a field-frequency lock signal for the NMR spectrometer.
5. *pH Adjustment:* The pH of the sample was adjusted to approximately 7.4 if necessary, as extreme pH values can affect the NMR spectra.

6. *Filtration:* The samples were filtered through a 0.22 µm filter to remove any particulate matter that could interfere with the NMR analysis.
7. *Transfer to NMR Tubes:* Transfer the prepared samples to NMR tubes (typically 5 mm in diameter) for analysis.

Specific Considerations

- *Volume:* A sufficient volume of the sample (usually 500–600 µL) is available for NMR analysis to obtain high-quality spectra.
- *Temperature Control:* Samples were kept at a consistent temperature (usually 25°C) during NMR measurements to ensure reproducibility.
- *Sample Handling:* All the samples were carefully handled to avoid contamination or degradation, especially for sensitive metabolites.

By following these preparation steps, the samples will be ready for high-resolution NMR spectroscopy, enabling the accurate characterization and quantification of metabolic components.

Sample Processing and NMR Experiments

To characterize and quantify the different metabolic components of samples, all tests are performed via an appropriate probe on an 800 MHz, 600 MHz, or 400 MHz spectrometer. Using a 1D CPMG (broad signals are subjected to spectral editing or filtering via a CPMG pulse experiment), we record NMR spectra for all the samples. This is followed by 2D NMR spectroscopic methods [both heteronuclear (HSQC/HMQC/HMBC) and homonuclear (COSY/TOCSY/NOESY/ROESY for ^{1}H-^{1}H correlation and INADEQUATE for ^{13}C—^{13}C correlation)] to create 2D spectra for metabolomics. The data were acquired, processed, and analyzed via TopSpin software (Bruker BioSpin 2020. TopSpin 4.0). A Chenomx (Chenomx Inc., Attn: David Tam, 1700 Enbridge Centre, 10175–101 Street, Edmonton, AB T5J 0H3, Canada) processor is used for baseline correction and calibration. A Chenomx profiler is used for peak fitting or binning.

Statistical Analysis

The software MetaboAnalyst (MetaboAnalyst 6.0) was used to carry out a series of univariate and multivariate analyses to compute potential markers of GBC. Normalization of the dataset is performed. Principal component analysis is performed on the binned sample spectra after the spectra have been processed to obtain a first understanding of the data. Later, supervised methods such as PLS-DA and OPLS-DA were used to gain a deeper understanding. ANOVA/t tests were used to identify significant metabolites.

Identification of Metabolites

On the basis of signal multiplicity and assignments published in the literature and online databases (BMRB, HMDB), metabolites are manually identified.

Analysis of Metabolic Pathways

Metabolic pathway analysis was carried out via MetaboAnalyst, a straightforward and freely accessible program that combines route enrichment analysis and topological analysis. The Kyoto Encyclopedia of Genes and Genomes database provides the metabolic pathways used by MetaboAnalyst. NMR-identified metabolites that differ significantly between serum samples from patients and healthy controls were loaded into MetaboAnalyst, and enrichment and pathway topology analysis algorithms were selected.

7.10 CONCLUSION

Metabolomics has emerged as a transformative tool for understanding the pathophysiology of gallbladder diseases, including chronic cholecystitis, xanthogranulomatous cholecystitis, and gallbladder carcinoma (25). Through comprehensive metabolic profiling via advanced techniques such as NMR spectroscopy, researchers can identify critical biomarkers that enable early detection of these conditions. This field has provided valuable insights into the metabolic changes associated with both benign and malignant gallstone diseases, highlighting distinct metabolic pathways and potential therapeutic targets (26). The integration of metabolomics into clinical research offers a promising pathway to improve diagnosis, prognosis, and patient outcomes, particularly in patients with hepatobiliary diseases (27). With ongoing advancements in analytical methods, metabolomics is expected to play a pivotal role in personalized medicine, guiding more effective treatment strategies for gallbladder carcinoma and other related conditions.

REFERENCES

1. Siddiqui MA, Pandey S, Azim A, Sinha N, Siddiqui MH. Metabolomics: An emerging potential approach to decipher critical illnesses. *Biophys Chem.* 2020 Dec 1;267:106462.
2. Lasica N, et al. Metabolomics as a potential tool for monitoring patients with aneurysmal subarachnoid hemorrhage. *Front Neurol.* 2022;13:1101524.
3. Nicholson JK, Lindon JC, Holmes E. 'Metabonomics': Understanding the metabolic responses of living systems to pathophysiological stimuli viamultivariate statistical analysis of biological NMR spectroscopic data. *Xenobiotica.* 1999;29:1181–9. doi: 10.1080/004982599238047.
4. Fermier B, et al. Specific metabolome profile of exhaled breath condensate in patients with shock and respiratory failure: A pilot study. *Metabolites.* 2016;6(3).
5. Fedoruk RP, Lee CH, Banoei MM, Winston BW. Metabolomics in severe traumatic brain injury: A scoping review. *BMC Neurosci.* 2023 Oct 16;24(1):54.
6. Bertini I, Luchinat C, Miniati M, Monti S, Tenori L. Phenotyping COPD by 1H NMR metabolomics of exhaled breath condensate. *Metabolomics.* 2014 Apr 1;10(2):302–11.
7. Lou D, et al. Quantitative metabolic analysis of plasma extracellular vesicles for the diagnosis of severe acute pancreatitis. *J Nanobiotechnol.* 2022 Jan 28;20(1):52.
8. Sonkar K, et al. 1 H NMR metabolic profiling of human serum associated with benign and malignant gallstone diseases. *Metabolomics.* 2013;9:515–28.
9. Dowling RH, et al. Role of intestinal transit in the pathogenesis of gallbladder stones. *Can J Gastroenterol Hepatol.* 1997;11:57–64.
10. Qiao T, et al. The systematic classification of gallbladder stones. *PLoS One.* 2013;8(10):e74887.
11. Hundal R, Shaffer EA. Gallbladder cancer: Epidemiology and outcome. *Clin Epidemiol.* 2014;6:99–109.
12. Stinton LM, Shaffer EA. Epidemiology of gallbladder disease: Cholelithiasis and cancer. *Gut Liver.* 2012;6:172–87.
13. Lazcano-Ponce EC, et al. Epidemiology and molecular pathology of gallbladder cancer. *CA Cancer J Clin.* 2001;51(6):349–64.
14. Randi G, Franceschi S, La Vecchia C. Gallbladder cancer worldwide: Geographical distribution and risk factors. *Int J Cancer.* 2006;118(7):1591–602.
15. Van Erpecum KJ. Pathogenesis of cholesterol and pigment gallstones: An update. *Clin Res Hepatol Gastroenterol.* 2011;35(4):281–7.
16. Van Erpecum KJ, et al. Gallbladder histopathology during murine gallstone formation: Relation to motility and concentrating function. *J Lipid Res.* 2006;47:32–41.
17. Sharma A, et al. Gallbladder cancer epidemiology, pathogenesis and molecular genetics: Recent update. *World J Gastroenterol.* 2017;23(22):3978.
18. Mishra SK, Kumari N, Krishnani N. Molecular pathogenesis of gallbladder cancer: An update. *Mutation Res Fundamental Mol Mech Mutagenesis.* 2019;816:111674.
19. Jayalakshmi K, et al. Lipid profiling of cancerous and benign gallbladder tissues by 1H NMR spectroscopy. *NMR Biomed.* 2011;24(4):335–42.
20. Sharma RK, et al. 1 H nuclear magnetic resonance (NMR)-based serum metabolomics of human gallbladder inflammation. *Inflamm Res.* 2017;66:97–105.

21. Sharma RK, et al. Gallstones: A worldwide multifaceted disease and its correlations with gallbladder carcinoma. *PLoS One*. 2016;11(11):e0166351.
22. Ko H, et al. Amphiphilic metabolites in gallbladder bile: Potential biomarkers for gallbladder diseases. *Appl Spectroscopy Rev*. 2016;51(7–9):706–17.
23. Du Y, Wijaya WA, Liu WH. Advancements in metabolomics research in benign gallbladder diseases: A review. *Medicine*. 2024;103(21):e38126.
24. Rawla P, et al. Epidemiology of gallbladder cancer. *Clin Exp Hepatol*. 2019;5(2):93–102.
25. Jain K, et al. Sequential occurrence of preneoplastic lesions and accumulation of loss of heterozygosity inpatients with gallbladder stones suggest causal association with gallbladder cancer. *Ann Surg*. 2014;260:1073–80.
26. Muszynska C, et al. Predictors of incidental gallbladder cancer in patients undergoing cholecystectomy for benign gallbladder disease: Results from a population based gallstone surgery registry. *Surgery*. 2017;162:256–63.
27. Diehl AK. Gallstone size and the risk of gallbladder cancer. *JAMA*. 1983;250:2323–6.

8 Metabolomics
Potential Biomarkers in Gallbladder Carcinogenesis

Ravi Bhushan,[2] *Raviranjan Kumar Gupta,*[1] *and Shyam Babu Prasad*[1*]
[1]Cancer Genomics and Therapeutics Laboratory, Department of Zoology, School of Life Sciences, Mahatma Gandhi Central University Bihar (MGCUB), India.
[2]Centre for Genetic Disorders, Institute of Science, Banaras Hindu University, Uttar Pradesh, India
*Corresponding author: shyambabuprasad@mgcub.ac.in

8.1 INTRODUCTION

Oncogenic transformation alters gross metabolic intracellular alterations that promote cancer progression. Gallbladder cancer (GBC) is an aggressive and rare type of cancer that does not show any signs and symptoms in the early stage of development, which represents significant challenges in its treatment. This chapter highlights the emerging role of metabolomics in GBC progression. It also focuses on how metabolic profiling can be utilized for early diagnosis, prognosis, and to devise targeted therapy. Metabolomics may reveal new insights about cancer progression and can be used to identify potential treatment options or predict responsiveness to cancer treatments. Further, its significance in cancer development, treatment, and potential applications in cancer development are summarized in this book chapter.

8.1.1 Overview of Gallbladder Cancer

GBC is an aggressive cancer marked by genetic mutations in oncogenes like KRAS, EGFR, and HER2, along with the inactivation of tumor suppressor genes such as TP53, disrupting normal cell signaling and promoting uncontrolled growth [1, 2]. A significant challenge in treating GBC is its asymptomatic nature in the early stages, resulting in diagnosis often occurring at a very advanced stage when symptoms like abdominal pain and jaundice appear [3]. Thus, late-stage identification contributes to a poor prognosis, with a 5-year survival rate under 5%, and only surgery offers a potentially curative option for a few patients. In addition, adjuvant therapies, including chemotherapy and radiotherapy, show limited effectiveness, highlighting the need for improved diagnostic methods and innovative treatment strategies at an early stage [4, 5].

Recent advancements in molecular biology and omics technologies provide hope for developing targeted therapies and personalized approaches by revealing metabolic alterations associated with GBC. Recent research highlights that metabolomics is a powerful tool that can identify cancer biomarkers and drivers of tumorigenesis, which may improve current and future opportunities for cancer diagnosis, monitoring, and treatment.

DOI: 10.1201/9781003506201-8

8.1.2 Current Challenges in Diagnosis and Treatment

Imaging techniques like ultrasonography, CT, and MRI are commonly used to diagnose GBC, but they often struggle with early-stage detection and distinguishing between benign and malignant lesions [6, 7]. Endoscopic ultrasound and positron emission tomography can provide better evaluations, though they may not be widely accessible and patient-friendly due to high costs [8]. In addition, reliable blood-based biomarkers for early GBC detection are lacking, although research into potential markers like CA19-9 and CEA is ongoing [9]. GBC treatment options are limited, with surgical intervention being the only curative approach, yet most patients are diagnosed too late for surgery. For those who do undergo surgery, recurrence rates are high, and palliative chemotherapy and radiotherapy offer limited efficacy, resulting in poor survival outcomes, typically not exceeding 12 months [10].

The molecular heterogeneity of GBC, involving mutations in pathways like KRAS, TP53, and EGFR, complicates treatment development [11]. Moreover, the treatment of GBC is hindered by a lack of targeted therapies, unlike those available for other cancers. Additionally, the tumor microenvironment's desmoplasia and immunosuppression may reduce the effectiveness of current immunotherapies [12–14]. While advances in precision medicine show promise for identifying new therapeutic targets, their clinical application is still in the early stages. Geographic disparities in GBC prevalence and limited access to advanced care in high-incidence areas further complicate diagnosis and treatment, alongside the high costs of newer therapies. Addressing these challenges requires focused efforts in research, clinical practice, and public health to enhance early detection, and exploring different treatment modalities is imperative.

8.1.3 Role of Metabolomics in Cancer Research

Metabolomics is a powerful tool in cancer research that examines small-molecule metabolites and their alterations in tumors. Cancer cells undergo "metabolic reprogramming" to meet growth demands, modifying key pathways like glycolysis, lipid metabolism, and amino acid biosynthesis [15, 16]. The study of metabolomics identifies cancer-specific metabolic signatures, such as the Warburg effect, where tumor cells favor aerobic glycolysis, leading to increased glucose consumption and lactate production [17]. Studies reveal how cancers exploit various metabolic pathways, with alterations in lipid and amino acid metabolism linked to tumor growth, invasion, and survival in nutrient-deprived environments.

Metabolomics plays a crucial role in cancer research by elucidating metabolic reprogramming mechanisms and aiding in the discovery of biomarkers for early cancer detection, diagnosis, and prognosis. The metabolic profiles of cancer patients differ significantly from healthy individuals, with changes in metabolites like lactate and pyruvate potentially indicating tumor development. These biomarkers can lead to noninvasive diagnostic tests and help personalize treatment approaches by stratifying patients according to their metabolic profiles [18]. Metabolomics enables the evaluation of treatment responses and the discovery of novel therapeutic strategies by targeting metabolic vulnerabilities in cancer cells. Integrating metabolomics with other omics technologies enhances our understanding of cancer's complexity, supporting advancements in diagnostics and therapies [19]. Despite challenges with standardization and clinical translation, the potential of metabolomics in cancer research continues to grow and has a high impact on cancer research.

8.2 UNDERSTANDING THE METABOLIC REPROGRAMMING IN CANCER

Metabolic reprogramming is a hallmark of cancer, enabling tumor cells to adapt to their high energy demands and thrive in the often-hostile microenvironment. Cancer cells undergo significant alterations in key metabolic pathways to support rapid proliferation, resist apoptosis, and survive in nutrient-poor conditions [20]. One of the most well-known examples is the Warburg effect, where cancer cells rely on aerobic glycolysis to produce energy, even in the presence of oxygen. This shift from oxidative phosphorylation to glycolysis, despite its inefficiency in ATP production, provides cancer cells with an abundant supply of metabolic intermediates required for biosynthesis, which supports cellular growth and division [17]. In addition to altered glucose metabolism, cancer cells exhibit dysregulated lipid and amino acid metabolism. Enhanced lipid biosynthesis, for instance, is critical for membrane formation in rapidly dividing cells and also contributes to signaling pathways that promote tumor growth and metastasis. Amino acid metabolism, particularly involving glutamine and serine, is often rewired to fuel anabolic processes, providing substrates for nucleotide and protein synthesis, and maintaining redox balance. These alterations allow cancer cells to manage oxidative stress and evade metabolic checkpoints that would normally limit proliferation [21]. The reprogramming of metabolism in cancer is driven by oncogenes (e.g., MYC and KRAS) and tumor suppressors (e.g., TP53), which influence key metabolic enzymes and pathways [22, 23]. The tumor microenvironment, including hypoxia and nutrient deprivation, plays a role in shaping the metabolic phenotype of cancer cells, driving adaptations that promote tumor progression and resistance to therapy [17]. Targeting metabolic reprogramming in cancer has emerged as a promising therapeutic strategy. By inhibiting key metabolic pathways that cancer cells depend on, such as glycolysis, glutaminolysis, or lipid biosynthesis, researchers aim to exploit the metabolic vulnerabilities of tumors and improve treatment outcomes (Figure 8.1).

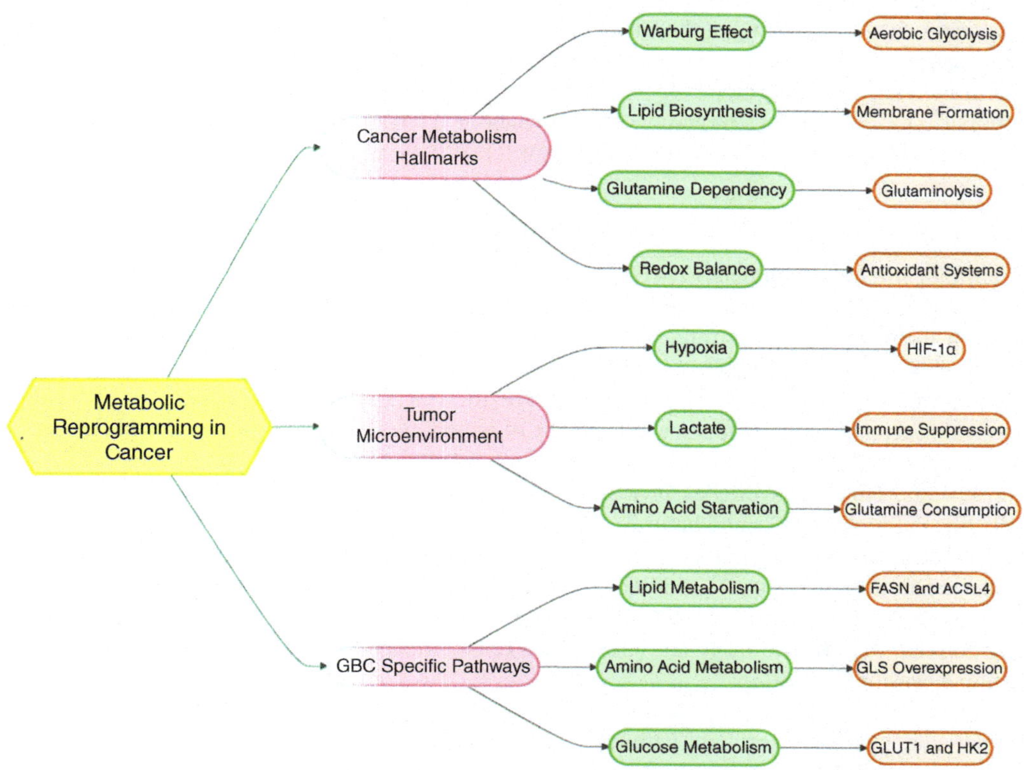

FIGURE 8.1 Metabolic reprogramming in cancer: Metabolic reprogramming in cancer involves the Warburg effect, where cells use aerobic glycolysis for growth and division, and altered lipid and amino acid metabolism supporting tumor progression. Targeting these metabolic changes offers potential therapeutic strategies by exploiting tumors' unique vulnerabilities.

8.2.1 Hallmarks of Cancer Metabolism

Cancer metabolism involves significant changes in cellular processes that support the rapid growth and survival of tumor cells using ATP production. Cancer cells favor glycolysis for ATP production even in the presence of oxygen, benefiting from the biosynthetic intermediates glycolysis provides [17], and increased lipid production is driven by oncogenes such as MYC and AKT, which activate enzymes like fatty acid synthase, highlighting the importance of altered glucose and lipid metabolism in cancer progression [24].

Amino acid metabolism is altered in cancer cells, with glutamine being critically important. Cancer cells are highly dependent on glutamine for energy, nucleotide and protein synthesis, and maintaining redox balance, a phenomenon known as "glutamine addiction," especially in aggressive tumors like glioblastomas and pancreatic cancers [21]. Glutaminolysis fuels the tricarboxylic acid (TCA) cycle, supporting rapid growth and survival of cancerous cells [25]. Very interestingly, cancer cells manage high levels of reactive oxygen species (ROS) from metabolic dysregulation by upregulating antioxidants like glutathione, balancing ROS for signaling while avoiding oxidative stress. Under hypoxic conditions, cancer cells rely more on glycolysis and activate hypoxia-inducible factor 1-alpha (HIF-1α), promoting angiogenesis and glucose transport to adapt and proliferate despite low oxygen [17, 26]. Cancer cachexia is a severe form of metabolic disorder characterized by loss of lean body mass or sarcopenia with or without loss of fat mass in advanced cancer stages, and it is controlled by numerous factors like cytokines and neuropeptides.

Together, these characteristics of cancer metabolism highlight how tumor cells adjust their metabolic processes to fulfill their growth and survival needs. Gaining insight into these changes enhances our understanding of the basic biology of cancer and reveals potential therapeutic targets. Many current cancer treatments aim to inhibit crucial metabolic pathways, including glycolysis, glutaminolysis, and lipid metabolism, in order to take advantage of the metabolic weaknesses of tumors [27]. Consequently, targeting cancer metabolism emerges as a promising strategy for creating more effective and targeted cancer therapies.

8.2.2 Metabolic Shifts in Tumor Microenvironment

The tumor microenvironment (TME) is a complex ecosystem comprised of cancer cells, stromal cells, immune cells, and extracellular matrix, characterized by unique metabolic shifts influenced by both intrinsic reprogramming of cancer cells and extrinsic factors like hypoxia and nutrient depletion [14]. Hypoxia, often present in rapidly growing tumors, leads to the stabilization of hypoxia-inducible factors (HIFs), particularly HIF-1α, which promotes anaerobic glycolysis and cell survival. Lactate accumulation is a key feature of metabolic reprogramming in the TME, as tumor cells produce excess lactate through aerobic glycolysis, which is then secreted via monocarboxylate transporters. This lactate creates an acidic environment that inhibits the activity of crucial immune cells like cytotoxic T cells and natural killer cells, while promoting the recruitment of immunosuppressive cells such as regulatory T cells (Tregs) and myeloid-derived suppressor cells, ultimately hindering antitumor responses. Additionally, tumor cells have an increased demand for amino acids like glutamine and arginine, essential for biosynthetic pathways and redox balance, but their high consumption can deplete these nutrients in the TME, impairing T-cell function and aiding tumor immune evasion [28, 29].

Lipid metabolism is reprogrammed in the TME to support cancer growth by increasing lipid uptake and synthesis through pathways like PI3K/AKT/mTOR. This accumulation of lipids benefits tumor cells by serving as building blocks, signaling molecules, and energy stores, while also inducing metabolic changes in surrounding stromal cells and immune cells. For instance, cancer-associated fibroblasts and tumor-associated macrophages shift their metabolism to provide nutrients and support pro-tumor functions. Ultimately, these metabolic shifts are crucial for tumor progression and present new therapeutic targets to improve cancer treatment outcomes [24, 30].

8.2.3 Specific Metabolic Pathways in GBC

In GBC, lipid metabolism is significantly altered to meet the high energy and biosynthetic needs of rapidly proliferating tumor cells, characterized by increased lipid biosynthesis and elevated levels of phosphatidylcholine (PC) and phosphatidylethanolamine (PE). Key enzymes like fatty acid synthase (FASN) and acyl-CoA synthetase long-chain family member 4 (ACSL4) are upregulated, supporting tumor growth and lipid accumulation. Similarly, amino acid metabolism, particularly glutamine and arginine metabolism, plays a critical role in GBC; glutamine serves as an essential energy source through its conversion to glutamate, while disrupted arginine metabolism affects polyamine and nitric oxide production, contributing to immune evasion and tumor proliferation [31].

GBC cells exhibit altered glucose metabolism, characterized by enhanced aerobic glycolysis, or the Warburg effect, which increases glucose uptake and lactate production even in the presence of oxygen. Upregulation of glucose transporters like GLUT1 and hexokinase 2 supports this metabolic shift, creating an acidic microenvironment that promotes tumor invasion while suppressing antitumor immunity. Additionally, GBC cells show increased activity in nucleotide metabolism, particularly with enzymes involved in purine and pyrimidine synthesis, facilitating rapid DNA replication and cell division. These altered metabolic pathways present potential targets for therapeutic intervention and underscore the need for further research into the metabolic vulnerabilities of GBC (Figure 8.1) [32].

8.3 LIPID METABOLISM IN GALLBLADDER CARCINOGENESIS

Lipid metabolism is crucial in gallbladder carcinogenesis, supporting the proliferation and survival of cancer cells. GBC cells show notable changes in lipid biosynthesis and uptake, particularly with the upregulation of FASN, which aids in producing fatty acids needed for membranes and energy. Increased levels of phospholipids like PC and PE are found in GBC tumors, promoting cell proliferation and invasion. Moreover, altered cholesterol metabolism also plays a role in enhancing tumor progression. Targeting lipid metabolic pathways, such as inhibiting FASN or ACSL4, could be a promising therapeutic strategy for GBC, especially in advanced cases (Figure 8.2) [33].

8.3.1 Alterations in Phospholipid Composition

Phospholipids, crucial components of cellular membranes, are vital for functions like signal transduction and vesicular transport. In cancer, particularly gallbladder carcinoma, changes in phospholipid composition are linked to tumor progression. Notably, alterations in PC and PE levels have been observed, indicating an increased need for membrane production in rapidly dividing cancer cells. The CDP-choline pathway, identified by P. Kennedy in 1956 and responsible for synthesizing PC, often shows overexpression of enzymes like choline kinase, which enhances tumor growth and metastasis [34].

PE and sphingolipids, particularly sphingomyelin (SM), are vital phospholipids in GBC biology. PE is crucial for maintaining membrane integrity and is linked to enhanced mitochondrial function and apoptosis resistance, helping cancer cells survive metabolic stress by preserving mitochondrial health. Additionally, the conversion between PC and PE supports the dynamic needs of tumor cell membranes [35]. Altered sphingomyelin levels affect cell signaling related to growth and apoptosis, contributing to increased cancer cell survival and tumor progression. Moreover, lysophosphatidylcholine (LPC), a breakdown product of PC, promotes tumor invasion and metastasis through pro-inflammatory signaling, influencing the tumor microenvironment. The metabolism of phospholipids is also interconnected with lipid signaling, where phospholipase A2 (PLA2) hydrolysis produces arachidonic acid (AA) and lysophospholipids. These lipid mediators promote inflammation, angiogenesis, and tumor growth, with increased PLA2 activity correlating with poor prognosis in GBC patients. In summary, changes in phospholipid composition—particularly in PC, PE, and sphingolipids—significantly contribute to GBC development, thereby highlighting phospholipid metabolism as a potential therapeutic target for improving patient outcomes [36, 37].

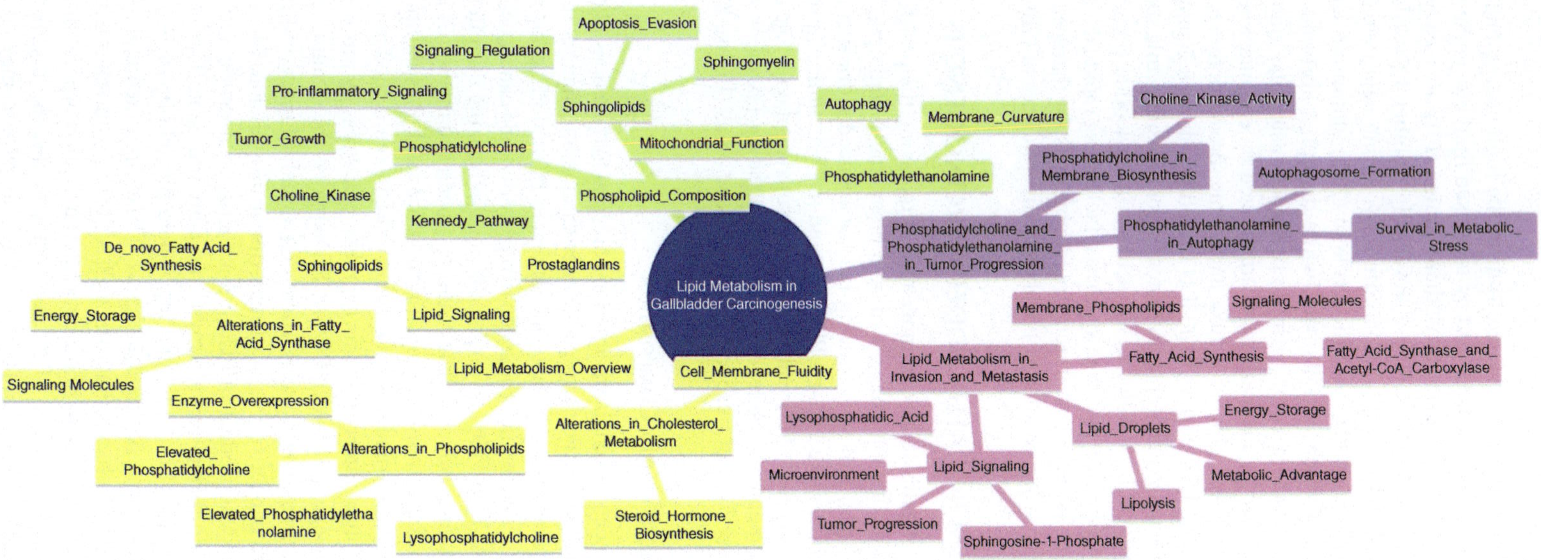

FIGURE 8.2 Lipid metabolism in gallbladder carcinogenesis: Lipid metabolism is crucial in gallbladder carcinogenesis, characterized by upregulation of fatty acid synthase (FASN) and increased levels of phosphatidylcholine (PC) and phosphatidylethanolamine (PE). Targeting lipid metabolic pathways, such as FASN and ACSL4 inhibition, may offer novel therapeutic strategies for gallbladder cancer.

8.3.2 Phosphatidylcholine and Phosphatidylethanolamine in Tumor Progression

PC and PE are major phospholipids in eukaryotic cell membranes, playing crucial roles in normal cellular function and cancer progression, particularly in GBC. The synthesis of PC through the Kennedy pathway involves the action of choline kinase, whose elevated expression is linked to increased tumor growth and metastasis due to enhanced membrane biogenesis and oncogenic signaling through pathways like MAPK and Akt. Additionally, PC metabolism produces LPC via phospholipase A2, which promotes cancer cell migration, supports pro-inflammatory signaling, and aids immune evasion, thereby contributing to tumor progression and metastatic potential [35, 38].

PE is a critical membrane phospholipid that plays a vital role in maintaining cellular structure and function, particularly in rapidly dividing cells. It is essential for membrane integrity and fluidity, and it is involved in various metabolic and signaling processes linked to cancer progression. A significant function of PE in cancer is its promotion of autophagy, a cellular process that enables cells to survive under metabolic stress by recycling cellular components—a necessity for cancer cells that often face nutrient deprivation and hypoxia in the tumor microenvironment. PE conjugates with microtubule-associated protein 1 light chain 3 (LC3), facilitating autophagosome formation and thereby supporting cancer cell survival during adverse conditions. In gallbladder cancer specifically, alterations in PE levels correlate with changes in mitochondrial function and resistance to apoptosis; elevated PE levels help sustain mitochondrial membrane potential, which is crucial for energy production and cell survival. The interconversion of PC and PE via phospholipid remodeling pathways also allows cancer cells to adapt to metabolic shifts, enhancing their survival further. Research indicates that the balance between the synthesis and degradation of PC and PE is essential for maintaining membrane homeostasis in cancer cells, and disruptions in this balance may promote tumorigenesis. Thus, the dysregulation of PE and phosphatidylcholine metabolism is pivotal in tumor progression, particularly in fostering cell proliferation, survival, and metastasis. Targeting the key enzymes involved in the synthesis and metabolism of these phospholipids offers a promising therapeutic strategy for cancers like gallbladder cancer, where metabolic reprogramming is a hallmark of disease progression [38].

8.3.3 Role of Lipid Metabolism in Tumor Cell Invasion and Metastasis

Lipid metabolism is crucial for the invasive and metastatic potential of tumor cells, enabling their spread from primary sites to distant organs. Cancer cells undergo metabolic reprogramming, particularly in lipid biosynthesis, which provides energy, structural components, and signaling molecules necessary for successful invasion and metastasis. Key enzymes such as FASN and ACC drive increased fatty acid synthesis, essential for building membrane phospholipids and producing signaling molecules that foster a supportive tumor microenvironment. In aggressive malignancies like GBC, the upregulation of lipid biosynthesis correlates with heightened metastatic potential, as cancer cells rely on increased lipids for energy and signaling during their migratory processes [39].

Lipid droplets serve as energy reservoirs, meeting the high energy demands of migrating cancer cells, particularly under metabolic stress. Additionally, bioactive lipids such as lysophosphatidic acid and sphingosine-1-phosphate (S1P) facilitate cancer cell motility and survival by activating key signaling pathways. These lipids also influence the tumor microenvironment, promoting pro-tumorigenic changes and immunosuppression through lipid metabolites. Overall, dysregulated lipid metabolism supports the invasive behavior of tumor cells, highlighting the potential for therapeutic strategies targeting these metabolic pathways to impede cancer progression and metastasis (Figure 8.3) [40].

8.4 AMINO ACID METABOLISM IN GALLBLADDER CANCER

Amino acid metabolism is integral to the progression of GBC, as tumor cells undergo metabolic reprogramming to facilitate rapid growth and survival in challenging environments. Cancer cells

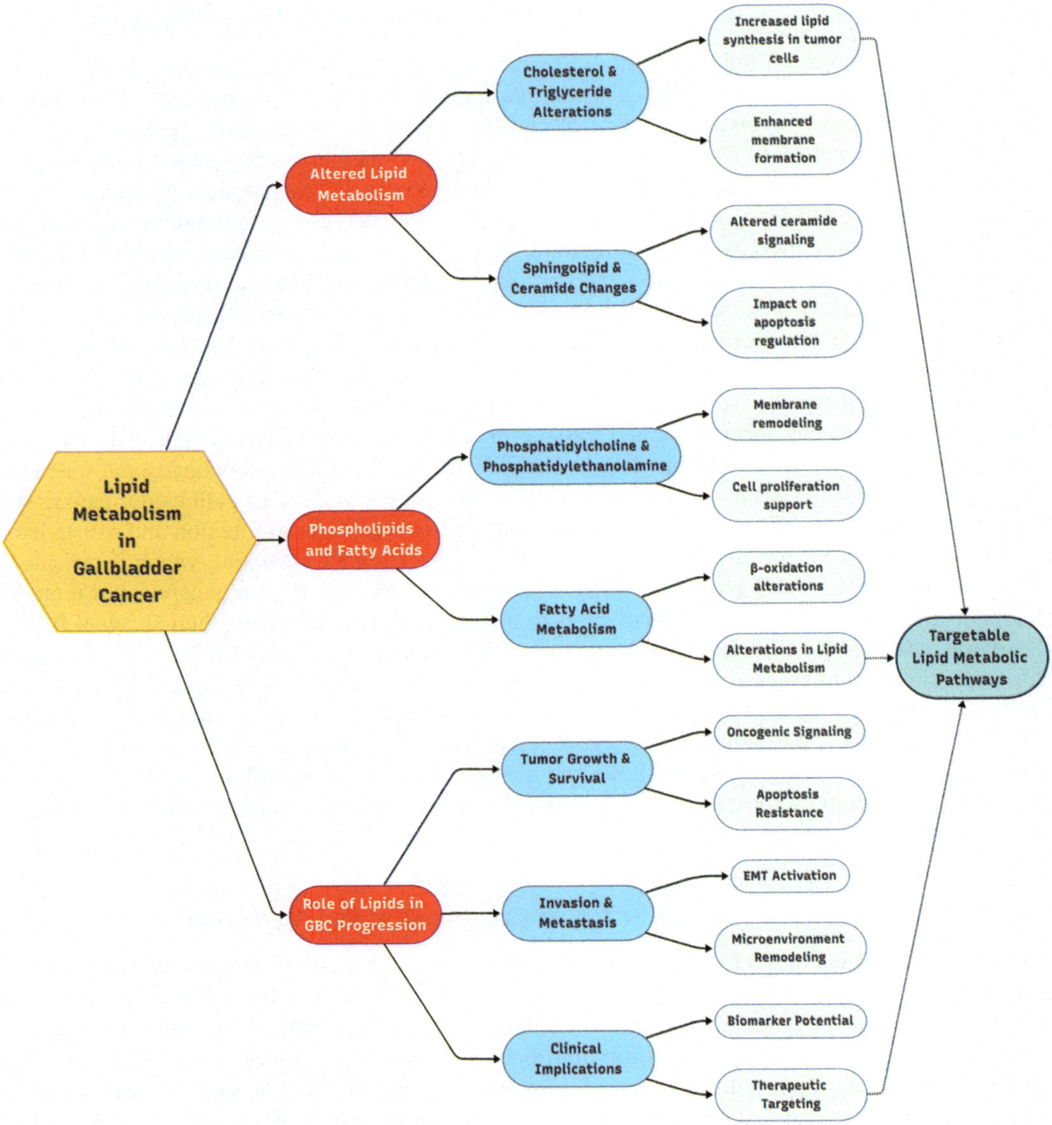

FIGURE 8.3 Phospholipids metabolism. Phospholipids, including PC and PE, are crucial for gallbladder cancer progression by supporting membrane biosynthesis, cell survival, and tumor invasion. Elevated PC and PE levels, driven by increased activity of choline kinase and alterations in phospholipid remodeling, enhance tumor growth and metastasis. Additionally, sphingolipids and LPC contribute to inflammation and metastatic potential, highlighting phospholipid metabolism as a potential therapeutic target.

demand an increase in amino acids to support anabolic processes, maintain biosynthesis, and uphold redox balance. Dysregulation of essential amino acids, particularly glutamine, serine, and arginine, has been linked to tumorigenesis by promoting growth, invasion, and metastasis in GBC [21, 27].

Glutamine is especially crucial in GBC progression, with cancer cells exhibiting what is termed "glutamine addiction." As a key carbon and nitrogen donor, glutamine supports the synthesis of nucleotides, proteins, and lipids while also fueling the TCA cycle for energy production. The metabolism of glutamine through glutaminolysis provides signals for energy and biosynthetic precursors essential for rapid cell proliferation. Inhibiting glutamine metabolism has been proposed as a potential therapeutic strategy, aiming to induce metabolic stress and apoptosis in GBC cells reliant on glutamine [21].

Arginine and serine metabolism also play important roles in GBC. Arginine, a semi-essential amino acid, is metabolized in cancer cells to support tumor growth, while high arginase activity can suppress immune responses, allowing the tumor to evade detection. Targeting arginine metabolism may help impair tumor growth and restore anti-tumor immune activity. Similarly, increased serine biosynthesis in cancer cells, driven by upregulation of specific enzymes, supports nucleotide synthesis critical for cell growth. Additionally, tryptophan metabolism through the kynurenine pathway is involved in immune evasion, further aiding GBC progression. Addressing these metabolic pathways provides promising therapeutic opportunities to disrupt the survival mechanisms of GBC cells (Figure 8.4) [41].

8.4.1 Dysregulation of Essential Amino Acids: Glutamine and Arginine

The dysregulation of essential amino acids, particularly glutamine and arginine, is a defining characteristic of cancer metabolism that significantly contributes to tumor progression, survival, and immune evasion. In GBC and other cancers, the altered metabolism of these amino acids meets the high proliferative demands of tumor cells and fosters a metabolic environment that supports cancer growth. A deeper understanding of the roles of glutamine and arginine in cancer could provide valuable opportunities for new therapeutic strategies [41].

Glutamine is crucial for cancer cell survival and growth, becoming an essential nutrient for rapidly proliferating tumor cells, a phenomenon referred to as "glutamine addiction." It serves as a primary source of nitrogen and carbon for critical biosynthetic processes, including nucleotide and protein synthesis, while also fueling the TCA cycle, essential for energy production. Key to this pathway is glutaminase (GLS), which converts glutamine into glutamate, an intermediate that supports high bioenergetics in cancer cells. Additionally, glutamine metabolism helps maintain redox balance by providing glutamate for glutathione synthesis, a critical antioxidant that protects cells from oxidative stress. Targeting glutaminolysis has emerged as a promising cancer therapy, with inhibitors like CB-839 showing potential in disrupting tumor metabolism and enhancing treatment efficacy [42, 43].

Arginine metabolism is another critical pathway altered in cancer, with implications for tumor growth, immune evasion, and angiogenesis. Arginine, a semi-essential amino acid, is converted into ornithine and urea by the enzyme arginase. This process increases the production of polyamines, which facilitate cancer cell proliferation and support the progression of tumors such as GBC. Furthermore, the dysregulation of arginine metabolism leads to immune evasion by depleting arginine in the tumor microenvironment, hindering T-cell activation necessary for anti-tumor immune responses. Developing strategies to target arginine metabolism may restore immune function and inhibit tumor growth, paving the way for new therapeutic approaches that target glutamine and arginine dysregulation in cancer treatment [14, 29, 37, 41].

8.4.2 Amino Acid Metabolism and Tumor Growth

Amino acid metabolism is crucial for tumor cell growth and proliferation, as cancer cells undergo metabolic reprogramming to support their increased demands for biosynthesis and energy. Essential and non-essential amino acids are utilized to meet the high metabolic needs of rapidly dividing tumor cells, allowing them to adapt and thrive in challenging microenvironments. Among these amino acids, glutamine stands out due to its diverse metabolic roles, including providing precursors for nucleotides, amino acids, and lipids, as well as supporting the TCA cycle for energy production. This reliance on glutamine, often referred to as "glutamine addiction," highlights its potential as a target for therapeutic interventions aimed at disrupting tumor growth [25, 43].

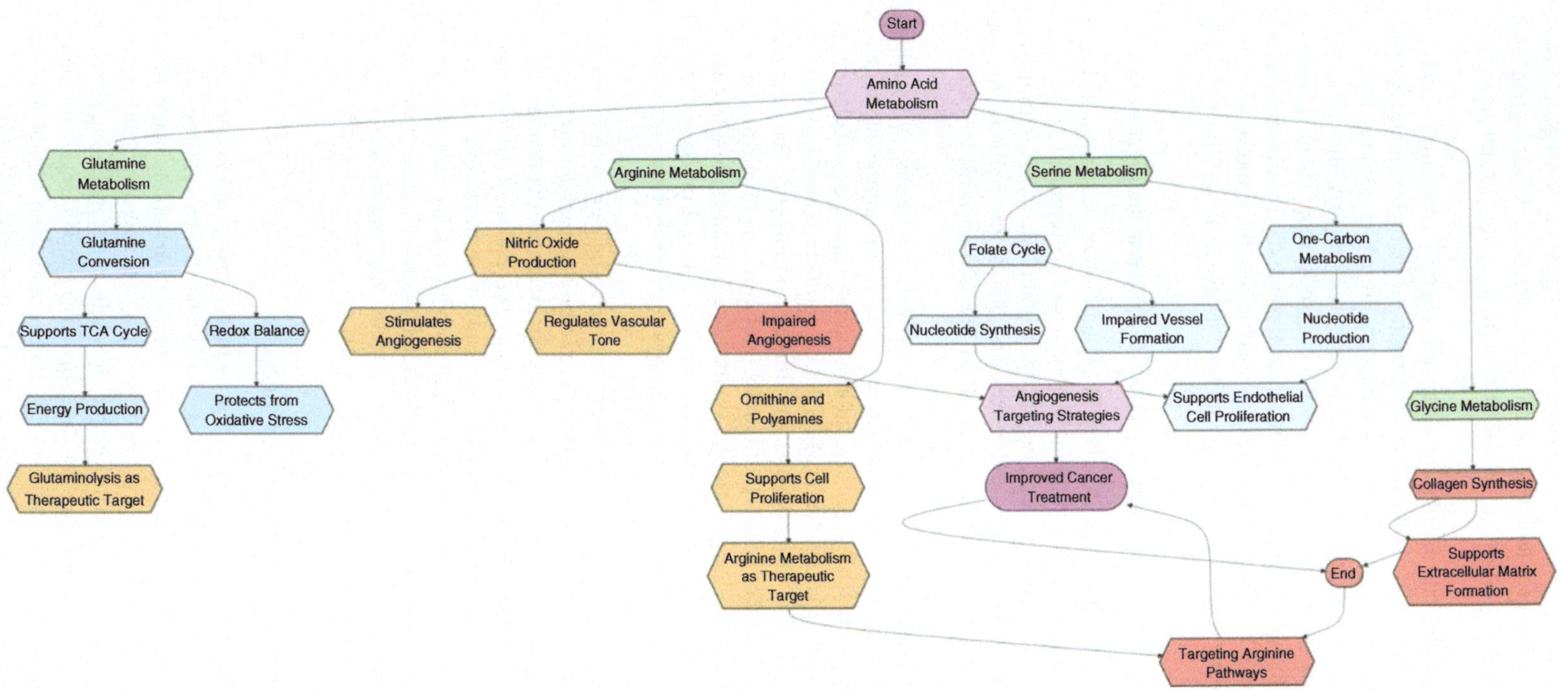

FIGURE 8.4 Amino acid metabolism. Amino acid metabolism is crucial in GBC progression, with key pathways including glutaminolysis, arginine metabolism, serine biosynthesis, and the tryptophan-kynurenine pathway driving tumor growth and immune evasion. Dysregulation of these pathways supports rapid proliferation, biosynthesis, and survival in GBC, highlighting them as potential therapeutic targets to disrupt cancer metabolism and enhance treatment outcomes.

In addition to their biosynthetic functions, amino acids significantly influence cell signaling pathways that regulate growth and proliferation. The mechanistic target of rapamycin (mTOR) pathway is particularly important, as it integrates amino acid availability with nutritional status to stimulate anabolic processes in cancer cells. Key amino acids, especially leucine and arginine, are pivotal for activating mTORC1, which in turn promotes protein synthesis and tumor growth. Consequently, the manipulation of amino acid availability and metabolism presents a promising strategy to target mTOR signaling and hinder cancer progression [44].

Furthermore, amino acids play a vital role in maintaining redox balance and facilitating immune evasion in tumors. Tumor cells often experience oxidative stress due to their rapid growth, with certain amino acids, like cysteine and glutamine, essential for producing glutathione, a major antioxidant that protects against ROS. This antioxidant defense helps tumor cells survive oxidative damage, contributing to their resilience. Additionally, tumors can alter amino acid metabolism to suppress immune function, with mechanisms such as tryptophan degradation inhibiting T-cell activity. Understanding these metabolic adaptations allows for the exploration of new therapeutic strategies targeting glutaminolysis, mTOR signaling, and immune escape to improve outcomes for cancer patients [26, 41, 44].

8.4.3 Role of Amino Acid Metabolism in Angiogenesis

Amino acid metabolism significantly influences angiogenesis, the process of forming new blood vessels from existing ones, which is essential for both normal physiological functions and pathological conditions like cancer development. Tumors, in particular, depend on the creation of new blood vessels to satisfy their rapid growth and nutrient needs. Recent studies show that amino acids are not only fundamental building blocks for proteins but also critical regulators of angiogenesis by affecting signaling pathways, generating metabolic intermediates, and providing precursors for angiogenic molecules [45].

Glutamine stands out as a crucial amino acid in promoting angiogenesis, particularly in tumor cells that heavily rely on its metabolism. Through glutaminolysis, glutamine is converted into glutamate and then into α-ketoglutarate, which is vital for energy production and biosynthesis. In endothelial cells (ECs), glutamine metabolism supports their dual roles in proliferating and migrating during blood vessel formation. Disruption of this metabolic pathway can hinder these processes, thereby reducing angiogenesis and limiting tumor growth [25, 43].

Another important amino acid involved in angiogenesis is arginine, which serves as a precursor for nitric oxide (NO) synthesis. NO is essential for regulating vascular tone and enhancing blood vessel formation. Produced by endothelial nitric oxide synthase (eNOS), NO aids in dilating blood vessels and promoting EC proliferation and migration. In the tumor environment, increased arginine metabolism and NO production encourage angiogenesis to meet higher oxygen and nutrient demands, making the regulation of this pathway vital for cancer progression [46].

Serine and glycine also play significant roles in one-carbon metabolism, essential for EC proliferation and nucleotide synthesis. Serine, synthesized from glucose, feeds into the folate cycle, vital for generating purines and pyrimidines necessary for DNA replication. Glycine, derived from serine, contributes to collagen synthesis, which supports the extracellular matrix of new vessels. Additionally, the mTOR signaling pathway is activated by amino acids like leucine, driving EC growth and proliferation. Targeting amino acid metabolism and these signaling pathways presents a promising therapeutic approach to inhibit angiogenesis and tumor development (Figure 8.4) [47].

8.5 METABOLOMICS IN EARLY DETECTION OF GBC

Metabolomics presents a significant advantage in the early detection of GBC by capturing the complex biochemical changes that occur during tumor development. Cancer cells undergo metabolic reprogramming, leading to specific alterations in metabolites related to energy production, lipid and amino acid metabolism, and nucleotide synthesis. These changes create unique metabolic profiles detectable in biofluids such as blood, urine, or bile, allowing for non-invasive early identification of GBC. Studies have shown that certain metabolites, including PC, PE, glutamine, and arginine, are altered in GBC patients, indicating their potential as biomarkers for early disease detection [8, 38, 41, 43].

The integration of metabolomics with other omics technologies, like genomics and proteomics, enhances the understanding of GBC pathogenesis and improves the specificity and sensitivity of potential biomarkers. Bile fluid analysis is particularly promising, as it contains metabolites relevant to GBC and can differentiate between malignant and benign conditions. While challenges, such as variability in metabolomic profiles due to diet and lifestyle, exist, advances in analytical techniques have significantly improved the reliability of metabolomics data. Future research should focus on validating these biomarkers in large-scale studies and developing non-invasive diagnostic tools that could transform GBC screening and improve patient outcomes (Figure 8.5).

8.5.1 Identifying Metabolic Biomarkers for Early-Stage GBC

Early detection of GBC is challenging due to its asymptomatic early stages and the absence of specific diagnostic markers. Most patients are diagnosed at advanced stages, limiting treatment options and resulting in poor prognosis. Therefore, finding reliable early-stage biomarkers is a critical goal in cancer research. Recently, metabolomics has emerged as a valuable method for identifying metabolic biomarkers linked to early-stage GBC by analyzing small-molecule metabolites in biological samples. This approach highlights metabolic changes associated with cancer, such as increased glycolytic activity (the Warburg effect), which can be detected in bodily fluids. These unique metabolites may provide non-invasive diagnostic tools for early detection of GBC before it becomes advanced and difficult to treat [17].

Metabolomics has shown promise in identifying key metabolic changes during the early stages of GBC, particularly in lipid metabolism [37, 39]. Alterations in phospholipids, like PC and PE, can indicate malignancy, serving as potential blood-based biomarkers for early detection. Additionally, elevated levels of amino acids, such as glutamine and arginine, and changes in bile metabolites reflect metabolic shifts in tumor development [38]. Bile analysis could help differentiate malignant GBC from benign conditions. Integrating metabolomics with genomics and proteomics enhances biomarker specificity and sensitivity, although challenges, like variability in metabolomic profiles and the need for standardization, remain. Overall, metabolomics holds significant potential for revolutionizing early GBC diagnosis and improving patient outcomes.

8.5.2 Screening and Diagnostic Potential of Metabolomics

Metabolomics is emerging as a powerful tool for disease screening and diagnosis, particularly in cancer. By analyzing metabolites in biological systems, scientists can capture the metabolic changes that occur during disease progression, offering crucial insights into an organism's physiological state. Specific metabolic profiles unique to diseases like GBC have been identified, which can serve as valuable biomarkers for early detection. Notably, metabolomics can uncover distinctive changes in lipid and amino acid metabolism in patients with GBC, enhancing prospects for identifying early-stage disease [19].

Metabolomics sheds light on the dysregulated biochemical pathways associated with diseases, aiding in the understanding of disease mechanisms and potential therapeutic targets. The integration of metabolomics with other omics technologies can enhance the identification of biomarkers and improve diagnostic assays. While challenges, like variability in metabolite levels and the need for standardization, remain, the potential for metabolomics to revolutionize early detection and personalized treatment is significant. By addressing these challenges, the field could greatly enhance clinical applications, ultimately improving patient outcomes and disease management strategies.

8.6 PROGNOSIS AND RISK ASSESSMENT USING METABOLOMICS

Metabolomics, the comprehensive analysis of metabolites in biological samples, has become a vital tool in cancer research, particularly for GBC, which is known for its aggressive nature and poor

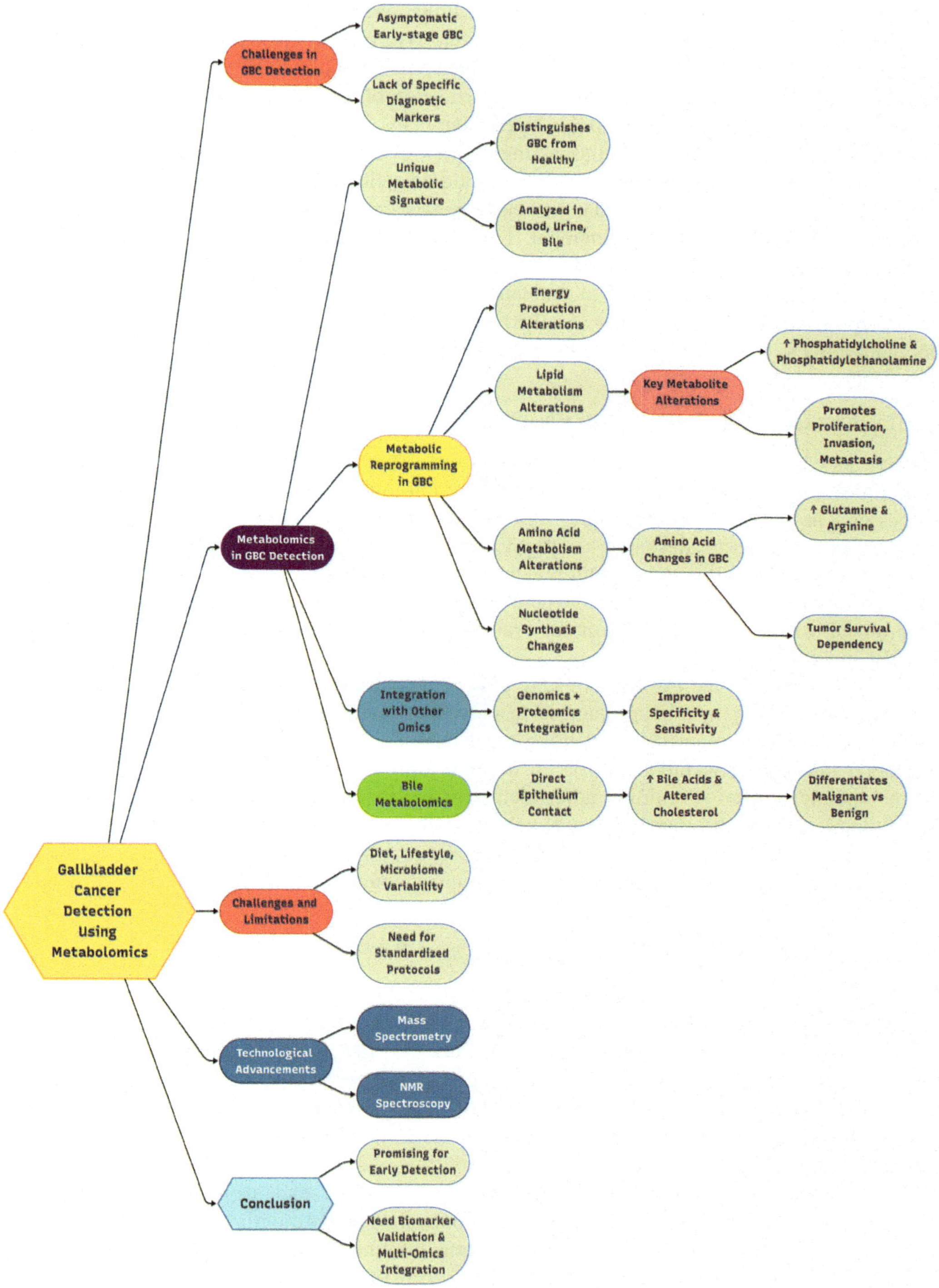

FIGURE 8.5 Gallbladder cancer detection using metabolomics. Metabolomics offers a promising approach for the early detection of gallbladder cancer (GBC) by identifying unique metabolic signatures in biofluids such as blood and bile. Key metabolic alterations, including changes in lipid and amino acid profiles, differentiate GBC patients from healthy individuals or those with benign conditions. Integrating metabolomics with genomics and proteomics enhances biomarker specificity and sensitivity, potentially revolutionizing early detection and improving patient outcomes.

survival rates. By offering insights into the metabolic state of an organism, metabolomics helps in understanding disease progression and individual risk factors. Research has identified specific metabolic alterations, such as elevated levels of phosphatidylcholine and various amino acids, that correlate with advanced GBC, enabling the stratification of patients based on their risk of disease progression and allowing for personalized treatment plans [38, 40].

Metabolomics is instrumental in assessing the risk of developing GBC in individuals with predisposing conditions, such as chronic inflammation or gallbladder diseases [19, 26]. By analyzing biofluids from these high-risk populations, early metabolic changes can be detected, acting as biomarkers for imminent cancer development. This approach facilitates early intervention, enhancing the likelihood of successful treatment. Additionally, monitoring metabolic profiles before, during, and after therapy allows clinicians to evaluate treatment effectiveness and make necessary adjustments, optimizing patient outcomes. As the field of metabolomics continues to evolve, its integration with clinical practice is set to improve prognosis and risk assessment in GBC significantly.

8.6.1 Metabolomic Profiles and Prognosis of GBC

Metabolomic profiles have become essential in understanding GBC, a highly aggressive cancer with low survival rates. Analyzing metabolites—small molecules involved in various biochemical processes—provides important insights into the metabolic changes that correlate with the disease's progression and severity. Research has highlighted specific metabolic signatures, such as elevated levels of certain lipids like PC and PE, which are associated with more aggressive tumor characteristics. Additionally, alterations in amino acid metabolism, particularly glutamine and arginine, have been linked to tumor growth and immune evasion, further underlining their potential as biomarkers for GBC prognosis [48].

Utilizing metabolomic profiles not only aids in risk stratification but also enhances treatment planning. Identifying metabolic changes in preoperative samples can help clinicians assess the likelihood of disease recurrence, enabling more targeted intervention for high-risk patients. Furthermore, monitoring metabolic shifts during and after treatment provides valuable information on therapeutic responses, guiding adjustments for optimal patient outcomes. The integration of metabolomic data with other omics technologies, like genomics and proteomics, creates a more comprehensive understanding of GBC and improves prognostic accuracy, paving the way for personalized treatment strategies and better overall management of the disease [4, 49].

8.6.2 Linking Metabolic Biomarkers to Patient Risk

Metabolomic profiles have become vital in understanding the prognosis of GBC, a disease known for its aggressive nature and low survival rates. Analyzing metabolites—small molecules integral to various biochemical processes—offers insights into the metabolic changes associated with GBC. These profiles help identify disease progression, treatment responses, and overall patient outcomes. Notably, distinct metabolic signatures linked to disease severity have been discovered, such as elevated levels of PC and PE, which are associated with more aggressive tumor characteristics [15].

The metabolism of amino acids, particularly glutamine and arginine, shows significant alterations in GBC patients. Elevated glutamine levels correlate with advanced disease stages, while arginine metabolism is connected to tumor immune evasion and blood vessel formation. The identification of these metabolic biomarkers aids in risk stratification, allowing clinicians to tailor treatment strategies based on predicted disease recurrence and patient survival. Furthermore, integrating metabolomic data with genomics and proteomics enhances prognostic capabilities, paving the way for more effective and personalized treatment approaches in GBC management [38, 40].

8.7 PERSONALIZED TREATMENT APPROACHES VIA METABOLOMICS

Metabolomics, the in-depth study of metabolites in biological systems, is transforming personalized medicine, particularly in the context of GBC. This field offers critical insights into the unique metabolic changes present in individual tumors, which can significantly impact a patient's response to therapies and overall prognosis. By analyzing these distinct metabolic profiles, clinicians can pinpoint dysregulated pathways in each patient's cancer, allowing for targeted treatment strategies that exploit these vulnerabilities. For instance, if a tumor shows a reliance on a specific lipid metabolism pathway, therapies that target this pathway can be prioritized, increasing the likelihood of effective treatment outcomes.

In addition to facilitating targeted therapies, metabolomics aids in identifying novel biomarkers that guide treatment decisions. These metabolic biomarkers, which reflect alterations associated with cancer, can help classify patients based on how likely they are to respond to particular therapies. This stratification enables clinicians to select the most suitable treatments aligned with an individual's metabolic landscape. Furthermore, real-time monitoring of a patient's metabolic profile during treatment allows for adaptive strategy modifications, ensuring that care is responsive to changes in tumor behavior, such as resistance to initial therapies.

The integration of metabolomics into clinical practice can enhance personalized treatment approaches for gallbladder cancer by predicting treatment outcomes and disease progression. By correlating specific metabolic characteristics with clinical responses, researchers can develop tailored therapeutic strategies designed to maximize efficacy and diminish the risk of relapse. The continuous advancement of metabolomics holds the potential to revolutionize cancer care, making treatments more precise, effective, and individualized for patients.

8.7.1 Targeting Metabolic Pathways in GBC Treatment

Targeting metabolic pathways in GBC offers a novel therapeutic strategy by focusing on the distinct alterations in lipid and amino acid metabolism that occur in tumor cells. GBC often exhibits abnormal lipid metabolism characterized by increased synthesis of phospholipids, which are crucial for cell membrane function and contribute to tumor growth and metastasis. Inhibiting key enzymes involved in lipid biosynthesis, such as choline kinase and lysophosphatidic acid acyltransferase, represents a promising approach to disrupt this altered metabolism. Similarly, amino acid metabolism, particularly the dysregulation of glutamine and arginine, plays a significant role in cancer cell proliferation. Targeting enzymes like glutaminase, which converts glutamine to glutamate, or arginase, which modulates arginine metabolism, could starve cancer cells of vital nutrients and impede their growth.

The tumor microenvironment significantly influences GBC progression, presenting further therapeutic opportunities. The presence of hypoxia and nutrient deprivation in the microenvironment drives adaptive metabolic changes in tumor cells. By targeting HIFs that mediate these responses, researchers may enhance the sensitivity of tumors to existing treatments [17]. Combining metabolic inhibitors with traditional therapies, like chemotherapy or immunotherapy, could create synergistic effects, improving overall treatment outcomes and potentially overcoming resistance mechanisms. As our understanding of GBC metabolism evolves, these targeted strategies may lead to more effective and tailored treatment options for patients battling this challenging cancer (Figure 8.6) [19, 50].

8.7.2 Identifying Metabolic Vulnerabilities in Tumor Cells

Identifying the metabolic vulnerabilities of tumor cells offers a groundbreaking method in cancer research, paving the way for targeted treatments that capitalize on the unique metabolic traits of

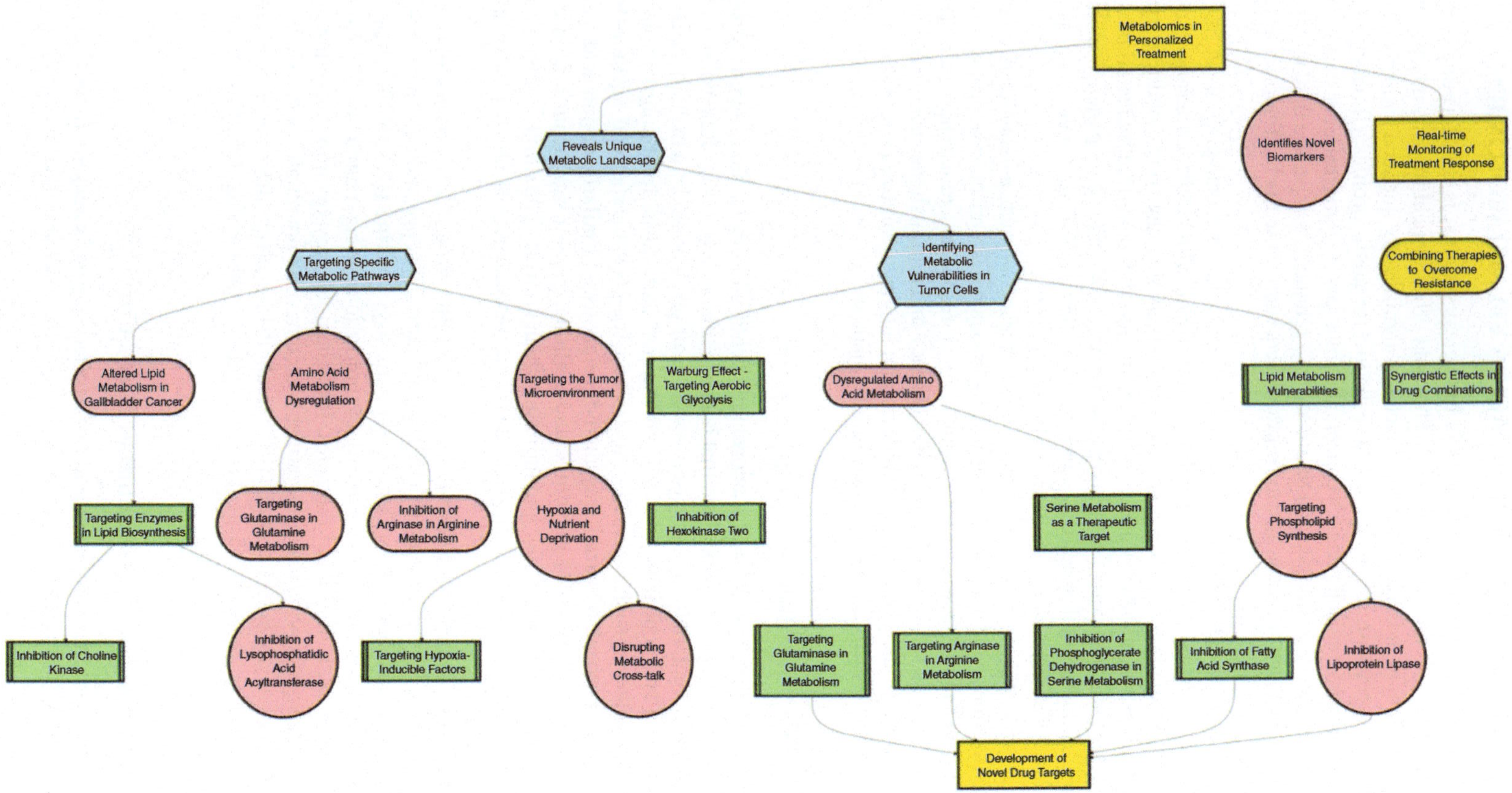

FIGURE 8.6 Metabolomics in personalized treatment. Metabolomics enhances personalized treatment in GBC by profiling unique metabolic alterations in individual tumors. This approach enables the identification of specific metabolic vulnerabilities and novel biomarkers, guiding the selection and optimization of targeted therapies. Real-time monitoring of metabolic changes during treatment further allows for adjustments to therapeutic regimens, improving efficacy and patient outcomes.

these cells. Tumor cells often display specific metabolic changes essential for their rapid growth and resistance to traditional therapies, primarily involving alterations in energy production, biosynthesis, and responses to cellular stress. A notable example is the Warburg effect, where many tumors prefer aerobic glycolysis, converting glucose to lactate even in the presence of oxygen. This metabolic pathway enables rapid proliferation and can be disrupted by targeting key glycolytic enzymes, like hexokinase, which has shown potential in preclinical studies to induce apoptosis in tumor cells [17].

Tumors often exhibit disruptions in amino acid metabolism, particularly with glutamine, which is vital for growth, survival, and synthesis of critical biomolecules. Targeting enzymes such as glutaminase can deprive cancer cells of necessary nutrients, while dysregulation of arginine metabolism can also be exploited through inhibitors of relevant enzymes [25, 43]. Moreover, altered lipid metabolism represents another critical vulnerability, with tumors often synthesizing specific lipids that facilitate cell proliferation and metastasis [37]. The tumor microenvironment, particularly under hypoxic conditions, further influences these metabolic pathways, providing additional targets for therapy. By concentrating on these metabolic alterations, cancer research aims to uncover new therapeutic strategies that can enhance treatment efficacy and offer personalized solutions for patients (Figure 8.6) [14, 29].

8.7.3 Novel Drug Targets from Lipid and Amino Acid Metabolism

Recent research into lipid and amino acid metabolism in cancer has identified several promising drug targets that may lead to innovative therapeutic interventions. Cancer cells often undergo significant metabolic changes that facilitate their unchecked growth, survival, and resistance to traditional treatments. By focusing on these altered metabolic pathways, scientists aim to develop therapies that can specifically inhibit tumor progression. Key targets in lipid metabolism include FASN and lipoprotein lipase, both of which are overexpressed in various cancers. Inhibitors of these enzymes have shown potential in preclinical studies to reduce tumor growth and induce cancer cell death [30].

Amino acid metabolism in cancer presents further targets for therapy. Enzymes such as glutaminase (GLS), which plays a vital role in glutamine metabolism, and arginase, involved in arginine metabolism, are notably dysregulated in tumors. Inhibitors of these enzymes have shown efficacy in clinical trials, reducing amino acid availability and impairing cancer cell proliferation. Additionally, the targeting of serine metabolism through enzymes like phosphoglycerate dehydrogenase (PHGDH) has emerged as another promising avenue for treatment. Overall, these approaches signify a significant step forward in personalized cancer therapy, holding the potential to improve patient outcomes [43, 44] (Figure 8.6).

8.8 INTEGRATING METABOLOMICS WITH OTHER OMICS TECHNOLOGIES

Integrating metabolomics with other omics technologies—such as genomics, transcriptomics, and proteomics—provides a comprehensive framework for understanding the complexities of biological systems and disease mechanisms, particularly in conditions like cancer. Metabolomics focuses on the analysis of small molecules or metabolites, revealing critical insights into the metabolic state of cells and tissues, while genomics identifies genetic variants and mutations linked to disease risk. However, genomics alone cannot explain how these changes manifest within biological systems, which is where metabolomics is essential. By examining metabolite profiles, researchers can explore how genetic variations influence metabolic pathways and contribute to disease, leading to the identification of biomarkers and therapeutic targets. Transcriptomics complements metabolomics by offering insights into gene expression, helping to clarify how changes in gene expression drive metabolic reprogramming in diseases. Meanwhile, proteomics examines protein levels and modifications, allowing researchers to connect changes in protein expression with metabolic alterations. Together, these technologies enable a systems biology approach that uncovers interactions among biological molecules, enhancing our understanding of complex diseases. For instance, integrating metabolomics with genomics has unveiled metabolic biomarkers tied to specific genetic mutations, and merging transcriptomics with metabolomics has highlighted how shifts in gene expression can drive metabolic changes in tumors. Overall, this multiomics integration provides valuable insights into disease mechanisms and supports the advancement of personalized medicine by improving biomarker identification and therapy development.

8.8.1 Synergistic Approaches: Genomics, Proteomics, and Metabolomics

The integration of genomics, proteomics, and metabolomics presents a powerful approach to understanding complex biological systems and diseases, significantly enhancing our ability to uncover mechanisms and develop targeted therapies. Genomics provides a detailed insight into an organism's genetic blueprint, identifying variants and mutations linked to disease. However, it often lacks the functional insights needed to relate genetic alterations to cellular processes. This limitation is addressed by proteomics, which offers a snapshot of the protein landscape, revealing expression levels, modifications, and interactions, thereby linking genomic data to protein function and cellular pathways. Proteomics captures changes in protein expression due to genetic and environmental factors, helping bridge the gap between genetic information and biological function. Furthermore, metabolomics studies small molecules and metabolites that reflect the metabolic state of cells; by integrating metabolomics with genomics and proteomics, researchers can connect genetic and proteomic changes to metabolic disruptions critical to disease development and progression. This holistic systems biology approach allows for the exploration of interactions between genetic, proteomic, and metabolic alterations, particularly in cancer research, where it elucidates how mutations drive protein expression changes and metabolic pathways leading to tumorigenesis. Ultimately, this synergistic framework has propelled advancements in personalized medicine, enabling the identification of patient-specific biomarkers and therapeutic targets that lead to tailored treatment strategies. By combining these technologies, researchers can deepen their understanding of disease mechanisms and improve precision medicine to enhance patient outcomes.

8.8.2 Comprehensive Multiomics Approaches to GBC

Genomics forms the foundation of multiomics approaches in GBC by identifying genetic mutations and variations that drive tumor initiation and progression, while proteomics sheds light on changes in protein expression and interactions that are critical for tumor development. Metabolomics further enriches our understanding by profiling small molecules and metabolites, revealing metabolic reprogramming associated with GBC. By integrating these layers of data, researchers can identify synergistic biomarkers and therapeutic targets that might otherwise be overlooked, such as novel metabolic signatures for early detection and critical molecular pathways for targeted therapies. This multifaceted approach supports personalized medicine, allowing clinicians to tailor treatment strategies to individual molecular profiles, ultimately enhancing patient outcomes and advancing cancer research [51].

8.9 FUTURE DIRECTIONS IN METABOLOMICS RESEARCH FOR GBC

Metabolomics research offers significant potential to enhance our understanding and treatment of GBC, an aggressive disease with limited therapeutic options. Future advancements in metabolomic profiling techniques, particularly through improvements in mass spectrometry and nuclear magnetic resonance spectroscopy [48], are set to increase the sensitivity and resolution of analyses. These developments will enable the detection of lower-abundance metabolites and a more comprehensive characterization of metabolic pathways in GBC. Furthermore, integrating high-throughput metabolomics with genomics, proteomics, and transcriptomics will provide a holistic view of GBC's molecular foundations, helping to uncover complex interactions and identify novel biomarkers and therapeutic targets.

Another promising direction lies in exploring the gut microbiome's influence on GBC development and progression. Research indicates that the gut microbiome can affect systemic metabolism and immune responses, potentially playing a role in cancer pathogenesis. Combining metabolomic and microbiome data could lead to insights for early detection and treatment strategies. Additionally, personalized medicine approaches, guided by individual metabolic profiles, could optimize treatment options to improve patient outcomes. Monitoring treatment responses with metabolic biomarkers will also be crucial for adapting strategies in real-time. Ultimately, targeted interventions addressing metabolic vulnerabilities identified through metabolomics may pave the way for new or repurposed therapies, offering more effective solutions for managing this challenging cancer type.

8.10 CONCLUSION

Metabolomics significantly enhances our understanding of GBC by providing insights into the metabolic changes associated with this aggressive disease. By analyzing small molecules in biological samples, metabolomics uncovers alterations in crucial metabolic pathways, including lipid and amino acid metabolism, which are essential to GBC development. This technology has successfully identified biomarkers like altered phosphatidylcholine and glutamine levels, directly linked to tumor progression and metastasis. Furthermore, metabolomics aids in early detection by highlighting unique metabolic profiles that differentiate early-stage GBC from benign conditions, potentially leading to timely interventions and improved patient outcomes.

The clinical implications of metabolomics in GBC extend beyond early detection; they include enhanced prognostic and therapeutic strategies. By identifying specific metabolic biomarkers, clinicians can better assess disease progression and patient risk, allowing for personalized treatment approaches. The integration of metabolomics with genomics and proteomics promises a comprehensive understanding of GBC, paving the way for new drug targets and more effective therapies. Continued advancements in metabolomics technologies and interdisciplinary collaboration are vital for translating these findings into clinical practice, ultimately improving diagnostic accuracy, treatment efficacy, and patient survival in gallbladder cancer.

REFERENCES

1. Awasthi, S., et al., Genomic landscape of gallbladder cancer: Insights from whole exome sequencing. *Int J Surg*, 2024. 110(11): p. 6883–6897.
2. Sharma, A., et al., Gallbladder cancer epidemiology, pathogenesis and molecular genetics: Recent update. *World J Gastroenterol*, 2017. 23(22): p. 3978–3998.
3. Okumura, K., et al., Gallbladder cancer: Historical treatment and new management options. *World J Gastrointest Oncol*, 2021. 13(10): p. 1317–1335.
4. Debela, D.T., et al., New approaches and procedures for cancer treatment: Current perspectives. *SAGE Open Med*, 2021. 9: p. 20503121211034366.
5. Liu, B., et al., Exploring treatment options in cancer: Tumor treatment strategies. *Signal Transduct Target Ther*, 2024. 9(1): p. 175.
6. Soundararajan, R., et al., Imaging patterns of wall thickening type of gallbladder cancer. *Clin Exp Hepatol*, 2022. 8(4): p. 255–266.
7. Vijayakumar, A., et al., Early diagnosis of gallbladder carcinoma: An algorithm approach. *ISRN Radiol*, 2013. 2013: p. 239424.
8. Amezcua-Hernandez, V., et al., Preoperative EUS vs. PET-CT evaluation of response to neoadjuvant therapy for esophagogastric cancer and its correlation with survival. *Cancers (Basel)*, 2023. 15(11).
9. Garcia, P., et al., Current and new biomarkers for early detection, prognostic stratification, and management of gallbladder cancer patients. *Cancers (Basel)*, 2020. 12(12).
10. Kanthan, R., et al., Gallbladder cancer in the 21st century. *J Oncol*, 2015. 2015: p. 967472.

11. Tang, Y., and Y. Fan, Combined KRAS and TP53 mutation in patients with colorectal cancer enhance chemoresistance to promote postoperative recurrence and metastasis. *BMC Cancer*, 2024. 24(1): p. 1155.
12. Tang, T., et al., Advantages of targeting the tumor immune microenvironment over blocking immune checkpoint in cancer immunotherapy. *Signal Transduct Target Ther*, 2021. 6(1): p. 72.
13. Shafqat, A., et al., Reprogramming the immunosuppressive tumor microenvironment: Exploiting angiogenesis and thrombosis to enhance immunotherapy. *Front Immunol*, 2023. 14: p. 1200941.
14. Sadeghi Rad, H., et al., Understanding the tumor microenvironment for effective immunotherapy. *Med Res Rev*, 2021. 41(3): p. 1474–1498.
15. Qiu, S., et al., Small molecule metabolites: Discovery of biomarkers and therapeutic targets. *Signal Transduct Target Ther*, 2023. 8(1): p. 132.
16. Schmidt, D.R., et al., Metabolomics in cancer research and emerging applications in clinical oncology. *CA Cancer J Clin*, 2021. 71(4): p. 333–358.
17. Mathew, M., et al., Metabolic signature of Warburg effect in cancer: An effective and obligatory interplay between nutrient transporters and catabolic/anabolic pathways to promote tumor growth. *Cancers (Basel)*, 2024. 16(3).
18. Wang, W., et al., Cancer metabolites: Promising biomarkers for cancer liquid biopsy. *Biomark Res*, 2023. 11(1): p. 66.
19. Danzi, F., et al., To metabolomics and beyond: A technological portfolio to investigate cancer metabolism. *Signal Transduct Target Ther*, 2023. 8(1): p. 137.
20. Yang, J., et al., Cancer metabolism and carcinogenesis. *Exp Hematol Oncol*, 2024. 13(1): p. 10.
21. Wei, Z., et al., Metabolism of amino acids in cancer. *Front Cell Dev Biol*, 2020. 8: p. 603837.
22. Nong, S., et al., Metabolic reprogramming in cancer: Mechanisms and therapeutics. *MedComm (2020)*, 2023. 4(2): p. e218.
23. Tarrado-Castellarnau, M., P. de Atauri, and M. Cascante, Oncogenic regulation of tumor metabolic reprogramming. *Oncotarget*, 2016. 7(38): p. 62726–62753.
24. Jin, H.R., et al., Lipid metabolic reprogramming in tumor microenvironment: From mechanisms to therapeutics. *J Hematol Oncol*, 2023. 16(1): p. 103.
25. Yoo, H.C., et al., Glutamine reliance in cell metabolism. *Exp Mol Med*, 2020. 52(9): p. 1496–1516.
26. Bhardwaj, V., and J. He, Reactive oxygen species, metabolic plasticity, and drug resistance in cancer. *Int J Mol Sci*, 2020. 21(10).
27. Tufail, M., C.H. Jiang, and N. Li, Altered metabolism in cancer: Insights into energy pathways and therapeutic targets. *Mol Cancer*, 2024. 23(1): p. 203.
28. Wang, Z.H., et al., Lactate in the tumour microenvironment: From immune modulation to therapy. *EBioMedicine*, 2021. 73: p. 103627.
29. Perez-Tomas, R., and I. Perez-Guillen, Lactate in the tumor microenvironment: An essential molecule in cancer progression and treatment. *Cancers (Basel)*, 2020. 12(11).
30. Wu, Y., et al., Reprogramming of lipid metabolism in the tumor microenvironment: A strategy for tumor immunotherapy. *Lipids Health Dis*, 2024. 23(1): p. 35.
31. Cheng, H., et al., Lipid metabolism and cancer. *Life (Basel)*, 2022. 12(6).
32. Feng, J., et al., Emerging roles and the regulation of aerobic glycolysis in hepatocellular carcinoma. *J Exp Clin Cancer Res*, 2020. 39(1): p. 126.
33. Xia, W., et al., The role of cholesterol metabolism in tumor therapy, from bench to bed. *Front Pharmacol*, 2023. 14: p. 928821.
34. Tavasoli, M., et al., Genetic diseases of the Kennedy pathways for membrane synthesis. *J Biol Chem*, 2020. 295(51): p. 17877–17886.
35. Stoica, C., et al., Bilayer forming phospholipids as targets for cancer therapy. *Int J Mol Sci*, 2022. 23(9).
36. Li, R.Z., et al., The key role of sphingolipid metabolism in cancer: New therapeutic targets, diagnostic and prognostic values, and anti-tumor immunotherapy resistance. *Front Oncol*, 2022. 12: p. 941643.
37. Broadfield, L.A., et al., Lipid metabolism in cancer: New perspectives and emerging mechanisms. *Dev Cell*, 2021. 56(10): p. 1363–1393.
38. Calzada, E., O. Onguka, and S.M. Claypool, phosphatidylethanolamine metabolism in health and disease. *Int Rev Cell Mol Biol*, 2016. 321: p. 29–88.
39. Fu, Y., et al., Lipid metabolism in cancer progression and therapeutic strategies. *MedComm (2020)*, 2021. 2(1): p. 27–59.
40. Khan, F., et al., Unraveling the intricate relationship between lipid metabolism and oncogenic signaling pathways. *Front Cell Dev Biol*, 2024. 12: p. 1399065.
41. Chen, C.L., et al., Arginine signaling and cancer metabolism. *Cancers (Basel)*, 2021. 13(14).

42. Jin, J., et al., Targeting glutamine metabolism as a therapeutic strategy for cancer. *Exp Mol Med*, 2023. 55(4): p. 706–715.
43. Cluntun, A.A., et al., Glutamine metabolism in cancer: Understanding the heterogeneity. *Trends Cancer*, 2017. 3(3): p. 169–180.
44. Chen, J., et al., Amino acid metabolism in tumor biology and therapy. *Cell Death Dis*, 2024. 15(1): p. 42.
45. Saman, H., et al., Inducing angiogenesis, a key step in cancer vascularization, and treatment approaches. *Cancers (Basel)*, 2020. 12(5).
46. Wu, G., et al., Role of L-arginine in nitric oxide synthesis and health in humans. *Adv Exp Med Biol*, 2021. 1332: p. 167–187.
47. Pan, S., et al., Serine, glycine and one-carbon metabolism in cancer (Review). *Int J Oncol*, 2021. 58(2): p. 158–170.
48. Ranjan, R., and N. Sinha, NMR-based metabolomics in gallbladder cancer research. *Methods Mol Biol*, 2019. 2037: p. 231–241.
49. Du, Y., W.A. Wijaya, and W.H. Liu, Advancements in metabolomics research in benign gallbladder diseases: A review. *Medicine (Baltimore)*, 2024. 103(21): p. e38126.
50. Chen, Y., et al., Metabolomic machine learning predictor for diagnosis and prognosis of gastric cancer. *Nat Commun*, 2024. 15(1): p. 1657.
51. Maan, K., et al., Metabolomics and transcriptomics based multi-omics integration reveals radiation-induced altered pathway networking and underlying mechanism. *NPJ Syst Biol Appl*, 2023. 9(1): p. 42.

9 Long Noncoding RNA Signatures in Gallbladder Cancer Tumorigenesis

Khushi Verma and Puneet*
Department of General Surgery, Institute of Medical Sciences, Banaras Hindu University, India
*Corresponding author: khushiimsbhu24@gmail.com

9.1 INTRODUCTION

Numerous mutations or copy number alterations in cancer are frequently found in noncoding DNA regions, according to the genomic mutation landscape of the disease (Wang et al., 2019). Only 20,000 proteins, or less than 2% of the overall genomic sequence, were found to be encoded after the human genome was sequenced, and the remaining 98% of the human genome was found to be devoid of any protein-encoding sequence (Wilusz et al., 2009). Regulatory and structural RNAs, such as ribosomal and transfer RNAs, belong to the nonprotein-coding RNA (ncRNA) family. Furthermore, three distinct categories of regulatory RNAs have been identified:

(a) Small noncoding RNAs are characterized as RNAs with lengths of 20–50 nucleotides. These include cis-regulatory RNAs (cisRNAs), piwi-interacting RNAs (piRNAs), telomere-specific small RNAs (telsRNAs), and interfering RNAs (siRNAs).
(b) Medium noncoding RNAs (mRNAs) include small nuclear RNAs (snRNAs), small cytoplasmic RNAs (scRNAs), transcription initiation RNAs (tiRNAs), and small nucleolar RNAs (snoRNAs). Their lengths vary from 50 to 200 nucleotides.
(c) Long noncoding RNAs (lncRNAs) are noncoding RNAs longer than 200 nucleotides that are either totally or partially incapable of coding proteins (Jurga and Barciszewski, 2020).

There is currently increasing evidence to suggest that the distinction between transcripts with and without coding potential may not be as clear. Traditionally, an mRNA's open reading frame (ORF) contains a start codon, which initiates protein translation. An ORF typically consists of codons spanning 50 amino acids in bacteria and at least 100 amino acids in eukaryotes, terminating protein synthesis at the stop codon when the ribosome disassembles. Nevertheless, a variety of noncoding RNAs with ORFs smaller than the classical level are misannotated because of these arbitrary standards for defining what constitutes a protein-coding transcript. There is increasing evidence that some putative noncoding RNAs (ncRNAs) encode small proteins, or micropeptides, which are a subclass of lncRNAs (Hartford and Lal, 2020).

9.2 DEFINITION AND NOMENCLATURE OF LNCRNAS

On the basis of current estimations, the human genome has over 16,000 lncRNA genes that encode over 28,000 different lncRNAs. These lncRNAs can be longer than two kilobases, but their coding potential is less than 100 amino acids, and their length normally ranges from 1000 to 10,000 nucleotides (Balas et al., 2018; Chen et al., 2019). Long noncoding RNA (lncRNA)-encoding genes

DOI: 10.1201/9781003506201-9

typically have unique DNA sequences, distinctive transcription factor (TF)-binding regions, and promoters of their own (Dhanoa et al., 2018).

Numerous distinguishing characteristics of mRNAs, such as the Kozak consensus sequence and ORFs of notable lengths, are absent from a broad class of produced RNA molecules known as long noncoding RNAs (lncRNAs) (Hu et al., 2018). Like mRNAs, RNA polymerase II (Pol II) transcribes most lncRNAs, which are then polyadenylated and capped. While lncRNAs were initially thought to be unstable, this is only true for a minority of lncRNAs. Polyadenylation stabilizes most lncRNAs; however, non-polyadenylated lncRNAs can also be stabilized by secondary structures, such as triple-helical structures at their 3' ends. These 3' sequence characteristics not only have a stabilizing effect but also promote effective nuclear export (Bridges et al., 2021).

Since most lncRNAs undergo significant alternative splicing, the number of possible isoforms increases substantially. According to a recent analysis, lncRNAs are significantly more alternatively spliced than mRNAs; even if their splicing efficiency is high, their specific intron splicing frequency appears to be lower than that of mRNAs (Bridges et al., 2021). Although the expression levels of lncRNAs are often lower than those of mRNAs, they exhibit more robust tissue-specific expression patterns, indicating crucial roles in processes particular to distinct cell types (Bridges et al., 2021).

9.3 CONSERVATION OF LNCRNAs

Furthermore, lncRNAs typically lack the high level of sequence conservation observed in mRNAs across species. However, a recent study revealed that although a significant number of lncRNAs exhibit sequence and/or positional conservation between human and mouse embryonic stem cells, mouse cells process these lncRNAs differently than human cells, leading to their localization to different subcellular compartments and ultimately serving different functions. Conserved mRNAs, however, presented identical patterns of localization in both species, which contrasts with previous findings (Bridges et al., 2021; Mattick et al., 2023). Many lncRNAs exhibit tissue-specific, temporal, and spatial regulation in addition to moderate levels of expression, suggesting that lncRNA expression is tightly controlled and organized during both development and disease (Hu et al., 2018).

It has also been demonstrated that lncRNAs and signaling receptors directly interact. The fact that lncRNAs interact with a variety of molecular species emphasizes the importance of comprehending lncRNA localization and local interactions to predict their function (Bridges et al., 2021). A lncRNA's subcellular location can provide valuable insight into how it functions. Long noncoding RNAs are often found in the mitochondria, cytoplasm, nucleolus, and nucleus (Fernandes et al., 2019). LncRNAs that are located in the nucleus can affect pre-mRNA splicing, regulate transcription (via chromatin remodeling or enhancer RNAs), and act as scaffolds for the formation of protein complexes and subnuclear domains. There is a chance that some lncRNA transcripts in the cytoplasm will be translated into micropeptides. A recent study revealed that various lncRNAs encode micropeptides that perform essential biological functions across a wide range of organisms, including humans, flies, and bacteria (Hartford and Lal, 2020). In addition, two categories of lncRNA function modes are distinguished as follows: trans-regulatory lncRNAs can control gene expression at a distance from their transcription site by modulating protein function or altering the state of chromatin, whereas cis-regulatory lncRNAs affect only neighboring genes on the same allele from which they are transcribed (Zampetaki et al., 2018).

9.4 BIOGENESIS OF LONG NONCODING RNAs

On the basis of their transcription direction with respect to other genes and their biogenesis position in the genome, lncRNAs are categorized into six types (Siddiqui et al., 2019):

A. Intergenic noncoding RNAs: These long noncoding RNAs are located in the genome between two genes that encode proteins.

B. Intronic lncRNAs: These are derived from an intron of a separate transcript.
C. Exonic or sense lncRNAs: These overlap with exons of a separate transcript on the same strand that codes for a distinct protein.
D. Antisense transcript lncRNAs: These are produced when a gene that codes for a protein has its transcription orientation reversed. The transcriptional orientation can also be categorized as either trans, which occurs on the opposite strand of a pseudogene, or cis, which occurs on the opposite strand of a protein-coding gene.
E. Bidirectional lncRNAs: These are expressed simultaneously with a neighboring transcript on the opposite strand.
F. Miscellaneous lncRNAs: Numerous other lncRNA subtypes, classified on the basis of the location of their production, have been discovered. Pseudo-lncRNAs, which originate from pseudogenes, are among them.

Pseudo-lncRNAs are obtained from pseudogenes; these include enhancer lncRNAs (eRNAs), which are transcribed from enhancer regions by RNAP II or III; transcribed-ultraconserved lncRNAs (T-UCR lncRNAs), which are produced from ultraconserved regions, such as Evf-2; phosphate and tensin homolog pseudogenes (pTENpg1); and promoter lncRNAs, such as promoter upstream transcripts (PROMPTS), which are produced from promoter sequences (Khandelwal et al., 2015).

9.5 DIVERGENT FUNCTIONS OF LNCRNAS

Despite their noncoding nature, most lncRNAs have a variety of functions in disease and biological developmental processes. Extensive research is necessary to clarify the specific roles and mechanisms of action of long noncoding RNAs. In general, lncRNAs have important regulatory functions in the expression of genes involved in a number of diseases, including cancer. Additionally, recent research has indicated that lncRNAs are involved in viral infection (Bridges et al., 2021). Four different categories can be used to classify the molecular roles of lncRNAs:

- *Signals*: Certain long noncoding RNAs (lncRNAs) are molecules that can respond to a variety of stimuli and are detected as markers of transcriptional activity. For example, the transcription factors p53 and MEG3 combine to produce growth differentiation factor 15 (GDF-15), which stops the proliferation of cancer cells. Notably, several cancer types have decreased MEG3 expression (Khandelwal et al., 2015).
- *Decoys*: Regulatory proteins can access RNA or DNA target-binding sites, and through their role as molecular decoys, lncRNAs are versatile regulators of transcription capable of both promoting and repressing gene expression. One notable example is growth arrest-specific transcript 5 (Gas5), a lncRNA with a distinctive hairpin structure that mimics the DNA-binding domain of glucocorticoid receptors. Gas5 uses its hairpin sequence motif to efficiently limit the transcription of metabolic genes during starvation by emulating the glucocorticoid receptor's DNA-binding site (Khandelwal et al., 2015).
- *Guides*: lncRNAs play crucial roles in orchestrating the spatial organization and precise localization of protein complexes at specific genomic regions, thereby influencing genome regulation. By interacting with key regulatory proteins, such as transcription factors and chromatin modifiers, lncRNAs guide these molecules to specific genomic sites. This targeting can occur in *cis* (affecting neighboring genes) or *trans* (impacting distant genes) relative to the lncRNA's transcription site. Homeobox antisense intergenic RNA (HOTAIR), a prominent example of a guide lncRNA, operates in *trans* to direct the chromatin modifier polycomb repressive complex 2 (PRC2) to target loci, such as the developmental HOXD region. Aberrant overexpression of HOTAIR redirects PRC2 to genes associated with cancer, leading to transcriptional repression. This dysregulation is strongly linked to tumor progression, invasion, and metastasis in cancers of the gastrointestinal tract, liver, pancreas, and breast (Balas et al., 2018).

- *Scaffolds:* Long noncoding RNAs (lncRNAs) serve as molecular scaffolds by facilitating the assembly of ribonucleoprotein (RNP) complexes. This involves recruiting and organizing two or more proteins into distinct, functional complexes, enabling precise regulation of various cellular processes. Through transcriptional activation or repression, these often ephemeral RNP complexes can selectively target certain genomic areas to control gene expression in time- and space-restricted ways. One well-known example of an RNA scaffold required for telomerase activity is telomerase RNA TERC (TERRA). The telomerase complex, which keeps telomere ends intact, is combined (Balas et al., 2018; Zhou et al., 2019).

9.6 EXPLORING LNCRNA INTERACTIONS WITH OTHER MOLECULES

Long noncoding RNAs (lncRNAs) interact with diverse molecules, such as proteins, DNA, RNA, and metal ions, and play pivotal roles in various biological processes, including chromatin remodeling, epigenetic modifications, gene transcription and translation, RNA stability, and genome defense. These functional roles are supported by the ability of lncRNAs to adopt specific tertiary structures tailored to their tasks. On the basis of their modes of interaction, lncRNA functionality can be categorized into five distinct models, each reflecting a unique mechanism of molecular interplay (Hu et al., 2018; Li et al., 2019).

- *Protein-lncRNAs:* Numerous binding sites for different molecules engaged in crucial cellular processes are provided by the complex secondary and tertiary structures of lncRNAs. LncRNAs work together as scaffolding molecules to interact with many proteins. Several theories explain how long noncoding RNAs (lncRNAs) are linked to proteins to impact gene expression. By attaching themselves to proteins and covering their ubiquitination sites, lncRNAs can prevent the accumulation of the desired protein. One characteristic of lncRNAs is protein binding, and the bound domain may change the function of the lncRNA. For example, certain PRC1 and PRC2 complex sites can directly interact with the antisense 3.8 kb lncRNA ANRIL to change the amount of chromatin modification and, consequently, the cis expression of protein-coding genes (Hu et al., 2018; Li et al., 2019).
- *lncRNA-DNAs*: LncRNAs regulate gene transcription via a variety of intricate processes, including binding to genomic DNA elements and localizing to particular DNA sequences, such as gene promoters. Triplexes of RNA–dsDNA are formed as a result of this process. These lncRNAs are also known as protector-associated lncRNAs or PARRs. Through hydrogen bonding, lncRNAs can enter the major groove of the DNA duplex and function as a third strand. In general, Hoogsteen hydrogen bonding via a third strand is weaker than Watson–Crick hydrogen bonding. Roberts and Crothers (1992) revealed that RNA is capable of forming triplex structures, which are more stable than DNA. Blume et al. (2003) first identified the *dhfr* minor transcript, a nucleus-enriched lncRNA transcribed from the upstream minor promoter of the *dihydrofolate reductase (DHFR)* gene. This lncRNA plays a regulatory role by modulating the transcriptional activity of the *dhfr* core promoter. Later studies by Martianov et al. confirmed that the *dhfr* minor transcript facilitates the formation of a stable RNA–DNA triplex structure at the primary promoter region of the *DHFR* gene. This structure blocks the binding of the general transcription factor TFIIB, thereby repressing *DHFR* gene expression (Hu et al., 2018; Li et al., 2019).
- *lncRNA-RNAs:* LncRNAs can demonstrate perfect or imperfect complementarity when interacting with other RNA molecules, such as mRNAs and miRNAs. Subcellular distribution, stability, and pre-mRNA splicing are regulated by lncRNAs. Certain antisense lncRNAs can bind to homologous mRNAs at splice sites, effectively masking these regions and preventing spliceosome assembly. Additionally, lncRNAs can stabilize and enhance the synthesis of mRNAs by interacting with their miRNA targets through miRNA response elements. These molecules are also capable of simultaneously binding both proteins and

mRNAs. An example of a noncoding RNA that functions as a miRNA sponge is PTENpg1, which is transcribed from the PTEN pseudogene. This pseudogene interacts with its antisense RNA β isoform via RNA–RNA pairing. The sponge activity of PTENpg1 is enhanced by this interaction, making it more effective at sequestering miRNAs. The translation and stability of PTEN mRNA are thereby increased (Hu et al., 2018; Li et al., 2019).

- *Peptide-coding:* Although most long noncoding RNAs (lncRNAs) do not produce proteins, a subset of their short ORFs can encode biologically active micropeptides in various organisms, including bacteria, flies, and humans (Hartford and Lal, 2020). An example is myoregulin (MLN), which encodes a conserved 46-amino-acid peptide. In humans and mice, the MLN gene consists of three exons spanning genomic regions of approximately 16.5 kb and 15.0 kb, respectively. The MLN peptide forms a single transmembrane alpha helix and integrates into the sarcoplasmic reticulum (SR) membrane. It binds within the SR to inhibit the SERCA pump, a key regulator of Ca^{2+} levels, as demonstrated through biochemical and animal studies. MLNs play a pivotal role in adult skeletal muscle function, as evidenced by studies in MLN-KO mice. Compared with WT mice, these mice exhibited potentially elevated SR Ca^{2+} levels in myoblasts and enhanced skeletal muscle performance. Similarly, the lncRNA DWORF (dwarf open reading frame) encodes a 34-amino-acid peptide localized to the SR membrane. Animal studies have demonstrated that increased DWORF expression increases the peak systolic Ca^{2+} transient amplitude and SR Ca^{2+} load. Conversely, DWORF KO mice displayed impaired late Ca^{2+} relaxation, reduced Ca^{2+} clearance, and diminished SERCA affinity for Ca^{2+}.

  Another notable lncRNA is LINC00961, which encodes a 90-amino-acid polypeptide expressed in the heart, lung, and skeletal muscle. This polypeptide interacts with lysosomal v-ATPase in late lysosomes or endosomes to repress mTORC1 activation. Interestingly, the absence of this polypeptide aggravated muscle regeneration deficits in animal studies.

  Additionally, the lncRNA HOXB-AS3 encodes a 53-amino-acid micropeptide that plays a regulatory role in colorectal cancer (CRC). This micropeptide inhibits CRC cell line proliferation by competitively binding hnRNP A1. These findings underscore that mRNAs and lncRNAs can act not only as noncoding RNAs but also as templates for encoding functional peptides critical for muscle homeostasis and other physiological processes (Hu et al., 2018; Hartford and Lal, 2020).

- *lncRNA catalysts:* Catalytic RNAs, or ribozymes, are RNA molecules with enzymatic activity that play critical roles in biological processes such as protein synthesis and RNA processing. Notable examples include self-cleaving group I introns, the ribosome, and RNase P RNA. Owing to the challenges in experimentally detecting RNA structures and their regulation by intricate secondary and tertiary structures, computational methods have been developed to predict RNA structures from primary sequences.

  Among experimental techniques, nucleotide analog interference mapping has proven invaluable for analyzing RNA substructures and functional groups at the atomic level. This approach has been instrumental in identifying several catalytic RNAs, including those associated with hepatitis delta virus, Varkud satellite ribozymes, and hairpin RNAs. However, uncovering the catalytic functions of long noncoding RNAs (lncRNAs) and understanding their mechanisms remain significant challenges, requiring a combination of advanced computational and experimental strategies (Hu et al., 2018).

The numerous functions of lncRNAs include the regulation of epigenetic modifications, post-transcription, translation, and gene transcription. An overview of dysregulated long noncoding RNAs implicated in gallbladder cancer and their biological roles is presented in Table 9.1. Furthermore, studies have demonstrated that during cancer development, lncRNAs regulate the immune system, migration, invasion, apoptosis, the cell cycle, and the maintenance of stemness (Jiang et al., 2019; Kumar et al., 2024).

Studies conducted on long noncoding RNAs (lncRNAs) have indicated that these molecules are associated with numerous disorders; however, the development of multiple

TABLE 9.1
Gallbladder Cancer-Associated Dysregulated Long Noncoding RNAs and Their Reported Biological Functions

S. No.	LncRNA	Genomic location	Regulation	Main effect	Biological significance	Gene/protein/pathways affected	References
1	HOTAIR	12q13.13	Up	Oncogenic	Promotes cell proliferation and invasion	↓miRNA-130a	Ma et al., 2014
2	LET	15q24.1	Down	Tumor suppressor	P21, Bax/Bcl-2, apoptosis proteins	HIF-1a/HDAC3/lncRNA-LET/NF-90 axis p21; caspase-3 and Bax/Bcl-2	Ma et al., 2014
3	CCAT1	8q24.21	Up	Oncogenic	Promotes cell proliferation and invasion	miRNA 218–5P, Bmi1	Ma et al., 2015
4	H19	11p15.5	Up	Oncogenic	Promotes cell proliferation, EMT, invasion and metastasis	H19/miR-194–5/AKT2axis; ↓E-cadherin, ↑Vimentin and ↑Twist1	Wang et al., 2016
5	MEG3	14q32.2	Down	Tumor Suppressor	Inhibits cell proliferation and induces apoptosis	↑p53gene; ↑caspase-3 and cyclin D	Liu et al., 2016
6	ANRIL	9p21.3	Up	Oncogenic	Promotes cell proliferation and inhibits apoptosis	↓ p53 and ↑ cyclin D1	Liu et al., 2016
7	GCASPC	6q25.3	Down	Tumor suppressor	Suppress cell proliferation	Sponge to miRNA 17–3p, negatively regulates pyruvate carboxylase	Ma et al., 2016
8	UCA1	19p13.12	Up	Oncogenic	Promote GBC cell proliferation, metastasis and tumor growth	↑ EZH2, recruit EZH2 to p21 and E-cadherin promoter	Cai et al., 2017
9	TUG1	22q12	Up	Oncogenic	Promote GBC cell proliferation and metastasis	Sponge to miR-300	Ma et al., 2017
10	HOXA-AS2	7p15.2	Up	Oncogenic	Promote proliferation, induce EMT		Zhang et al., 2017
11	DILC	13q34	Up	Oncogene	Proliferation and metastasis	↑GBC cells progression by activating Wnt/β-catenin signaling.	Liang et al., 2019
12	MALAT1	11q13.1	Up	Oncogenic	Promotes cell proliferation, migration, and metastasis	↑ MMP 9; ↑ MEK1/2, ERK1/2, MAPK, JNK1/2/3 in ERK/MAPK pathway	Lin et al., 2019
13	PAGBC	1q23.2	Up	Oncogenic	Promote tumor growth and metastasis	Sponge to miR-1–3b and MIR-511 and activate the AKTmTOR pathway	Wu et al., 2017
14	Loc344887	Chro3	Up	Oncogenic	Increase cell proliferation, migration, invasion and promote EMT	↓ expression of Twist, Vimentin, N-cadherin and ↑ increased the expression of E-cadherin	Wu et al., 2017
15	AFAP-ASI	-	Up	Oncogenic	Proliferation, invasion and EMT	MET protein	Ma et al., 2016
16	Linc00261	20p11.21	Down	Tumor suppressor	-	-	Niu et al., 2020

tumors across the body is linked to their aberrant expression (Balas et al., 2018). Recent studies suggest that lncRNAs may be involved in altering the tumor microenvironment and in tumor dissemination (Jiang et al., 2019).

9.7 LONG NONCODING RNAs IN GALLBLADDER CANCER

The discovery of the first lncRNA H19 by Brannan et al. in 1990 marked a significant milestone in RNA biology. Since then, numerous lncRNAs have been identified, with extensive online resources cataloging their molecular features and biological functions. Recent estimates revealed the presence of more than 6,700 lncRNA genes in the human genome. These RNAs, typically longer than 100 kilobases, are characterized by the absence of large open reading frames.

LncRNAs are transcribed by RNA polymerase II or III and present diverse features; they can be nuclear or cytoplasmic, polyadenylated or non-polyadenylated, and spliced or unspliced. The expression levels of these genes are generally lower than those of protein-coding genes, and they often display tissue-specific expression patterns.

In the context of gallbladder cancer, several lncRNAs have been implicated in deregulation, highlighting their potential role in disease pathology. Below is a summary of lncRNAs associated with this deregulation, providing insights into their relevance in gallbladder carcinogenesis.

Several long noncoding RNAs (lncRNAs) have been identified as key regulators in gallbladder cancer (GBC) pathogenesis, displaying diverse genomic locations, regulatory patterns, and functional roles. Among these, a significant number are upregulated and exhibit oncogenic properties, promoting tumor growth, invasion, metastasis, and epithelial-to-mesenchymal transition (EMT). For instance, HOTAIR (12q13.13) enhances cell proliferation and invasion by downregulating miR-130a, while CCAT1 (8q24.21) acts through miR-218–5p and Bmi1 to drive similar oncogenic processes. H19 (11p15.5) contributes to GBC progression by facilitating EMT and metastasis via the H19/miR-194–5p/AKT2 axis and by modulating key EMT markers such as E-cadherin, Vimentin, and Twist1. Similarly, ANRIL (9p21.3) promotes cell proliferation and inhibits apoptosis by downregulating p53 and upregulating cyclin D1, whereas UCA1 (19p13.12) facilitates tumor growth and metastasis through recruitment of EZH2 to the p21 and E-cadherin promoters. Other upregulated lncRNAs, such as TUG1, HOXA-AS2, DILC, MALAT1, PAGBC, Loc344887, and AFAP-ASI, have also been linked to GBC aggressiveness, acting through diverse mechanisms including sponging specific microRNAs, activating oncogenic signaling pathways (e.g., AKT/mTOR, Wnt/β-catenin, and ERK/MAPK), and influencing proteins like MET and MMP-9.

In contrast, several lncRNAs exhibit downregulated expression and function as tumor suppressors. LET (15q24.1) inhibits tumor progression by regulating the HIF-1α/HDAC3/lncRNA-LET/NF-90 axis and apoptosis-related proteins such as p21, caspase-3, and Bax/Bcl-2. MEG3 (14q32.2) suppresses proliferation and induces apoptosis through upregulation of p53, caspase-3, and cyclin D. GCASPC (6q25.3) acts as a sponge for miR-17–3p, thereby negatively regulating pyruvate carboxylase and suppressing cell proliferation. Additionally, Linc00261 (20p11.21) has been identified as a tumor suppressor, although its exact mechanisms remain under investigation.

Together, these findings highlight the critical roles of lncRNAs in GBC, where oncogenic lncRNAs drive malignant behaviors and tumor-suppressive lncRNAs lose their protective effects. Understanding these regulatory networks provides promising avenues for novel diagnostic biomarkers and therapeutic targets in gallbladder cancer management.

9.8 CONCLUSION

The discovery of long noncoding RNA molecules offers promising opportunities for improving the diagnosis and treatment of gallbladder cancer. It is essential to investigate the critical roles and potential applications of long noncoding RNAs in regulating GBC initiation, progression, chemoresistance, and prognosis. Furthermore, unraveling the complex cross-regulatory networks among lncRNA axes and their downstream targets or signaling pathways will provide deeper insights into GBC pathogenesis and identify novel therapeutic strategies.

REFERENCES

Balas MM, Johnson AM. Exploring the mechanisms behind long noncoding RNAs and cancer. *Noncoding RNA Res*. 2018;3(3):108–117.

Blume SW, Meng Z, Shrestha K, Snyder RC, Emanuel PD. The 5'-untranslated RNA of the human dhfr minor transcript alters transcription pre-initiation complex assembly at the major (core) promoter. *J Cell Biochem*. 2003;88:165–180.

Bridges MC, Daulagala AC, Kourtidis A. LNCcation: LncRNA localization and function. *J Cell Biol*. 2021;220(2):1–17.

Cai Q, Jin L, Wang S, et al. Long noncoding RNA UCA1 promotes gallbladder cancer progression by epigenetically repressing p21 and E-cadherin expression. Oncotarget. 2017;8(29):47957–47968.

Chen D, Lu T, Tan J, Li H, Wang Q, Wei L. Long noncoding RNAs as communicators and mediators between the tumor microenvironment and cancer cells. *Front Oncol*. 2019;9:1–10.

Dhanoa JK, Sethi RS, VeRma R, Arora JS, Mukhopadhyay CS. Long noncoding RNA: Its evolutionary relics and biological implications in mammals: A review. *J Anim Sci Technol*. 2018;60(25):1–10.

Fernandes JCR, Acuña SM, Aoki JI, Floeter-Winter LM, Muxel SM. Long non-coding RNAs in the regulation of gene expression: Physiology and disease. *Noncoding RNA*. 2019;5(1):1–25.

Hartford CCR, Lal A. When long noncoding becomes protein coding. *Mol Cell Biol*. 2020;40(6):1–11.

Hu G, Niu F, Humburg BA, Liao K, Bendi S, Callen S, Fox HS, Buch S. Molecular mechanisms of long noncoding RNAs and their role in disease pathogenesis. Oncotarget. 2018;9(26):18648–18663.

Jiang MC, Ni JJ, Cui W Y, Wang BY, Zhuo W. Emerging roles of lncRNA in cancer and therapeutic opportunities. *Am J Cancer Res*. 2019;9(7):1354–1366.

Jurga S, Barciszewski J. RNA technologies. In *The Chemical Biology of Long Noncoding RNAs*. Springer, 2020.

Khandelwal A, Bacolla A, Vasquez KM, Jain A. Long noncoding RNA: A new paradigm for lung cancer. *Mol Carcinog*. 2015;54(11):1235–1251.

Kumar D, Gurrapu S, Wang Y, et al. LncRNA Malat1 suppresses pyroptosis and T-cell-mediated killing of incipient metastatic cells. *Nat Cancer*. 2024;5(2):262–282.

Li Z, Zhao W, Wang M, Zhou X. The role of long noncoding RNAs in gene expression regulation. In *Gene Expression Profiling in Cancer*, Ed Dinitrios Vlachakis. Intech Open, 2019, 1–17.

Liang C, Yang P, Han T, et al. Long noncoding RNA DILC promotes the progression of gallbladder carcinoma. *Gene*. 2019;694:102–110.

Lin N, Yao Z, Xu M, et al. Long noncoding RNA MALAT1 potentiates growth and inhibits senescence by antagonizing ABI3BP in gallbladder cancer cells. *J Exp Clin Cancer Res*. 2019;38(1):1–15.

Liu B, Shen ED, Liao MM, Hu YB, Wu K, Yang P, Zhou L, Chen WD. Expression and mechanisms of long noncoding RNA genes MEG3 and ANRIL in gallbladder Cancer. *Tumor Biol*. 2016;37(7):9875–9886.

Ma F, Wang SH, Cai Q, Jin L, Zhou D, Ding J, Quan Z. Long noncoding RNA TUG1 promotes cell proliferation and metastasis by negatively regulating miR-300 in gallbladder carcinoma. *Biomed Pharmacother*. 2017;88:863–869.

Ma F, Wang SH, Cai Q, Zhang MD, Yang Y, Ding J. Overexpression of LncRNA AFAP1-AS1 predicts poor prognosis and promotes cells proliferation and invasion in gallbladder cancer. *Biomed Pharmacother*. 2016;84:1249–1255.

Ma MZ, Kong X, Weng MZ, et al. Long noncoding RNA-LET is a positive prognostic factor and exhibits tumor-suppressive activity in gallbladder cancer. *Mol Carcinog*. 2015;54(11):1397–1406.

Ma MZ, Li CX, Zhang Y, Weng MZ, Zhang MD, Qin YY, Gong W, Quan ZW. Long noncoding RNA HOTAIR, a c-Myc activated driver of malignancy, negatively regulates miRNA-130a in gallbladder cancer. *Mol Cancer*. 2014;13(156):1–14.

Mattick JS, Amaral PP, Carninci P, et al. Long non-coding RNAs: Definitions, functions, challenges and recommendations. *Nat Rev Mol Cell Biol*. 2023;24:430–447.

Niu JZ, Liang XC, Xu ZW, et al. Long noncoding RNA Linc00261 as a novel potential diagnostic and prognostic biomarker for gallbladder cancer. *Transl Cancer Res*. 2020;9(10):6078–6085.

Roberts RW, Crothers DM. Stability and properties of double and triple helices: Dramatic effects of RNA or DNA backbone composition. *Science*. 1992;258(5087):1463–1466.

Siddiqui H, Ghafari AA, Choudhary H, Doghaither HA. Roles of long non-coding RNAs in colorectal cancer tumorigenesis: A review. *Mol Clin Oncol*. 2019;11:167–172.

Wang SH, Ma F, Tang ZH, Wu XC, Cai Q, Zhang MD, Weng MZ, Zhou D, Wang JD, Quan ZW. Long non-coding RNA H19 regulates FOXM1 expression by competitively binding endogenous miR-342–3p in gallbladder cancer. *J Exp Clin Cancer Res*. 2016 Oct 3;35(1):1–12.

Wang WT, Han C, Sun YM, Chen TQ, Chen YQ. Noncoding RNAs in cancer therapy resistance and targeted drug development. *J Hematol Oncol*. 2019;12(1):1–15.

Wilusz JE, Sunwoo H, Spector DL. Long noncoding RNAs: Functional surprises from the RNA world. *Genes Dev*. 2009;23(13):1494–1504.

Wu XS, Wang F, Li HF, et al. LncRNA-PAGBC acts as a microRNA sponge and promotes gallbladder tumorigenesis. *EMBO Rep*. 2017;18(10):1837–1853.

Zampetaki A, Albrecht A, Steinhofel K. Long Noncoding RNA Structure and Function: Is There a Link?. *Front Physiol*. 2018;9:1–8.

Zhang P, Cao P, Zhu X, et al. Upregulation of long non-coding RNA HOXA-AS2 promotes proliferation and induces epithelial-mesenchymal transition in gallbladder carcinoma. *Oncotarget*. 2017;8(20):33137–33143.

Zhou H, Wang B, Yang YX, et al. Long noncoding RNAs in pathological cardiac remodeling: A review of the update literature. *Biomed Res Int*. 2019:7159592.Abstract

10 Metabolic Reprogramming in Gallbladder Cancer

*Sana Ahuja and Sufian Zaheer**
Department of Pathology, Vardhman Mahavir Medical College and Safdarjung Hospital, New Delhi, India
*Corresponding author: sufianzaheer@gmail.com

10.1 INTRODUCTION

Gallbladder cancer (GBC) is a rare but highly aggressive malignancy that is often detected at advanced stages because of its asymptomatic nature in the early phases. Although it accounts for only 1.2% of all cases of cancer globally, it is responsible for almost 1.7% of all cancer deaths.[1] It is the most common malignancy of the biliary system and ranks fifth among gastrointestinal malignancies. GBC patients have a mean overall survival of only 6 months and very low 5-year survival rates.[2]

Although genetic mutations are vital in cancer development, alterations in energy and resource processing, known as metabolic reprogramming, are also critical for tumor development and progression. This chapter thoroughly examines the ways in which gallbladder cancer cells modify their metabolic pathways to promote rapid growth, survival, and invasiveness, highlighting distinctions from normal cell metabolism.

10.2 OVERVIEW OF CELLULAR METABOLISM IN NORMAL GALLBLADDER CELLS

In healthy cells, metabolism is tightly regulated to maintain a balance between energy production and use. Glucose serves as the main energy source and undergoes a series of reactions to generate ATP (adenosine triphosphate), which fuels cellular functions. This process involves two primary pathways: glycolysis and oxidative phosphorylation.

Glycolysis takes place in the cytoplasm, where glucose is broken down into pyruvate, yielding a modest amount of ATP. Under normal oxygen conditions, pyruvate is transported to the mitochondria, where it undergoes complete oxidation through the tricarboxylic acid (TCA) cycle and oxidative phosphorylation, producing a substantially greater amount of ATP.

Oxidative phosphorylation is the primary mode of energy production in cells with an adequate oxygen supply. This process is highly efficient and occurs in the mitochondria, where electrons from nutrients such as glucose are ultimately transferred through a series of protein complexes to produce ATP.

In healthy cells, the breakdown of proteins, lipids, and carbohydrates results in the production of ATP, the cellular energy currency. Glucose, obtained from carbohydrates, is converted into pyruvate and then into acetyl-coenzyme A (CoA). Within the tricarboxylic acid cycle, also called the "Krebs cycle," acetyl-CoA is further processed into various intermediate compounds. This process releases ATP and high-energy carriers, such as nicotinamide adenine dinucleotide phosphate, which in turn produce more ATPs. ATPs are released as energy when aerobic glucose is converted to carbon dioxide and water. When proteins are broken down, amino acids are converted into molecules such as pyruvate and acetyl-CoA. Similarly, the breakdown of fats yields glycerol,

DOI: 10.1201/9781003506201-10

a three-carbon molecule, along with fatty acids, both of which contribute to the formation of pyruvate and acetyl-CoA. Numerous hormones, such as glucagon, glucocorticoids, insulin, and growth hormone, control pathways leading to anabolism and catabolism. Healthy gallbladder cells preserve this equilibrium, guaranteeing that they have sufficient energy to carry out essential functions such as bile production and secretion, cellular repair, and detoxification.[3]

10.3 GLUCOSE METABOLISM

In contrast to normal cells, cancer cells—including those in gallbladder cancer—undergo significant metabolic shifts to meet the demands of rapid growth and proliferation. One of the hallmarks of cancer metabolism is increased reliance on glycolysis, even when oxygen is sufficient, a phenomenon known as the Warburg effect. This shift enables cancer cells to thrive under conditions that would typically be detrimental to normal cells.

The Warburg effect was proposed by Otto Warburg in the 1920s. The cancer cell uses carbon from glucose to fuel biosynthetic processes, which allows for biomass production and fast cell division. The main features of the Warburg effect in cancer cells are glutamine metabolism, glycolysis, lactate fermentation, the TCA cycle, the pentose pyruvate route, and the utilization of TCA cycle intermediates in the synthesis of lipids, amino acids, and nucleotides. Furthermore, many metabolic pathways are deregulated by oncogenic mutations affecting growth factor signaling pathways and other important molecules, such as MYC. Cancer cells gain a growth advantage from metabolic reprogramming, which promotes tumor development and progression. This accelerated glucose metabolism not only provides quick energy but also generates intermediate products that are essential for the synthesis of nucleotides, lipids, and proteins, all of which are necessary for rapid cell division.[4]

Although glycolysis generates ATP more efficiently and at a slower pace than oxidative phosphorylation (OXPHOS), its faster rate than that of mitochondrial oxidation indicates that glycolysis serves cancer cells in ways beyond just energy production. A variety of metabolic intermediates needed for cell development and proliferation are provided by the metabolic intermediates of both the TCA cycle and glycolysis. For example, the pentose phosphate pathway, which supplies ribose sugars for nucleotide biosynthesis, involves glucose-6-phosphate (Figure 10.1).[5]

Recent studies have proposed the concept of a two-compartment model. The reverse Warburg effect involves metabolic interplay between cancer cells and adjacent stromal cells, particularly cancer-associated fibroblasts (CAFs). Unlike the classical Warburg effect, in which cancer cells rely primarily on aerobic glycolysis for ATP production, in the reverse Warburg effect, stromal cells undergo metabolic reprogramming.

This concept proposes that hydrogen peroxide released by cancer cells into the TME causes oxidative damage in nearby CAFs. Oxidative stress results from the activation of NF-κB and HIF-1a. In CAFs, HIF-1a induces autophagy and lysosomal degradation, which leads to the loss of stromal Cav-1 and hence aerobic glycolysis. High-energy substrates such as lactate and pyruvate are then produced and enter cancer cells through monocarboxylate transporters (MCTs) to produce ATP through OXPHOS. It is known as a "lactate shuttle" conveyance. This mutualistic metabolic relationship between tumor cells and CAFs is the foundation for the reverse Warburg effect.[6,7]

In GBC, circβ-catenin is vital for facilitating the Warburg effect, a metabolic characteristic observed in numerous cancers, where cells favor glycolysis instead of oxidative phosphorylation even when there is adequate oxygen available. This shift toward aerobic glycolysis provides cancer cells with the metabolic flexibility needed to support rapid proliferation and survival in a hypoxic tumor microenvironment. Understanding how circβ-catenin contributes to this metabolic reprogramming requires a closer examination of the molecular pathways involved.

CircRNAs, including circβ-catenin, are increasingly recognized as important regulators of cancer metabolism.[8] Specifically, circβ-catenin has been shown to increase glycolytic flux in GBC cells. When circβ-catenin is upregulated, there is a notable increase in essential markers of increased

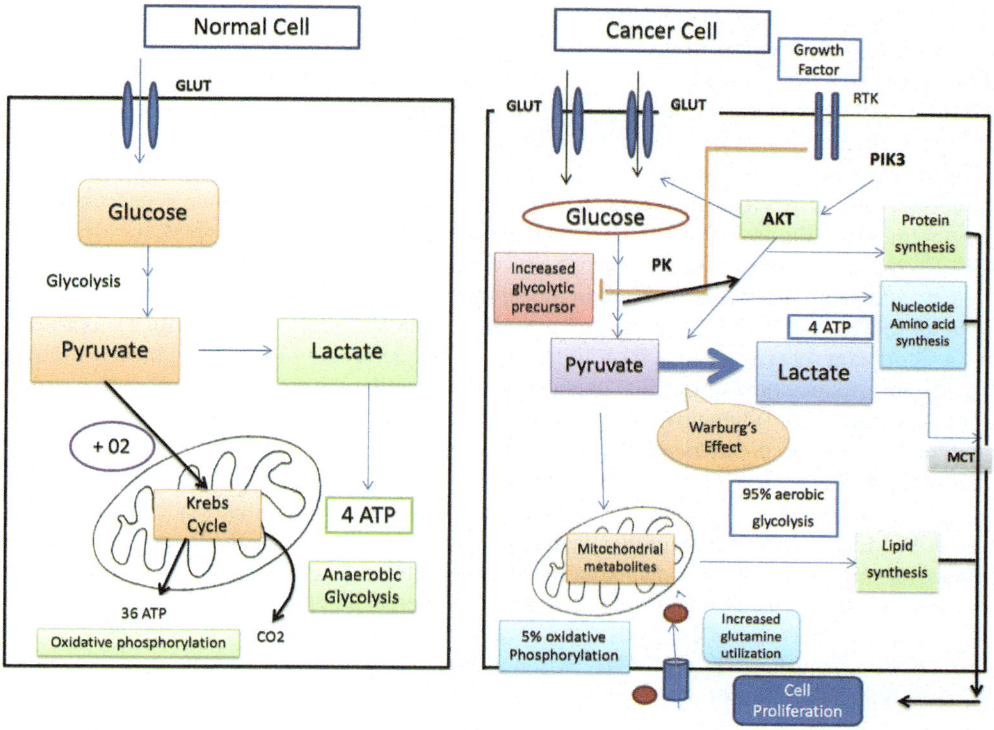

FIGURE 10.1 This illustration compares glucose metabolism in healthy cells with that in cancer cells, emphasizing the unique metabolic alterations associated with the Warburg effect. In healthy cells, glucose is primarily broken down through oxidative phosphorylation within the mitochondria. In contrast, cancer cells exhibit increased glycolysis, leading to substantial lactate production even when oxygen is present. Enhanced glucose uptake in cancer cells is also displayed, facilitated by overexpressed glucose transporters such as GLUT1. This uptake of glucose is directed toward lactate production instead of entering the TCA cycle, a hallmark of cancer cell metabolism.

glycolysis, including elevated production of lactate and pyruvate. This change is additionally supported by an increase in the extracellular acidification rate (ECAR), which indicates increased lactate secretion—a characteristic feature of glycolysis. In contrast, knockdown of circβ-catenin results in a decrease in these glycolytic markers, suggesting that circβ-catenin is pivotal for sustaining the glycolytic state of GBC cells.

In parallel, oxygen consumption rate (OCR) analysis revealed that circβ-catenin negatively influences oxidative phosphorylation (OXPHOS). When circβ-catenin is overexpressed, a significant reduction in the OCR is noted, indicating that mitochondrial respiration is suppressed in favor of glycolysis. Conversely, silencing circβ-catenin enhances the OCR, suggesting reactivation of mitochondrial respiration and partial reversal of the Warburg effect.[9] Furthermore, ATP production, a crucial marker of the cellular energy status, is correlated with circβ-catenin expression. The overexpression of circβ-catenin promotes ATP generation via glycolysis, whereas its downregulation decreases ATP levels, indicating a shift in energy production from glycolysis to oxidative phosphorylation. These findings underscore the central role of circβ-catenin in regulating the metabolic phenotype of GBC cells, steering them toward a glycolytic mode of energy production that supports tumor growth and survival.

The molecular mechanism underlying the regulation of the Warburg effect by circβ-catenin involves its role as a competing endogenous RNA (ceRNA). CircRNAs can act as sponges for microRNAs (miRNAs), thereby modulating the availability of miRNAs to regulate their target mRNAs. Circβ-catenin exerts its proglycolytic effects by sponging miR-223, a microRNA implicated in the regulation of various oncogenic processes.

Circβ-catenin contains binding sites that are complementary to those of miR-223, allowing it to sequester this miRNA and prevent it from binding to its target mRNAs. When circβ-catenin is downregulated, the level of miR-223 increases, leading to increased suppression of its target genes. Conversely, overexpression of circβ-catenin decreases the availability of miR-223, thereby derepressing its target genes, which are involved in the promotion of glycolysis and tumor growth.[9]

One of the key targets of the circβ-catenin/miR-223 axis is STMN1 (Stathmin 1), a protein that plays a significant role in cancer cell proliferation, cell cycle progression, and metabolic reprogramming. Bioinformatics predictions and experimental validation have shown that STMN1 is directly regulated by miR-223. When miR-223 is sequestered by circβ-catenin, STMN1 expression is upregulated, contributing to the metabolic switch toward glycolysis.

STMN1 has been shown to be upregulated in GBC tissues compared with normal adjacent tissues, and its expression is correlated with increased glycolytic activity and tumor progression. Inhibition of STMN1 leads to a reduction in lactate and pyruvate production, the ECAR, and ATP levels while increasing the OCR, thus reversing the Warburg effect. These findings indicate that STMN1 is a downstream effector of the circβ-catenin/miR-223 axis and plays a crucial role in maintaining the glycolytic phenotype in GBC cells.[9]

The circβ-catenin/miR-223/STMN1 axis not only supports the metabolic reprogramming of GBC cells but also promotes their proliferation and survival. By driving the Warburg effect, circβ-catenin enables GBC cells to thrive in the nutrient- and oxygen-deprived conditions typical of solid tumors. This metabolic adaptation is a key factor in tumor aggressiveness and resistance to therapy.

Targeting the circβ-catenin/miR-223/STMN1 pathway represents a potential therapeutic strategy for GBC. Disruption of this axis may reverse the Warburg effect, reactivate oxidative phosphorylation, and reduce the proliferative capacity of GBC cells. For example, restoring miR-223 levels or inhibiting STMN1 function could reduce glycolysis and impair the ability of tumors to generate the energy and biosynthetic precursors needed for rapid growth.

In GBC, the upregulation of circFOXP1 significantly contributes to the Warburg effect, a hallmark of cancer metabolism, where cells predominantly rely on glycolysis for energy production even in the presence of oxygen. This metabolic reprogramming enables cancer cells to meet their high energy and biosynthetic demands, promoting rapid proliferation and survival.

Studies have shown that circFOXP1 upregulation leads to increased glycolytic activity in GBC cells. When circFOXP1 expression is increased, key metabolic outputs associated with glycolysis, such as lactate and pyruvate production, increase significantly. This finding indicates a shift toward glycolysis as the primary mode of energy production, which is consistent with the Warburg effect. Conversely, knocking down circFOXP1 results in a substantial decrease in these metabolites, reflecting impaired glycolytic activity. These findings underscore the role of circFOXP1 in driving the metabolic switch toward glycolysis in GBC.

Additionally, the extracellular acidification rate (ECAR), a measure of lactate secretion and overall glycolytic flux, is markedly increased with circFOXP1 overexpression, further confirming its role in promoting glycolysis. On the other hand, circFOXP1 knockdown leads to a reduction in the ECAR, demonstrating a reversal of the glycolytic phenotype.[10]

CircFOXP1 also has a significant effect on oxidative phosphorylation (OXPHOS), the more efficient but slower process of ATP production that occurs in the mitochondria. In cells where circFOXP1 is upregulated, the OCR, a marker of mitochondrial respiration, is decreased. These findings indicate that circFOXP1 suppresses oxidative phosphorylation in favor of glycolysis, reinforcing the Warburg effect. Conversely, when circFOXP1 is silenced, the OCR increases, signaling enhanced mitochondrial activity and shifting back toward oxidative phosphorylation. These findings suggest that circFOXP1 actively downregulates mitochondrial respiration to promote a glycolytic phenotype.[10]

The regulation of ATP production by circFOXP1 further illustrates its role in promoting the Warburg effect. In cells exhibiting elevated circFOXP1 expression, ATP generation occurs mainly through glycolysis instead of oxidative phosphorylation. While glycolysis is less efficient than oxidative phosphorylation, it offers cancer cells a quick way to produce energy, as well as the biosynthetic precursors required for cell growth and division. However, in cells in which circFOXP1 is knocked down, there is a noticeable shift toward ATP production through oxidative phosphorylation, reflecting a more typical metabolic state akin to that of noncancerous cells.[10]

CircFOXP1 further promotes the Warburg effect in GBC cells by interacting with the RNA-binding protein PTBP1. This interaction enhances the expression and cytoplasmic localization of PTBP1, which plays a key role in regulating the alternative splicing of pyruvate kinase isoforms, particularly the oncogenic PKM2 isoform. Additionally, PTBP1 binds to the mRNA of PKLR, another key enzyme involved in glycolysis, protecting it from degradation and increasing its protein levels. The increased expression of PKLR, driven by circFOXP1-mediated PTBP1 activity, further promotes glycolysis and contributes to tumor progression in GBC cells.[10,11]

The ability of circFOXP1 to modulate the balance between glycolysis and oxidative phosphorylation gives GBC cells a significant growth advantage. By promoting the Warburg effect, circFOXP1 enables cancer cells to rapidly generate energy and molecular precursors needed for proliferation. This metabolic reprogramming also allows cancer cells to thrive in the nutrient- and oxygen-poor conditions that are often present within tumors, increasing their adaptability and resistance to stress.

In summary, circFOXP1 upregulation in GBC cells promotes the Warburg effect by enhancing glycolysis and suppressing oxidative phosphorylation. This shift in metabolism supports tumor growth and progression, making circFOXP1 a potential target for therapeutic strategies aimed at disrupting the metabolic adaptations of cancer cells. By targeting circFOXP1, reversing the Warburg effect, restoring normal metabolic processes, and reducing the aggressive growth of GBC cells may be possible.

10.4 LIPID METABOLISM

In addition to glucose metabolism, lipid metabolism plays a crucial role in supporting the growth and survival of GBC cells. Lipids are essential for cell membrane formation, energy storage, and signaling. Cancer cells, including those in the gallbladder, often exhibit increased lipid uptake, synthesis, and storage.

Lipid Biosynthesis: Cancer cells frequently activate lipid biosynthesis pathways, even under nutrient-deprived conditions. This allows the cells to generate the necessary components for new cell membranes, a critical requirement for rapidly dividing cells. Enzymes involved in fatty acid synthesis, such as fatty acid synthase (FASN), are often upregulated, contributing to increased lipid production (Figure 10.2).[12]

Lipid Storage and Utilization: Cancer cells also store lipids in lipid droplets, which serve as reservoirs of energy and building blocks. These stored lipids can be mobilized when needed to support energy production or the synthesis of new membranes. Additionally, lipid droplets help cancer cells manage oxidative stress, a common challenge for rapidly growing tumors.[13]

In GBC, circRNAs have been implicated in the reprogramming of lipid metabolism, which is crucial for tumor development and metastasis.[14] Among these, circEZH2 (hsa_circ_0006357) has emerged as a key player. Research has demonstrated that circEZH2 is significantly upregulated in GBC tissues and is correlated with advanced tumor-node-metastasis stage, suggesting its role in promoting tumor aggressiveness.[15]

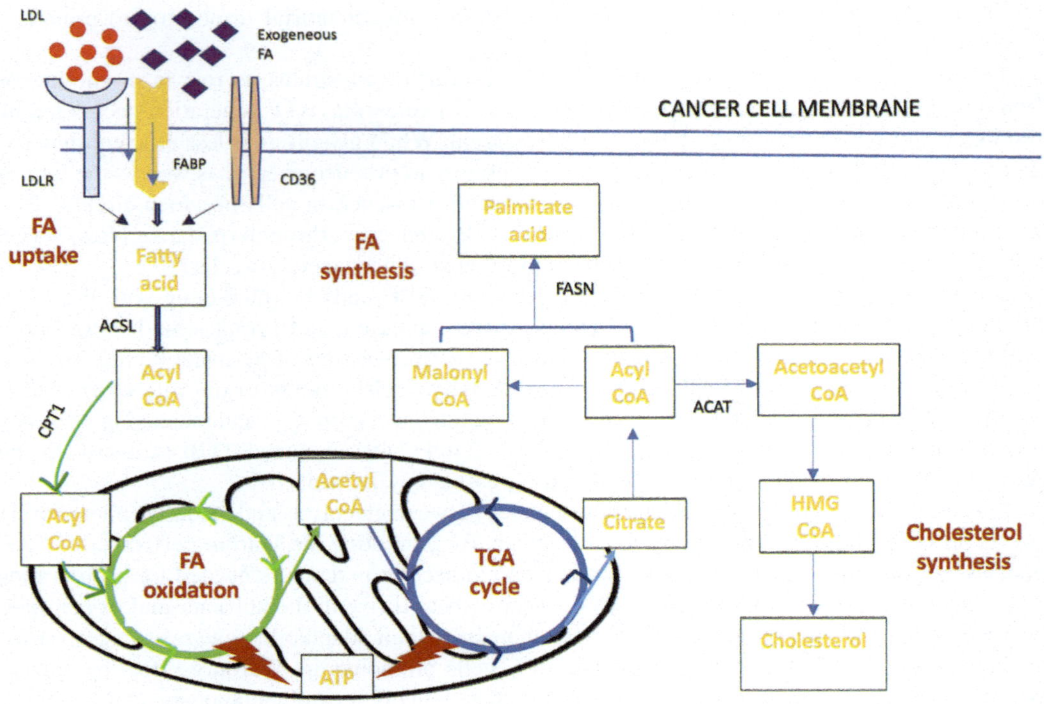

FIGURE 10.2 This diagram outlines essential pathways for fatty acid (FA) utilization and cholesterol biosynthesis in cancer cells. External fatty acids enter the cell via transporters, including the LDL receptor, fatty acid–binding protein (FABP), and CD36, and are subsequently transformed into acyl-CoA by acyl-CoA synthetase (ACSL) for use in multiple metabolic pathways. Additionally, fatty acids can be synthesized internally, with fatty acid synthase (FASN) catalyzing the formation of palmitate from malonyl-CoA and acyl-CoA. Within mitochondria, acyl-CoA is imported through carnitine palmitoyltransferase 1 (CPT1) for β-oxidation, yielding acetyl-CoA, which then fuels the TCA cycle for ATP production, thus supporting cancer cell growth. Acyl-CoA also contributes to cholesterol production, converting it into acetoacetyl-CoA and then HMG-CoA, the precursor for cholesterol. These lipid metabolic adaptations provide the energy and structural components essential for the rapid proliferation of cancer cells.

Lipid metabolism reprogramming is a hallmark of cancers, including GBC. circEZH2 contributes to this reprogramming by modulating the balance between monounsaturated fatty acids (MUFAs) and saturated fatty acids (SFAs). Mechanistically, circEZH2 acts as a molecular sponge for miR-556–5p, a microRNA that negatively regulates the expression of stearoyl-CoA desaturase 1 (SCD1), an enzyme involved in the synthesis of MUFAs from SFAs. By sponging miR-556–5p, circEZH2 upregulates SCD1 expression, leading to an increase in MUFA levels and a decrease in SFA levels in GBC cells.[15] This alteration in lipid composition supports the malignant progression of GBC by enhancing membrane fluidity, which is essential for rapid cell proliferation and metastasis.[16] The stability of circEZH2 in GBC cells is further enhanced by N6-methyladenosine (m6A) modification, a common RNA modification that affects RNA stability and function. The m6A reader protein IGF2BP2 recognizes and binds to m6A-modified circEZH2, thereby increasing its stability. Interestingly, circEZH2 also interacts with IGF2BP2 to prevent its ubiquitination-mediated degradation, creating a positive feedback loop that further upregulates circEZH2 expression. This feedback loop amplifies the effects of circEZH2 on lipid metabolism reprogramming, contributing to the aggressive nature of GBC.[15]

The reprogramming of lipid metabolism by circEZH2 has several functional implications for GBC progression. The increased MUFA levels resulting from SCD1 upregulation not only support rapid cell proliferation but also influence other metabolic pathways that are crucial for tumor survival and growth. For example, alterations in lipid composition can affect membrane dynamics and signal transduction, promoting cancer cell invasion and metastasis. Additionally, the changes in lipid metabolism induced by circEZH2 can lead to the activation of ferroptosis, a form of cell death driven by lipid peroxidation, which may further influence the response of GBC cells to various therapies.[15]

Given its central role in lipid metabolism reprogramming and GBC progression, circEZH2 represents a promising therapeutic target. Inhibiting circEZH2 or disrupting its interaction with miR-556–5p could reverse the lipid metabolic changes that promote tumor growth, increasing the susceptibility of GBC cells to ferroptosis and other forms of cell death. Moreover, targeting the circEZH2–IGF2BP2 feedback loop could reduce the stability of circEZH2, thereby diminishing its oncogenic effects. Further research is needed to explore the therapeutic potential of circEZH2 inhibitors and their effectiveness in combination with existing GBC treatments.[15]

10.5 PROTEIN METABOLISM AND AMINO ACID UTILIZATION

Protein metabolism is another area where cancer cells diverge from normal cells. Proteins are constantly synthesized, degraded, and recycled in cells, and this balance is tightly regulated to meet the needs of the cell. However, in cancer cells, this balance is often disrupted to support their growth and survival. Proteins are essential not only as structural and functional components of cells but also as key regulators of processes such as signal transduction and stress response.[17] Owing to the rapid and uncontrolled growth typical of gallbladder cancer, these cells must continuously produce large amounts of proteins, creating a significant demand on their metabolic systems.

In gallbladder cancer, pathways that drive protein synthesis are often upregulated. The mammalian target of rapamycin (mTOR) pathway, a central regulator of protein synthesis, is frequently overactive in various cancers, including GBC. mTOR responds to growth signals and nutrient availability, leading to enhanced protein synthesis that fuels tumor growth.[18] The elevated activity of mTOR in gallbladder cancer supports the production of proteins needed for cell proliferation, angiogenesis, and survival, contributing to the aggressive nature of the disease.

This increased protein synthesis creates additional stress on the cellular machinery responsible for protein folding and quality control. In response to this stress, cancer cells may activate the unfolded protein response (UPR). The UPR helps cells cope with the increased demand for protein folding and degradation by expanding the capacity of the endoplasmic reticulum (ER) and managing the protein load. This adaptation allows cancer cells to survive the metabolic stress associated with increased protein production.[19] To sustain their high rate of protein synthesis, these cells often modify their amino acid metabolism. Amino acids are crucial building blocks for proteins, and cancer cells may become dependent on specific amino acids to fuel both their biosynthetic and energy-producing processes. Glutamine plays a vital role in metabolic reprogramming and serves as a critical substrate for the TCA cycle, supporting both energy production and the synthesis of key biomolecules.[20, 21]

Cancer cells may also rely on increased uptake and metabolism of other amino acids, such as serine and glycine, which are necessary for nucleotide synthesis and other metabolic pathways. By increasing the availability and utilization of these amino acids, they can meet the demands of their increased protein synthesis and growth. This dependence on amino acids extends to the increased expression of amino acid transporters, which facilitate the uptake of extracellular amino acids. These transporters allow cancer cells to maintain a sufficient supply of amino acids to fuel protein synthesis and other critical metabolic processes.[22]

Cancer cells not only increase protein synthesis but also actively engage in protein degradation pathways to recycle amino acids and manage cellular homeostasis. The ubiquitin–proteasome system (UPS) and autophagy are two major mechanisms involved in protein degradation. In cancer, the UPS may be upregulated to manage the increased production and turnover of proteins. By degrading misfolded or damaged proteins, the UPS helps gallbladder cancer cells maintain protein quality control and recycle amino acids for continued protein synthesis. This recycling process is essential to sustain the high rate of protein production necessary for tumor growth.[23] Autophagy, a cellular process that breaks down proteins and other macromolecules in the lysosome, also plays a critical role in cancer. Under nutrient-deprived conditions, which are common in the tumor microenvironment due to poor vascularization, autophagy provides an alternative source of amino acids and energy. This allows cancer cells to survive and continue proliferating even when external nutrients are limited.[24] Targeting protein metabolism offers potential therapeutic opportunities in gallbladder cancer. Since the mTOR pathway is frequently overactive in these tumors, the use of mTOR inhibitors is being explored as a strategy to reduce protein synthesis and slow tumor growth. By inhibiting mTOR, it may be possible to disrupt the metabolic processes that gallbladder cancer cells rely on for survival.[25]

In addition to targeting protein synthesis, therapies that interfere with amino acid metabolism are under investigation. For example, glutaminase inhibitors, which block glutamine metabolism, can cut off a critical fuel source for gallbladder cancer cells, leading to their death. Similarly, inhibitors of amino acid transporters can reduce the availability of essential amino acids, disrupting protein synthesis and tumor growth.[26]

Furthermore, strategies to modulate protein degradation pathways, such as proteasome inhibitors or autophagy modulators, are being explored in cancer therapy. These approaches aim to disrupt the balance between protein synthesis and degradation in gallbladder cancer cells, leading to increased stress and potentially triggering cell death.[27]

REFERENCES

1. Bray F, Ferlay J, Soerjomataram I, Siegel RL, Torre LA, Jemal A. Global cancer statistics 2018: GLOBOCAN estimates of incidence and mortality worldwide for 36 cancers in 185 countries. *CA Cancer J Clin*. 2018;68:394–424. doi: 10.3322/caac.21492.
2. Hundal R, Shaffer EA. Gallbladder cancer: Epidemiology and outcome. *Clin Epidemiol*. 2019;6:99–109. doi: 10.2147/CLEP.S37357.
3. Kalyanaraman B. Teaching the basics of cancer metabolism: Developing antitumor strategies by exploiting the differences between normal and cancer cell metabolism. *Redox Biol*. 2017;12:833–842. doi: 10.1016/j.redox.2017.04.018.
4. Ediriweera MK, Jayasena S. The role of reprogrammed glucose metabolism in cancer. *Metabolites*. 2023;13:345. doi: 10.3390/metabo13030345.
5. Bose S, Zhang C, Le A. Glucose metabolism in cancer: The Warburg effect and beyond. In: Le A, editor. *The Heterogeneity of Cancer Metabolism* [Internet]. 2nd edition. Cham, CH: Springer, May 21, 2021. doi: 10.1007/978-3-030-65768-0_1. www.ncbi.nlm.nih.gov/books/NBK573693/
6. Liang L, Li W, Li X, Jin X, Liao Q, Li Y, et al. 'Reverse Warburg effect' of cancer-associated fibroblasts (review). *Int J Oncol*. 2022;60:67. doi: 10.3892/ijo.2022.5357.
7. Fu Y, Liu S, Yin S, Niu W, Xiong W, Tan M, et al. The reverse Warburg effect is likely to be an Achilles' heel of cancer that can be exploited for cancer therapy. *Oncotarget*. 2017;8:57813–57825. doi: 10.18632/oncotarget.18175.
8. Liang WC, Wong CW, Liang PP, Shi M, Cao Y, Rao ST, et al. Translation of the circular RNA circβ-catenin promotes liver cancer cell growth through activation of the Wnt pathway. *Genome Biol*. 2019;20:84. doi: 10.1186/s13059-019-1685-4.
9. Wang S, Su T, Tong H, Zhou D, Ma F, Ding J, et al. Circβ-catenin promotes tumor growth and Warburg effect of gallbladder cancer by regulating STMN1 expression. *Cell Death Discov*. 2021;7:233. doi: 10.1038/s41420-021-00626-6.
10. Wang S, Zhang Y, Cai Q, Ma M, Jin LY, Weng M, et al. Circular RNA FOXP1 promotes tumor progression and Warburg effect in gallbladder cancer by regulating PKLR expression. *Mol Cancer*. 2019;18:145. doi: 10.1186/s12943-019-1078-z.

11. Lu W, Cao Y, Zhang Y, Li S, Gao J, Wang XA, et al. Upregulation of PKM2 promote malignancy and related to adverse prognostic risk factor in human gallbladder cancer. *Sci Rep.* 2016;6:26351. doi: 10.1038/srep26351.
12. Baenke F, Peck B, Miess H, Schulze A. Hooked on fat: The role of lipid synthesis in cancer metabolism and tumor development. *Dis Model Mech.* 2013;6:1353–1363. doi: 10.1242/dmm.011338.
13. Olzmann JA, Carvalho P. Dynamics and functions of lipid droplets. *Nat Rev Mol Cell Biol.* 2019;20:137–155. doi: 10.1038/s41580-018-0085-z.
14. Bach DH, Lee SK, Sood AK. Circular RNAs in cancer. *Mol Ther Nucleic Acids.* 2019;16:118–129. doi: 10.1016/j.omtn.2019.02.005.
15. Tong H, Yu X, Zhou D, Shen Z, Chen J, Si Y, et al. CircEZH2 promotes gallbladder cancer progression and lipid metabolism reprogramming through the miR-556–5p/SCD1 axis. *iScience.* 2024;27:110428. doi: 10.1016/j.isci.2024.110428.
16. Yang K, Wang X, Song C, He Z, Wang R, Xu Y, et al. The role of lipid metabolic reprogramming in tumor microenvironment. *Theranostics.* 2023;13:1774–1808. doi: 10.7150/thno.82920.
17. Ling ZN, Jiang YF, Ru JN, Lu JH, Ding B, Wu J. Amino acid metabolism in health and disease. *Signal Transduct Target Ther.* 2023;8:345. doi: 10.1038/s41392-023-01569-3.
18. Panwar V, Singh A, Bhatt M, Tonk RK, Azizov S, Raza AS, et al. Multifaceted role of mTOR (mammalian target of rapamycin) signaling pathway in human health and disease. *Signal Transduct Target Ther.* 2023;8:375. doi: 10.1038/s41392-023-01608-z.
19. Read A, Schröder M. The unfolded protein response: An overview. *Biology (Basel).* 2021;10:384. doi: 10.3390/biology10050384.
20. Muhammad N, Lee HM, Kim J. Oncology therapeutics targeting the metabolism of amino acids. *Cells.* 2020;9:1904. doi: 10.3390/cells9081904.
21. Wang B, Pei J, Xu S, Liu J, Yu J. Correction: A glutamine tug-of-war between cancer and immune cells: Recent advances in unraveling the ongoing battle. *J Exp Clin Cancer Res.* 2024;43:93. doi: 10.1186/s13046-024-03018-7. Erratum for: *J Exp Clin Cancer Res.* 2024;43:74. doi: 10.1186/s13046-024-02994-0.
22. Pan S, Fan M, Liu Z, Li X, Wang H. Serine, glycine and one--carbon metabolism in cancer (Review)-. *Int J Oncol.* 2021;58:158–170. doi: 10.3892/ijo.2020.5158.
23. Wang J, Xiang Y, Fan M, Fang S, Hua Q. The ubiquitin–proteasome system in tumor metabolism. *Cancers (Basel).* 2023;15:2385. doi: 10.3390/cancers15082385.
24. Yun CW, Lee SH. The roles of autophagy in cancer. *Int J Mol Sci.* 2018;19:3466. doi: 10.3390/ijms19113466.
25. Magaway C, Kim E, Jacinto E. Targeting mTOR and metabolism in cancer: Lessons and innovations. *Cells.* 2019;8:1584. doi: 10.3390/cells8121584.
26. Jin J, Byun JK, Choi YK, Park KG. Targeting glutamine metabolism as a therapeutic strategy for cancer. *Exp Mol Med.* 2023;55:706–715. doi: 10.1038/s12276-023-00971-9.
27. Tundo GR, Sbardella D, Santoro AM, Coletta A, Oddone F, Grasso G, et al. The proteasome as a druggable target with multiple therapeutic potentialities: Cutting and noncutting edges. *Pharmacol Ther.* 2020;213:107579. doi: 10.1016/j.pharmthera.2020.107579.

11 Gallbladder Carcinogenesis
Current Insights and Future Directions

*Amit Yadav and Usha Dutta**
Department of Gastroenterology, PGIMER, Chandigarh, India
*Corresponding author: ushadutta@gmail.com

11.1 INTRODUCTION

Gallbladder cancer (GBC) is the most common biliary tract cancer. GBC imposes a substantial impact on financial and economic handling, health-related quality of life, and social life of the patient. The 5-year survival rate for GBC is often less than 5% (1). GBC originates from the epithelium of either the gallbladder (GB) or the cystic duct. The GB lies on the undersurface of the liver, which is an anatomically busy area. The GB structure consists of a discontinuous muscle layer with no muscularis mucosa, and about 30% of its surface is attached to the liver; the rest is covered by peritoneum. GBC has a varying spectrum of presentation, including incidental diagnosis during cholecystectomy, non-specific symptoms, and advanced-stage disease with jaundice, ascites, or duodenal infiltration. Often, delay in diagnosis due to non-specific presenting symptoms and the aggressive biological nature of the tumor results in a dismal prognosis (2). Screening for GBC remains a challenge because of the lack of specific symptoms in early stages of disease, and the GB mucosa is not amenable to direct endoscopic visualization in contrast to other luminal organs. Adenocarcinoma is the most common histological subtype, representing >90% of cases (3). The tumor is usually unresectable at presentation and inherently resistant to chemotherapy and radiotherapy (4). Immunotherapy and targeted therapy may be emerging treatment strategies to extend survival and quality of life. Advancement in research over the past few decades has provided important insights into the genetic alterations and complex molecular pathways involved in GBC. These innovations in the understanding of GBC have the potential to transform clinical practices, improve early detection, and lead to the development of more effective treatments. Recent advances in diagnostics have provided a better understanding of disease pathogenesis, discovering mechanisms associated with inflammation-driven tumor initiation and progression, as well as distinctive genomic alterations from precancerous lesions to invasive GBC (5).

This chapter delves into the latest research on the molecular pathogenesis of GBC, including genetic mutations, epigenetic alterations, and the influence of environmental factors. Additionally, it explores novel diagnostic tools, emerging therapeutic strategies, and the challenges and opportunities that lie ahead in the fight against GBC.

11.2 GALLBLADDER CANCER: EPIDEMIOLOGY AND RISK FACTORS

North India and south-central Chile are the highest incidence areas of GBC in the world, with incidence ranges from 10 to 20 per 100,000 population. Western populations have low prevalence, ranging from 1 to 2 per 100,000 population (6). The incidence of GBC is characterized by marked geographical and ethnic variations. Even though gallstones (7)-related chronic inflammation is the most important risk factor for GBC, other local and/or systemic inflammatory conditions,

 DOI: 10.1201/9781003506201-11

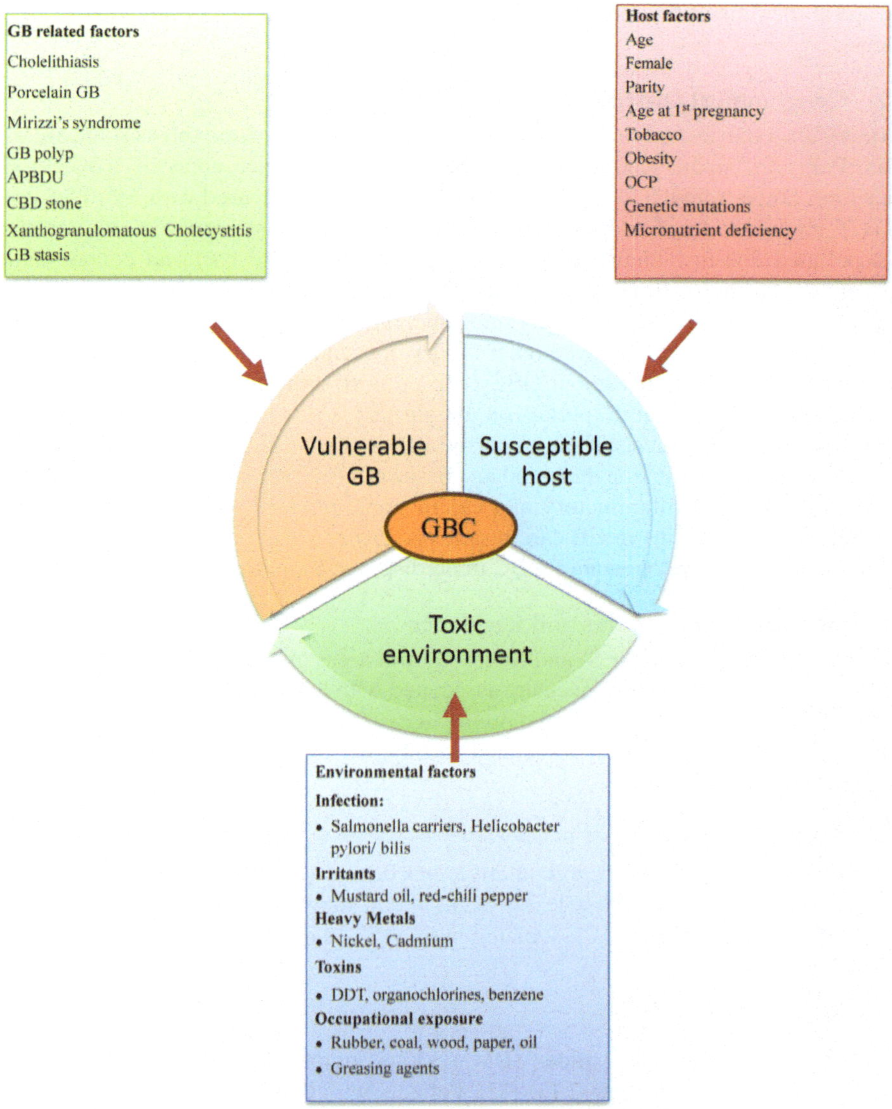

Figure 11.1 Epidemiological triad depicting risk factors of GBC.

environmental factors, and individual genetic susceptibility also play an important role in GB carcinogenesis (Figure 11.1 and Table 11.1) (1).

11.2.1 Host Factors

11.2.1.1 Age and Gender

Increasing age is associated with increasing risk for GBC (8). Gender differences exist across geographic regions. Women are at 2–6 times higher risk for developing GBC (9). The GB mucosa has been found to have estrogen and progesterone receptors, which may promote GB stasis, stone formation, and, in turn, increase exposure time of the GB mucosa to bacterial and chemical toxins (10). Furthermore, the presence of other risk factors, like younger age at the time of marriage, younger age at the time of first pregnancy, and having a higher number of pregnancies, accentuates this process.

11.2.1.2 Geographical Distribution

South American countries, particularly Chile, Bolivia, and Ecuador, as well as some areas of northern India, Pakistan, Japan, Korea, and Poland, are high-incidence areas of GBC. In Chile, the incidence rates range from 8.3 to 18.8 per 100,000 in women, compared with 3.2 to 5.8 per 100,000 in men (11). High-risk areas in India have water sources along major rivers, the water of which becomes polluted due to human waste, industrial pollutants, and bacterial contamination while flowing from the northern Himalayas toward the west and east. High levels of pesticides and heavy metals and nitrates, both of which are carcinogenic, have been identified as pollutants in the Ganges (1). Also, patients with GBC have higher biliary concentrations of these pesticides and nitrates compared with those with GS without GBC (12). In north-eastern India, mustard oil is the staple cooking oil, which has irritant properties on the gut and is often adulterated with butter yellow, a known carcinogen. Higher levels of sanguinarine and diethyl nitrosamine, carcinogenic adulterants in mustard oil, have been found in the blood and tissue of GBC patients compared with patients with GS. Mustard oil has pro-inflammatory and carcinogenic properties. Various studies have shown the potential role of aflatoxins in GB carcinogenesis. Prolonged exposure to these toxins acts in an additive manner in the development of GBC in predisposed individuals (13).

11.2.1.3 Socio-Economic Status and Residence

Poorer socio-economic status (SES) and living in rural areas are associated with increased risk for GBC (14). Low SES and rural residence are associated with lower literacy rates, overcrowding, poor access to health care, poor sanitation, and poorer access to clean drinking water than those belonging to upper SES. They are, hence, associated with higher exposure to fecal-oral infections like *S. typhi* and *H. pylori* (15).

11.2.1.4 Family History

A history of GBC or GS in first-degree relatives has been associated with a fivefold increased risk for GBC (summary RR: 4.8; 95% CI: 2.6–8.9) (14). A study from Varanasi demonstrated that a family history of biliary disease is associated with an increased risk for GBC (OR: 3.48; 95% CI: 1.38–8.98) (16).

11.2.1.5 Obesity

Obesity, defined as a body mass index (BMI) >30, is associated with a twofold increased risk for GBC (RR: 1.88; 95% CI: 1.66–2.13) (17). Patients with GBC have lower BMI, which reflects weight loss secondary to malignancy. Most patients are not reported to be obese initially; if we add expected weight loss to present BMI, most patients will not fall in the obesity category. Undernutrition, however, may be associated with suboptimal immune status and a pro-inflammatory state secondary to micronutrient and antioxidant deficiency, both of which, in turn, can promote malignancy.

11.2.1.6 Medications

Some medications have been implicated in biliary carcinogenesis, including methyldopa, oral contraceptives/hormone replacement therapy, and isoniazid (1).

11.2.2 Gallbladder-Related Risk Factors

11.2.2.1 Gallstones

Are GS a risk factor for GBC? GS are present in up to 90% of patients with GBC, especially in high-risk populations. GS are strongly associated with increased GBC risk and are diagnosed years before

GBC development. Cholesterol GS seem to be more common than pigment GS in patients with GBC. There is enough evidence to suggest that GS are causative rather than just incidental, as all the Bradford Hill criteria for causation are fulfilled (1). Physical, chemical, and biological characteristics of GS may influence chronic inflammatory processes and, therefore, may be associated with initiation of GB carcinogenesis. GS may promote GBC by causing direct mechanical injury of the GB mucosa during contraction, especially when the stones are large, irregular, and occupy a larger volume. Also, the GS surface may provide a surface for bacterial biofilm formation, which allows persistence of bacteria. The chemical composition of bile may also be altered due to deconjugation of bile acids, conjugated toxins, and accumulation of heavy metals and metabolites in the GB, which may be compounded by biliary stasis often found in patients with GS (3). Despite this evidence for causality, only ~1% of individuals with GS develop GBC, suggesting the involvement of additional contributing factors. GS alone, in the absence of cofactors, may result only in mechanical injury, which the GB mucosa may recover from by natural healing processes. A combination of multiple repeated insults of varying nature may overwhelm the tissue repair mechanism, giving way to chronic inflammation, mutagenesis, and carcinogenesis (18). Thus, removal of the GB harboring GS, even if the patient is asymptomatic in high-risk areas, may be a worthwhile step in some selected cases to prevent GBC-related mortality. Preventive cholecystectomy is advocated in high-incidence areas such as Chile (19) and Bolivia, but not in low-incidence areas such as North America and western Europe.

11.2.2.2 Structural Biliary Abnormalities

Anomalous pancreaticobiliary junction (APBJ) has been associated with an increased risk of GBC and warrants prophylactic cholecystectomy. The reflux of pancreatic juice causes chemical irritation in the GB mucosa, K-RAS mutation, and resultant papillary adenocarcinoma (20).

GB polyps are present in around 5% of adult patients. Most of them (95%) are non-neoplastic in nature. Two-thirds of polyps are composed of cholesterol esters; the others include adenomas, leiomyomas, or inflammatory polyps. Benign adenomas constitute <5% of all GB polyps, and their size ranges from 0.5–2 cm (21). There are salient features that predict malignancy: large polyps (>15 mm), a solitary or sessile mass, focal wall thickness greater than 4 mm (not attributed to wall edema, mucosal fold, adenomyomatosis, or GB under-distention), and rapid polyp growth (increase in size of >4 mm on annual follow-up imaging). Endoscopic ultrasound is useful in differentiating benign and malignant polyps. Presence of a hypoechoic, heterogeneous lesion with a height/width ratio of 0.8 and increased vascularity suggests increased risk for neoplasia. Prophylactic cholecystectomy is warranted in patients with polyps that possess such malignant-appearing features (22).

Chronic inflammation from any cause may lead to calcium deposition in the GB wall, termed the "porcelain GB" because of its bluish color and fragile, brittle consistency. This entity is often rare, being identified pathologically in less than 1% of GB specimens. Controversy exists over whether the porcelain GB is truly associated with an increased risk of cancer. Focal calcification, stippled or multiple punctate calcifications in the glandular spaces of the GB mucosa, are premalignant and, therefore, should be removed prophylactically. Those with a broad continuous band of calcification in the muscularis appear not to be harbingers of GBC, but they need close follow-up (23). Primary sclerosing cholangitis is associated with increased risk for biliary cancers, cholangiocarcinoma, as well as GBC. These patients need annual surveillance for GBC and are likely to benefit from prophylactic cholecystectomy.

11.2.2.3 Congenital Biliary Cysts

Biliary cysts are cystic dilatations that may occur singly or in multiples throughout the bile ducts. Biliary cysts may be congenital or acquired, and they are associated with a variety of anatomic abnormalities. APBJ is present in approximately 70% of patients with biliary cysts. Biliary cysts are associated with an increased risk of cancer, particularly cholangiocarcinoma, and to some extent GBC as well (24).

11.2.3 Environmental Risk Factors

11.2.3.1 Infections

H. pylori is a gram-negative, spiral-shaped bacterium, which usually colonizes the gastrointestinal tract. Virulent H. pylori strains induce a pro-inflammatory state, result in detrimental immune alterations in the host, and hence promote carcinogenesis along the same lines as gastric cancer pathogenesis (25). *H. bilis* is also implicated in GBC. *S. typhi* carriers have an 8- to 12-fold increased risk of developing GBC. Recent meta-analyses show an OR of 3.5 (95% CI: 2.48–5) using serology and 4.1 (95% CI: 2.41–7.12) using culture techniques (26).

11.2.3.2 Carcinogen Exposure

Smoking has been seen to be associated with an increased risk of GBC (27). In large population-based cohort studies, soil and water analyses showed higher levels of heavy metals, including nickel, arsenic, cadmium, chromium, and dichloro diphenyl trichloroethane (DDT), in regions with a higher prevalence of GBC. Exposure to mining, radon gas, rubber industries, paint, chemicals, paper, and wood industries has been associated with increased risk for GBC (3). Exposure to aflatoxin B1, a potent liver carcinogen, has also been associated with GBC, with ORs of 2.7 (95% CI: 1.7–4.3) in China and 9.4 (95% CI: 2.8–37.2) in Chile for detection of aflatoxin B1 in individuals with GBC compared with individuals with GS but not GBC (7, 28). In a study from India by Yadav et al., detectable serum aflatoxin adducts (OR = 6.8 [CI: 1.3–35.7]; p = 0.024) were identified as independent risk factors for GBC (29).

TABLE 11.1
Risk Factors Identified for Gallbladder Cancer

Factor	High risk subset	Remarks
Geographical (Incidence in women per lac population)	Chile, Ecuador, Columbia,	15.5, 12.9, 9.5
	North India, Pakistan	21.5, 13.8
Ethnicity (Incidence in women per lac population)	Hispanics	8.2
	American Indians	7
Age	5th decade	Younger age if multiple risk factors and high incidence areas
Gender	Female	2–6 times higher compared to men
Pregnancy	Higher parity (>5)	RR 4.2 (95% CI: 1.5–12.8)
Family history of GBC	First degree relatives	Summary RR 4.8 (95% CI:2.6–8.9)
Socioeconomic status	Poor SES	OR 3 (95% CI: 0.9–10)
Gall stone	Large GS >3 cm,	RR 10.1
	GS volume >6 mL	RR 5
	Multiple GS	Higher the number, higher the risk
	Long standing GS	Time (stone to cancer): 20 years
	Mirrizi's syndrome, Empyema GB	Increased likelihood of GBC
GB polyp	Sessile, focal wall thickness >4 mm, rapid polyp growth (>4 mm/year annual, associated GS	Increasing size, higher risk
Porcelain GB	Associated chronic cholecystitis and gallstones	Incidence of GBC variable: 10–61%

(Continued)

TABLE 11.1 (*Continued*)
Risk Factors Identified for Gallbladder Cancer

Factor	High risk subset	Remarks
Anomalous pancreaticobiliary junction	Common channel >8 mm especially in Japan, Korea, China	Younger, papillary, ↓invasive, no GS
Salmonella carriers	Especially in Chile, India	Summary RR 10.2 (95% CI:2–51)
Helicobacter species	Few and small studies only	Summary RR 9.9 (95% CI:1.4–70.5) for *H. pylori*
Obesity	BMI >30–35 versus BMI 18.5–25	Summary RR 1.88 (95% CI: 1.66–2.13)
Smoking	>10 cigarettes for 5 years	RR 11 (CI: 1.7–71)
Dietary intake	Low in vegetable, fiber, fruit, vitamin B6, C and E; High in calorie, carbohydrate, red meat, red chilli pepper	Increased risk
Serum lipids	High triglycerides and low LDL	RR 1.9 (95% CI: 1.3–2.8)
Chemical exposure	Oil, chemical, miner, textile and paper industry	Increased risk
Medications	Methyldopa, Isoniazid	Increased risk
	Oral contraceptives	Summary RR 0.9 (CI: 0.5–1.3)
	Hormone replacement therapy	Summary RR 0.9 (CI: 0.3–3.3)

11.3 GALLBLADDER CANCER: GENETIC AND MOLECULAR ALTERATIONS

The existing literature regarding the genetic and molecular alterations in GBC is still limited. Like other neoplasms, GBC is a multifactorial disorder involving multiple alterations at the genetic and molecular levels. Abnormalities in tumor suppressor genes (TSG), oncogenes, and DNA repair genes, the presence of microsatellite instability (MSI), and epigenetic alterations are some of the well-known factors reported to date (30).

11.3.1 Tumor Suppressor Genes

TSG play an essential role in regulating cell growth and ensuring genomic stability (31). Mutations or inactivation of these genes are often a hallmark of various cancers. Several TSG have been implicated in the causation of GBC.

a) *TP53:* The TP53 gene is one of the most frequently mutated genes in GBC, with mutations found in approximately 30–50% of cases (32). TP53 encodes a 53-kD nuclear phosphoprotein, which acts as a "Guardian of the genome," regulates the cell cycle, induces apoptosis in response to DNA damage, and prevents the propagation of damaged cells. Most of the TP53 mutations associated with GBC are missense mutations that produce a non-functional protein with an increased half-life (30). Mutated p53 proteins have a longer half-life compared to their wild-type p53 product and have been shown in regions with a high occurrence of GBC. Loss of TP53 function is associated with poor prognosis, as it enables the accumulation of genetic mutations and resistance to cell death, promoting tumor progression. TP53 mutations are relatively more common in advanced stages of the disease. Functional molecular studies have discovered that mutations in exons 5 and 8 of the TP53 gene cause deregulation of this gene and promote carcinogenesis (33).

b) *CDKN2A*: The cyclin-dependent kinase inhibitor 2A (CDKN2A) gene, which encodes p16INK4a, is another TSG frequently altered in GBC. The inactivation of CDKN2A leads to uncontrolled cell cycle progression, allowing for unchecked cell proliferation. Loss of CDKN2A expression has been observed in both primary GBC tumors and pre-cancerous lesions, suggesting its role in early carcinogenesis (34).
c) *SMAD4*: The SMAD4 gene is involved in the transforming growth factor-beta (TGF-β) signaling pathway, which regulates cell growth, differentiation, and apoptosis. Mutations or deletions in SMAD4 have been observed in GBC and are associated with aggressive disease, metastasis, and poor survival outcomes (35).
d) *Fragile histidine triad (FHIT)*: FHIT is a TSG that is lost during GB carcinogenesis and is reported on chromosome 3p. A high incidence of loss of heterozygosity in the FHIT region is seen in dysplastic and carcinomatous GB. The loss of FHIT prompts global genome instability, mostly through nucleotide imbalance, spontaneous replication stress, and DNA breaks (36).

11.3.2 Oncogenes

Oncogenes are genes that, when mutated or overexpressed, drive cancerous cell growth. Oncogenes are mutated forms of normal cellular genes (proto-oncogenes) that become altered in the cancer signaling cascade. Several oncogenes have been implicated in GBC pathogenesis, contributing to the dysregulation of key signaling pathways that control cell proliferation and survival.

a) *KRAS*: The KRAS gene encodes a small GTPase involved in cell signaling and is frequently mutated in GBC, with mutations found in approximately 15–20% of cases. KRAS mutations activate the mitogen-activated protein kinase (MAPK) pathway, leading to enhanced cell proliferation, survival, and migration. KRAS mutations are also associated with poor prognosis and resistance to chemotherapy in GBC. KRAS acts as an initial key player in numerous signal transduction mechanisms and associated pathways. KRAS gene mutations identified in GBC mostly affect codons 12, 13, and 61 (30, 37). GBC associated with an APBJ has a high KRAS gene mutation rate of 50–83% of cases (38).
b) *The epidermal growth factor receptor (EGFR):* EGFR is a transmembrane receptor belonging to the ErbB family of receptor tyrosine kinases. EGFR is often overexpressed in GBC, contributing to tumor growth and metastasis. EGFR mutations, including activating and resistant mutations, frequently occur in exons 18–21 of the EGFR gene. Studies examining the expression of EGFR in GBC have indicated highly variable receptor expression (39). Overexpression of EGFR activates several downstream signaling pathways, including the PI3K/AKT and MAPK/ERK pathways, which promote cell proliferation, survival, and angiogenesis.
c) *PIK3CA*: Mutations in the PIK3CA gene, which encodes the catalytic subunit of the PI3-kinase enzyme, have been found in GBC (40). These mutations activate the PI3K/AKT pathway, leading to increased cell survival, resistance to apoptosis, and enhanced tumor growth. The PI3K/AKT/mTOR signaling pathway is a potential therapeutic target in GBC.

11.3.3 DNA Repair Mechanisms

DNA repair mechanisms are critical for maintaining genomic stability. Defects in DNA repair pathways can lead to the accumulation of mutations and the development of cancer. In GBC, mutations in genes involved in DNA repair have been observed.

a) *BRCA1 and BRCA2*: Although more commonly associated with breast and ovarian cancers, mutations in the BRCA1 and BRCA2 genes have been reported in a subset of GBC patients (41). These genes are involved in homologous recombination repair, a critical mechanism for repairing double-strand DNA breaks. Loss of BRCA function in GBC may lead to genomic instability and increased sensitivity to DNA-damaging agents, such as platinum-based chemotherapy.

b) *MLH1 and MSH2*: Mutations in mismatch repair (MMR) genes, such as MLH1 and MSH2, have been implicated in GBC, particularly in patients with Lynch syndrome. These mutations result in microsatellite instability (MSI), which can drive carcinogenesis through the accumulation of mutations in critical genes involved in cell cycle regulation, apoptosis, and DNA repair. A study published by Goldin et al. demonstrated a high degree of microsatellite instability (MSI) in 10% of GBC patients (42).

11.3.4 Epigenetic Modifications

Epigenetic alterations, including DNA methylation, histone modifications, and non-coding RNAs, also play a crucial role in the development of GBC. These changes can silence tumor suppressor genes and activate oncogenes without altering the underlying DNA sequence, contributing to cancer initiation and progression (43).

a) *DNA Methylation*: Aberrant DNA methylation, particularly hypermethylation of TSG, is a common feature of GBC. For example, the promoter regions of genes such as p16INK4a, E-cadherin, and MLH1 are often hypermethylated in GBC, leading to their silencing and contributing to tumorigenesis. Acquisition of hypermethylation at multiple TSG-promoter sites may contribute to tumor formation and progression within the chronically inflamed GB (44). DNA methylation markers have also been proposed as potential biomarkers for early detection and prognosis in GBC.
b) *Histone Modifications*: Post-translational modifications of histones, such as acetylation, methylation, and phosphorylation, can influence gene expression and chromatin structure. These modifications have been observed in patients with GBC and may promote oncogene activation and TSG silencing. These modifications are reversible, making them potential targets for therapeutic interventions (45).
c) **Noncoding RNAs**: MicroRNAs (miRNAs) and long non-coding RNAs (lncRNAs) are important regulators of gene expression in cancer. In GBC, altered expression of several miRNAs, such as miR-21, miR-155, and miR-34a, has been reported. These miRNAs can regulate key genes involved in cell proliferation, invasion, and apoptosis. The potential use of miRNAs as biomarkers for diagnosis, prognosis, and therapy in GBC is an area of active research. Elucidating the role of miRNAs in the biology of GBC may provide novel therapeutic strategies for the management of GBC and identify effective biomarkers for early diagnosis (46).

11.3.5 Cell Adhesion Molecules

Cell adhesion molecules (CAMs) are proteins positioned on the cell surface and are involved in intercellular and cell–extracellular matrix interactions. Some adhesion molecules (immunoglobulin-like CAMs, cadherins, selectins, and integrins) play a crucial role in the development of repeated, invasive, and distant metastatic cancer conditions (47). Mucins are high–molecular-weight glycoproteins that provide a protective layer on epithelial surfaces, serve lubricating roles in the normal epithelium of gastrointestinal organs, and are significantly altered during neoplastic and paraneoplastic conditions. They are involved in cell–cell interactions, signaling, and metastasis. Normal mucosal cells secrete a variety of different mucins, such as MUC1, MUC2, MUC3, MUC4, MUC5A, and MUC6 (48). The expression of MUC1 is absent in normal GB mucosa, whereas MUC5AC and MUC6 are more diffusely expressed. MUC1 levels are significantly increased in GBC, while MUC2 and MUC5AC expression levels are reduced in GBC (49).

E-cadherin (CDH1) is crucial for the development and Ca^{2+}-dependent maintenance of cell adhesion, polarity, and differentiation of epithelial cells. Significant differences in the protein expression of E-cadherin have been observed between normal, inflamed, and cancerous GB tissues. Loss of E-cadherin expression is a prominent event in GBC compared to chronic cholecystitis, xanthogranulomatous cholecystitis, and normal GB (50).

11.4 GALLBLADDER CANCER: TUMOR MICROENVIRONMENT

The tumor microenvironment (TME) is an important factor in the progression and aggressiveness of GBC. The TME is composed of tumor cells, stromal cells, immune cells, blood vessels, and extracellular matrix components, all of which interact to promote tumor growth and metastasis (51, 52).

11.4.1 Chronic Inflammation

Chronic inflammation, often associated with conditions such as GS and chronic cholecystitis, is a key factor in the development of GBC (53). Inflammatory mediators, such as cytokines and chemokines, can promote tumorigenesis by stimulating cell proliferation, survival, and angiogenesis, as well as by suppressing the anti-tumor immune response.

11.4.2 Immune Evasion

GBC is characterized by immune evasion mechanisms that prevent effective anti-tumor immune responses. Tumor-associated macrophages (TAMs), regulatory T cells (Tregs), and myeloid-derived suppressor cells (MDSCs) contribute to the immunosuppressive TME, inhibiting the activation of cytotoxic T cells and natural killer cells (52). Targeting the immunosuppressive components of the TME is an area of interest for novel immunotherapies in GBC.

11.4.3 Angiogenesis

Angiogenesis, the formation of new blood vessels, is critical for tumor growth and metastasis. In GBC, overexpression of pro-angiogenic factors, such as vascular endothelial growth factor (VEGF), has been observed, contributing to tumor vascularization and the supply of nutrients and oxygen to the growing tumor mass. VEGF-A is highly expressed in GBC and correlates with poor prognosis, suggesting that VEGF-A expression might be used as a biomarker for predicting malignant behavior and for identifying a subset of patients (54).

11.5 GALLBLADDER CANCER: MOLECULAR PATHWAYS

Intracellular signaling pathways play a key role in carcinogenesis and cancer maintenance. The role of abnormal activation of signaling pathways has also been extensively studied as a way of directing rational cancer therapy in chemotherapy-resistant cancers.

11.5.1 Hedgehog Signaling Pathway

The Hedgehog (Hh) pathway is a key developmental signaling pathway involved in numerous developmental processes, including determination of cell fate, patterning, and proliferation, as well as its role in cancer initiation and progression. While this pathway is silenced in most adult tissues, aberrant activation has been documented in different types of cancer and contributes to the induction of malignant phenotypes and the regulation of cancer stem cells, fostering tumorigenesis (55). Recent studies have confirmed that aberrant activation of Sonic hedgehog signaling transduction is present in GBC and is associated with poor survival. Several proteins in the Hh pathway have been identified as potential drug targets for inhibiting aberrant Hh signaling (56).

11.5.2 PI3K/AKT/mTOR Signaling Pathway

The PI3K/AKT/mTOR pathway is involved in the pathogenesis of several human cancers, with pivotal effects on cell cycle progression, cell proliferation, and angiogenesis (57). Transmembrane growth factor receptors, including VEGFR, EGFR, insulin growth factor receptor 1, G protein-coupled receptors, and RAS proteins, are capable of activating PI3K, which phosphorylates phosphatidylinositol-4,5-bisphosphate (PIP2) to generate phosphatidylinositol-3,4,5-trisphosphate (PIP3). PIP3 subsequently binds to phosphoinositide-dependent kinase 1 (PDK1), thus phosphorylating and activating the serine/threonine kinase AKT (58). PTEN (phosphatase and tensin homolog), a tumor suppressor, is a phosphatase that dephosphorylates PIP3 into inactive PIP2, thereby dampening AKT and PDK1 activity (59). The PIK3CA gene encodes the p110α subunit of PI3K, an activator of AKT and a downstream effector of the mammalian target of rapamycin (mTOR) pathway. PI3K signaling deregulation as a consequence of PIK3CA mutations or PTEN loss has been described in 5–12% and 50% of GBC, respectively (60, 61). Studies have reported loss of PTEN expression in 51.8% of GBC, whereas most chronic cholecystitis and GB polyp cases (93.3%) expressed PTEN (62). Loss of PTEN expression is significantly more common in poorly differentiated adenocarcinomas with lymph node metastasis, invasion, and poor prognosis. The activation of mTOR occurs very early during GBC development, contributing to the malignant transformation process. The cumulative evidence provides a rationale for the potential use of mTOR inhibitors as a therapeutic strategy for GBC (63). Use of inhibitors against the mTOR signaling pathway has been proposed as an attractive therapeutic strategy, based on the fact that this pathway seems to be activated in GBC.

11.5.3 Notch Pathway

The Notch pathway is normally activated when membrane receptors interact with ligands on an adjacent cell, such as delta-like and jagged. Then, two consecutive proteolytic cleavages of the Notch receptor release its intracellular portion (termed ICN), which translocates into the nucleus to mediate target gene activation. Alterations of the Notch pathway in hematopoietic and solid tumors can lead either to activation or repression of the pathway, depending on the context and the activation status of other potentially oncogenic pathways (64). Notch signaling regulates cholangiocarcinoma carcinogenesis, since constitutive activation of NOTCH1 and overexpression of the intracellular domain of Notch (Notch-ICD) were sufficient to initiate cholangiocarcinoma development in mice (64). Studies on GBC are far more limited.

11.5.4 ErbB Signaling Pathway

ErbB receptors are cell-surface receptors that harbor a transmembrane tyrosine kinase domain capable of activating multiple downstream signaling pathways upon binding with epidermal growth factor (EGF). They play important roles in normal physiology and have been involved in the development of different cancers. ErbB signaling is the most extensively studied mutated pathway. Activating mutations in ERBB2 and ERBB3 have been implicated in oncogenic potential, and ErbB alteration highlights the importance of the ErbB2 receptor in GBC development and progression. Studies on GBC have focused on EGFR and HER2 tyrosine kinases, which appear to be attractive targets in GBC (65).

11.5.5 MAPK/ERK Signaling Pathway

The mitogen-activated extracellular signal-regulated kinase (MEK) pathway, also known as the RAS/RAF/MEK/ERK pathway, is one of the best-characterized kinase cascades in cancer cell biology (66). It is activated by a variety of growth signals, including EGF, PDGF, and cytokines, leading ultimately to MEK1 and MEK2 phosphorylation and the subsequent phosphorylation and nuclear localization of ERK1 and ERK2 (67). Dysregulation of the MEK/ERK pathway is observed in biliary cancers, probably due to mutations of the oncogenic kinases involved in this pathway (RAS, RAF, MEK) (68).

11.5.6 Angiogenesis Pathway

Angiogenesis is the biological process that drives the formation of new blood vessels from pre-existing vasculature. It has been described as a hallmark of cancer, necessary for tumor progression, during which an "angiogenic switch" is activated, causing the secretion of signals that recruit and expand the vascular network that sustains neoplastic growth. Some angiogenic regulators are signaling proteins that bind to stimulatory or inhibitory cell-surface receptors displayed by vascular endothelial cells. The well-known prototypes of angiogenesis inducers and inhibitors are vascular endothelial growth factor-A (VEGF-A) and thrombospondin-1 (TSP-1), respectively. VEGF-A is a diffusible endothelial cell-specific mitogen that induces endothelial cell proliferation and increases vascular permeability (69). VEGF expression is observed in 50–80% of GBC and correlates with metastatic disease and poor prognosis (70). Tian et al. investigated the involvement of the TP53 gene in regulating tumor angiogenesis and observed that abnormal TP53 accumulation was closely associated with a progressive increase of VEGF expression. Thus, the TP53–VEGF pathway can regulate tumor angiogenesis in GBC, and the combined analysis of mutant TP53 and VEGF expression might be useful for predicting tumor vascularity in GBC (71).

11.6 GALLBLADDER CANCER: PATHOGENESIS

Chronic inflammation is the key process that contributes to the development of GBC. The GB is already a physiologically vulnerable organ of the body because it has to empty its content against gravity back into the gut lumen. The GB tends to concentrate bile, which allows increased mucosal contact with concentrated toxic substances. The fundus has higher mucosal contact time (3). A diseased GB also tends to be static, enhancing mucosal exposure time. The GB has a weak muscular structure, which further disadvantages the emptying process. Moreover, obstruction along the pathway for bile flow due to CBD stones, strictures, or sphincter of Oddi abnormalities can impede the emptying process. Once bacteria enter the GB, it is difficult for the body to eradicate them. These bacteria tend to cause chronic inflammation, deconjugate conjugated bile acids and toxins, and thus locally release toxic metabolites (1). Inflammation provides a favorable environment that initiates malignant transformation and tumor progression. Current understanding of GBC pathogenesis recognizes two key pathways to developing GBC, both of which are driven by long-standing inflammation of either lithiasic or non-lithiasic origin. The most common histogenic pathway is the metaplasia–dysplasia–carcinoma sequence, which is seen in up to 85–90% of GBC, more frequently in India. Nearly 5–10% of GBCs seem to develop via an adenoma–carcinoma sequence, which is less frequently associated with GS but is also characterized by a chronic inflammatory process that favors adenoma formation and accumulation of molecular alterations, leading to the initiation and progression of GBC (Figure 11.2) (72).

11.6.1 Metaplasia–Dysplasia–Carcinoma Sequence

Most GBC (70–90%) arise in the setting of GS, which is the major risk factor associated with GBC. Cholelithiasis is often asymptomatic, but around a quarter of cases can manifest over decades by repeated acute episodes with total or partial occlusion of the cystic duct and massive inflammatory destruction of the epithelium. These continuous cycles of mucosal damage and repair and/or regeneration, alongside the pro-inflammatory lithogenic bile–related chemical injury, induce a

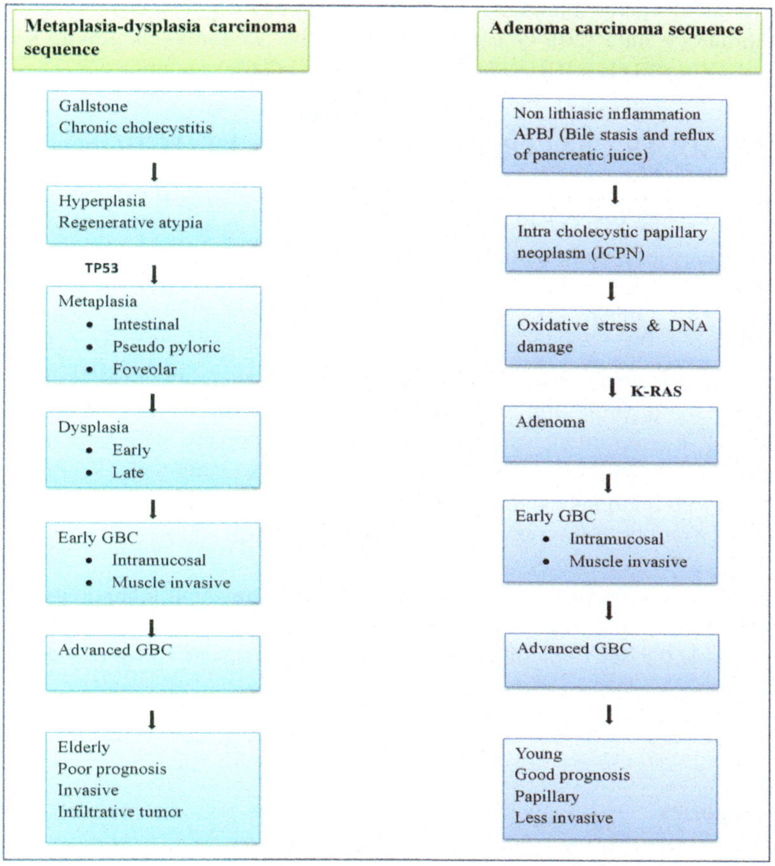

FIGURE 11.2 Pathogenesis pathways of GBC (metaplasia–dysplasia–carcinoma sequence and adenoma–carcinoma sequence).

long-standing chronic inflammatory response that disrupts cellular homeostasis. It is further characterized by overexpression of cyclooxygenase 2 (COX2) and infiltration by T cells, B cells, COX2/inducible nitric oxide synthase (iNOS)-positive macrophages, iNOS-positive granulocytes, and mast cells (53). Histological changes of metaplasia and dysplasia are also common in the background of chronic inflammation. Chronic inflammation induces DNA damage through oxidative stress generated by elevated levels of reactive oxygen species (73) and reactive nitrogen species (RNS). DNA damage activates various response pathways, including the p53-mediated tumor suppression pathway, which acts as a major barrier against tumorigenesis by controlling several cellular responses, such as apoptosis, senescence, and cell-cycle arrest in damaged cells (72).

11.6.2 Adenoma Carcinoma Sequence

GBC can also develop in the background of non-lithiasic inflammation from adenomatous precursor lesions. Intracholecystic papillary neoplasm (ICPN) is one of the precursor lesions of GBC, presenting as exophytic preinvasive growth. It is often associated with APBJ, where bile stasis and chronic reflux of pancreatic juice into the biliary tract and GB lead to activation of proteolytic pancreatic enzymes, which modify bile composition and generate strong cytotoxic substances (such as lysolecithin) that ultimately damage the GB epithelium and trigger a constant and excessive non-lithiasic inflammatory state (20). Genetic changes in the adenoma–carcinoma sequence include K-RAS (50%), EGFR (20.6%), RB1 (17.6%), and ERBB2 (17.6%) mutations (72).

11.7 GALLBLADDER CANCER: INNOVATIONS IN DETECTION

GBC is a highly lethal malignancy often diagnosed at an advanced stage due to its asymptomatic nature in the early stages. The challenge of early detection has spurred significant innovations in diagnostic technology and methodologies.

11.7.1 Imaging Advancements

Imaging techniques remain a cornerstone for the detection and staging of gallbladder cancer. Recent innovations have significantly improved the sensitivity and specificity of imaging modalities. Detailed radiological assessment using GB-RADS preoperatively will help in the detection of incidental GBC. Patients with asymptomatic GS with high-risk features (GS > 3 cm, non-functioning GB, family history of malignancy, associated GB polyps) can be considered for either preventive cholecystectomy or regular follow-up using GB-RADS (Table 11.2) (74).

a) *High-Resolution Ultrasound (HRUS)*: Advances in ultrasound technology, such as high-frequency probes and real-time elastography using acoustic radiation force impulse (ARFI), enable the detection of subtle changes in the GB wall and surrounding structures.

TABLE 11.2
GB-RADS Categories

GB-RADS Category	Description	Findings
0	Inadequate evaluation due to technical or patient factors or gallbladder-related factors	Few examples Morbid obesity Wall-echo-shadow complex Porcelain gallbladder Gas in the gallbladder lumen
1	Normal	Adequate gallbladder distension Wall thickness ≤ 3 mm
2	Benign	Symmetric circumferential thickening (>3 mm) with or without intramural changes or focal thickening with intramural changes Layered appearance Preserved interface with liver
3	Equivocal	Circumferential thickening without layered appearance Focal thickening without intramural features (cysts or echogenic foci) or layered appearance Distinct interface with liver
4	Malignancy is likely	Circumferential or focal thickening (>3mm) without layered appearance and with loss of interface with liver
5	Malignancy is highly likely	Same as GB-RADS 4 with definite extramural invasion as suggested by one of the following: Biliary or vascular involvement by direct extension of mural thickening Liver mass in contiguity with the mural thickening

Elastography, in particular, provides insights into tissue stiffness, helping differentiate between benign and malignant lesions (75).

b) *Endoscopic Ultrasound*: EUS has become an important tool in the diagnosis of gallbladder malignancies, as it allows for high-resolution imaging of the gallbladder wall and surrounding structures. EUS-guided fine-needle aspiration (FNA) can also help in obtaining tissue samples for histological analysis, improving diagnostic accuracy (76).
c) *Contrast-Enhanced Imaging*: Contrast-enhanced ultrasound (CEUS) and magnetic resonance imaging (MRI) have proven effective in identifying vascular patterns and enhancing lesion characterization (77). MRI techniques, like diffusion-weighted imaging and magnetic resonance cholangiopancreatography (MRCP), offer non-invasive approaches to assess tumor extent and biliary tree involvement.
d) *Positron Emission Tomography*: PET-CT using fluorodeoxyglucose (FDG) has emerged as a valuable tool for detecting metastatic disease and distinguishing between benign and malignant GB abnormalities (78).

11.7.2 Biomarkers

The identification of specific biomarkers has opened new avenues for the early detection of GBC. These biomarkers include:

a) *Serum Markers*: Current data do not support the application of serum cancer antigens as reliable biomarkers in the diagnosis of GBC, although these markers may be associated with prognosis (79).
    1. *CA 19–9 and CEA*: The diagnostic performance of these biomarkers remains suboptimal due to low sensitivity and specificity, as well as their non-specificity in other biliary and pancreatic diseases. CA 19–9 is valuable for detecting recurrence, with a sensitivity of 79.1%, specificity of 97.2%, positive predictive value of 95%, and negative predictive value of 87.5% (80).
    2. *CA-242*: The performance of CA-242 as a diagnostic test is superior compared to CEA and CA 19–9, which showed a sensitivity, specificity, PPV, and negative predictive value (NPV) of 64%, 83%, 88%, and 53%, respectively (81).
    3. *CYFRA 21–1 and MMP-7*: Studies have shown that CYFRA 21–1 is a promising biomarker for GBC, with a sensitivity of 93.7% and specificity of 96.2%. Serum levels of CYFRA 21–1 also correlate with tumor aggressiveness, prognosis, and recurrence-free survival (82).

b) *Circulating Tumor DNA (ctDNA):* Liquid biopsies that detect ctDNA in blood offer a non-invasive method for identifying genetic mutations and alterations specific to GBC. Kumari et al. evaluated the diagnostic value of circulating serum-free DNA (cfDNA) in GBC using quantitative PCR (qPCR) and found that GBC patients had substantially higher cfDNA levels than cholecystitis patients and the control group. In addition, a cfDNA cutoff value of >218.55 ng/mL distinguished cancer patients from healthy controls with 100% sensitivity (95% confidence interval [CI]: 89.6–100; $p < 0.001$) and 100% specificity (95% CI: 80.3–100; $p < 0.001$) (83). In another study, the use of ALU 247 (long DNA fragments derived from tumor necrosis) in the diagnosis of GBC was reported to have a sensitivity and specificity of 80.0% and 86.1%, respectively. The combination of ALU 247 and cfDNA can be a novel diagnostic tool to discriminate GBC from controls with good diagnostic accuracy (84). Qualitative studies on cfDNA have shown mutations in various genes, such as KRAS and p53, MSI, loss of heterozygosity (LOH), aberrant methylation of several genes, and tumor-specific cell-free mRNA.

c) *Circulating Tumor Cells (CTCs):* CTCs are shed from tumors and enter circulation, which can be extracted and quantified via flow cytometry protocols. Awasthi et al. demonstrated that EpCAM-positive CTCs might be used to discriminate GBC from healthy controls with a sensitivity of 92.6% and specificity of 91.7%. At a cutoff >6, CTCs were able to discriminate metastatic versus non-metastatic GBCs with a sensitivity, specificity, and diagnostic accuracy of 55.6%, 100.0%, and 85.2%, respectively (85).
d) *MicroRNAs (miRNAs):* MicroRNAs with tumor suppressor activity (miR-135–5p, miR-335, miR-34a, miR-26a, miR-146b-5p, miR-218–5p, miR-1, miR-145, miR-130a) are often downregulated, while those with oncogenic properties (miR-20a, miR-182, miR-155) are upregulated in patients with GBC. MicroRNA signatures in blood samples represent a future non-invasive diagnostic biomarker for GBC. The high stability of miRNAs when compared with mRNA molecules in both blood samples and formalin-fixed, paraffin-embedded tissue samples offers a great advantage, suggesting miRNAs as a promising approach to ameliorate GBC (86).

11.7.3 Molecular Diagnostics

Molecular profiling has provided a deeper understanding of the genetic and epigenetic landscape of GBC. Innovations in this area include:

a) *Next-Generation Sequencing (NGS):* NGS enables comprehensive analysis of genetic mutations, aiding in the identification of actionable targets and early detection markers. NGS is widely used to identify biomarkers to guide the targeted treatment approach in cancer, both in tissue and cfDNA (87).
b) *Proteomics and Metabolomics:* Advanced techniques, like mass spectrometry, are used to analyze protein and metabolite profiles in GBC, uncovering unique signatures that distinguish cancerous tissues from benign conditions (88). Liquid chromatography–mass spectrometry (LC–MS)–based proteomics is a novel, robust technique being utilized for protein profiling to assess molecular changes in a diseased state and to identify plausible biomarkers.
c) *Epigenetic Markers:* The use of methylation in the search for new biomarkers in GBC is a promising alternative, since this epigenetic modification is an early, progressive, and cumulative event in GBC. Determining DNA methylation patterns would allow them to be used as biomarkers for early detection, diagnosis, prognosis, and/or therapeutic selection in GBC (89).

11.7.4 Artificial Intelligence and Machine Learning

AI-driven tools have transformed diagnostic workflows by enhancing accuracy and efficiency. Deep learning AI algorithms have demonstrated promising results, with improved efficacy and accuracy at early diagnosis of GBC, achieving accuracy rates of 98.35% using deep neural networks and MobileNet architectures (90). Conventional neural network approaches have shown moderate efficacy, with an area under the curve of 0.81 (90). Despite their considerable promise, several barriers to clinical implementation persist, including regulatory compliance requirements, ethical considerations, and the need for validation and integration within existing clinical workflows. Despite these challenges, the integration of AI in CT scan analyses presents a promising method for the early diagnosis of GBC (90).

a) *Radiomics:* Radiomics is a rapidly growing field of radiology research that allows extraction of quantitative data from routinely acquired radiological images. These data can be used to decode tumor phenotype and heterogeneity, which are surrogate markers of biological aggressiveness, allowing prognostication, therapy, and response prediction. Studies have evaluated the clinical utility of radiomic features in biliary tract cancer and other abdominal malignancies. In a preliminary study, results highlight the potential utility of

CT texture-based radiomics analysis in patients with GBC. Medium texture scale parameters, including both mean and kurtosis, or kurtosis alone, may help predict histological grade and survival, respectively (91).

b) *Predictive Models:* AI-based models integrate clinical, imaging, and molecular data to stratify patients based on their risk of developing GBC. In a recent study, a deep learning-based approach showed high sensitivity (89.8–93%) and AUC (0.810–0.890) for detecting GBC in the presence of GS, contracted GB, lesion size <10 mm, and neck lesions, which was comparable to both radiologists ($p = 0.052–0.738$ for sensitivity and $p = 0.061–0.745$ for AUC). The sensitivity for deep learning-based detection of mural thickening–type GBC was significantly greater than that of one of the radiologists (87.8% versus 72.8%, $p = 0.012$), despite a reduced specificity (92).

c) *Computer-Aided Diagnosis (CAD):* CAD systems assist radiologists in identifying subtle abnormalities on imaging scans, reducing the likelihood of missed diagnoses. CAD systems can be used as a non-invasive diagnostic tool for early and accurate detection and grading of GBC (93).

11.7.5 Immunohistochemistry

Immunohistochemistry (IHC) has become an essential tool in the diagnosis, prognosis, and treatment of GBC, a malignancy with a typically poor prognosis due to its late-stage presentation and lack of early detection methods. IHC has proven to be a valuable adjunct in differentiating GBC from other biliary tract lesions and metastatic malignancies (94). A range of markers is employed to improve the accuracy of diagnosis. IHC staining for CK7 is typically positive in GBC, while CK20 expression is more common in colorectal cancers. The profile of CK7 and CK20 can help distinguish GBC from other primary or metastatic tumors (95). TP53 overexpression is often seen in GBC, and Ki-67 is a marker of proliferation that can help assess the tumor's aggressiveness. Overexpression of mucins is often found in GBC and can help confirm the diagnosis of carcinoma when evaluated alongside other histopathological findings. IHC not only assists in diagnosis but also offers prognostic insights. Several markers have been studied for their role in predicting outcomes and survival in GBC. The presence of TP53 mutations in GBC correlates with poor prognosis and may serve as an indicator of tumor progression (73). Loss of E-cadherin expression, a marker for epithelial-to-mesenchymal transition (EMT), has been linked to increased metastatic potential and poorer survival outcomes (96). Overexpression of VEGF, which promotes angiogenesis, can indicate more aggressive tumor behavior and a higher likelihood of metastasis, thus influencing treatment strategies (97). IHC can inform therapeutic decisions by identifying molecular targets for targeted therapies. One of the most promising areas is the use of HER2/neu expression. Studies suggest that HER2 overexpression may indicate a subset of GBC patients who could benefit from HER2-targeted therapies (98). Moreover, IHC for mismatch repair proteins (MLH1, MSH2, MSH6, PMS2) can help identify patients with microsatellite instability, who may respond to immunotherapies like PD-1 inhibitors (99). Despite challenges related to variability and interpretation, IHC remains a cornerstone in the pathological workup of GBC, and ongoing research into additional markers and targeted therapies continues to enhance its clinical utility.

11.8 GBC: EMERGING THERAPEUTIC STRATEGIES

Despite the challenges in treating GBC, significant progress has been made in the development of novel therapeutic strategies. Surgical resection remains the primary treatment for localized GBC, but most patients present with advanced disease, limiting the effectiveness of surgery. As a result, adjuvant therapies such as chemotherapy and radiation therapy are commonly used, but their success has been limited. Chemotherapy agents induce non-specific inhibition of tumor cell proliferation, usually via blockade of DNA synthesis, which has been extensively engaged in the treatment of a variety of cancers (100). The National Comprehensive Cancer Network (NCCN) suggests two options for GBC treatment: single-agent therapy, which is fluoropyrimidine- or gemcitabine-based treatment, and multiagent regimens, which include oxaliplatin, cisplatin, and capecitabine (101, 102). In addition to the intervention of most chemotherapeutic agents on DNA synthesis, over the past decade, there has been growing research work focusing on RNA molecules that control oncogene or TSG expression. It is emerging that microRNAs (miRNAs) and long non-coding RNAs (lncRNAs) act as indispensable factors to coordinately manipulate multiple gene expressions underlying carcinogenesis, in contrast to the traditional paradigm that non-coding RNAs are noisy and non-functional in the regulation of gene expression (103). Multiple studies have reported that miRNAs and lncRNAs mediate the proliferation, invasion, and chemotherapy resistance of GBC and serve as new therapeutic targets for the treatment of advanced GBC, including miR-125b-5p, miR-122, miR-223, miR-31, miR-30a-5p, lncRNA-HGBC, lncRNA-PAGBC, lncRNA PVT1, and lncRNA GBCDRlncl (103). Concomitantly, advanced nanotechnology employing various modified materials that offer optimal cargos to efficiently deliver RNA molecules has also given rise to promising benefits in the chemotherapy of GBC (104).

Albeit the benefit of adjuvant radiotherapy is evident in many other types of cancers, it remains unclear in GBC, since there is a lack of strong evidence of improved endpoints.

11.8.1 Targeted Therapy for GBC

Targeted therapy has been investigated in GBC to specifically block a number of molecular targets that are keenly associated with tumor cell proliferation, differentiation, migration, cancer stemness, vascular angiogenesis, and antitumor immune responses (105). Drugs for targeted therapy consist of small molecules and immunized antibodies. Small molecules with molecular weight <900 Da can be readily transported into cells to inactivate specific proteins or enzymes, thus inhibiting tumor cell growth (106), while therapeutic antibodies specifically bind to cell membrane receptors or their ligands to regulate cell proliferation or apoptosis (107). Some drugs target extracellular molecules that mediate angiogenesis or immune reaction in the tumor microenvironment, resulting in inhibition of tumor growth, angiogenesis, and metastasis (108). There are numerous targeted therapies on individual signaling pathways, including human epidermal growth factor receptor 2 (HER2), growth factor receptor tyrosine kinases (RTKs) (EGFR, vascular endothelial growth factor receptor [VEGFR]), programmed death receptor 1 (PD-1)/programmed death ligand 1 (PD-L1), TP53, KIT, CDKN2A/B, phosphatidylinositol 3-kinase (PI3K)/AKT/mammalian target of rapamycin (mTOR), and RAS/BRAF/MEK/MAK (103).

HER2, also known as ErbB2, is a member of the HER family, which includes HER1, HER2, HER3, and HER4, also called ErbB1 (EGFR), ErbB2, ErbB3, and ErbB4, respectively. HERs are cell-surface receptors that harbor a transmembrane tyrosine kinase domain capable of activating multiple downstream signaling pathways upon binding with epidermal growth factor (EGF). HER2/neu overexpression has been seen in 10–15% of advanced GBC (98). Preclinical and clinical therapeutic trials in GBC found that HER2+ SNU-2670 and SNU-2773 GBC cell lines are more sensitive to trastuzumab, dacomitinib, and afatinib than HER2– BTC cell lines (109). Zanidatamab, a bispecific HER2-targeted antibody, has demonstrated anti-tumor efficacy and safety in a range of HER2-overexpressing cancers (110). Clinical studies and case reports suggest that HER2-directed targeted therapy represents a novel and tolerable treatment approach for advanced BTCs that express somatic HER2 mutations (111–113).

The VEGF/VEGFR axis plays a key role in both physiological and pathological vascularization in diseases such as tumor angiogenesis. VEGFR1 is expressed on many types of cells, including endothelial cells, epithelial cells, inflammatory cells, and cancer cells (114). A multicentric phase II study of the VEGF antibody bevacizumab (NCT00361231) in combination with gemcitabine and oxaliplatin in advanced BTC, with a single-arm trial, demonstrated that the response rate was 40%, modified progression-free survival was 7 months, and overall survival was 12.7 months (115).

Longitudinal, case–control studies with large cohorts may be essential to establish the benefits of anti-angiogenic blockers. Nevertheless, at present, the combination regimen with angiogenic-targeted antibodies and other chemotherapeutic agents offers the optimal means to treat GBC.

EGFR, also commonly known as ErbB1 or HER1, similar to HER2, triggers multiple intracellular downstream signaling cascades, including ERK/MAPK, PI3K-AKT, SRC, PLC-γ1-PKC, JNK, and JAK-STAT pathways, mediating cancer proliferation, angiogenesis, cell motility, adhesion, and metastasis (116). Typical antibodies are cetuximab and panitumumab, which are able to prevent EGFR from activated dimerization, thus inhibiting downstream signaling (117). Tyrosine kinase inhibitors (TKIs) can bind to the ATP-binding pockets on the intracellular catalytic kinase domain of RTKs, committing to the disruption of downstream signaling. Three generations of EGFR-TKI drugs have been evolutionarily devised to fight against mutation activity of EGFR, including erlotinib and gefitinib (first generation), afatinib and dacomitinib (second generation), and osimertinib and olmutinib (third generation) (118). Somatic mutations of EGFR are quite low, ranging from 2.5 to 3.9% in patients with GBC (119). Therapeutic trials targeting EGFR in GBC patients have shown varied results. Nonetheless, more clinical settings with combined regimens targeting EGFR are essential to offer a powerful tool for clinical practice.

MAPK signaling pathway is a crucial intracellular signal transduction pathway that regulates varied cellular activities and is frequently dysfunctional in cancer. RAS/RAF/MEK/ERK is the most common pathway in MAPK signaling by which a variety of cancerous cells promote cell proliferation, death, differentiation, cell cycle progression, apoptosis, survival, metastasis, metabolism, and angiogenesis (66). A number of divergent therapeutic approaches in treatment with MAPK pathway blockers have been employed to unveil potential targets for GBC patients. In a phase II study of trametinib (GSK1120212, JTP-74057), the first generation of MEK1/2 inhibitor approved by the Food and Drug Administration, safer and more effective drug responses were shown than single gemcitabine treatment (120). A recent multi-institutional phase II study (NCT00553332) of selumetinib demonstrated that the drug response is of acceptable tolerability in patients with metastatic BTC (120).

PI3K/AKT/mTOR signaling pathway is well known to participate in various biological and physiological cellular processes, including cell growth, mobility, differentiation, metabolic activity, and apoptosis (58). Several inhibitors targeting the PI3K/AKT/mTOR pathway, including A66, Wortmannin, and LY294002, have been demonstrated to inhibit GBC cell proliferation, migration, and invasion both in vitro and in vivo. In addition, rapamycin, RAD001, and AZD8055 block mTOR and inhibit the growth and migration of GBC cells in vitro (103).

Overexpression and high mutation frequency of TP53 protein, which has a tumor-promoting signature rather than cell apoptotic activity, are correlated with poor survival in a broad type of cancers, including GBC (121). Although TP53 has the highest mutation rate in GBC, very limited clinical trials targeting mutated p53 have been reported to evaluate potential therapeutic benefit.

In recent years, intense research evidence focusing on antitumor immunity has established proof of concept that blockade of the interaction between tumor-derived checkpoint ligands and their corresponding binding receptors expressed by T cells elicits T cell immunity against tumors. PD-L1, also known as B7-H1 or CD274, is a membrane-associated protein and specifically binds to PD-1 expressed on T cells. Interaction of these immune checkpoint proteins leads to disruption of the major histocompatibility complex coupled with T cell receptors, the identical interaction of membrane proteins by which cytotoxic T cells recognize and eliminate tumor cells (122). Lin (123) and Weinberg et al. (124) showed that 18% of GBC (n = 66) and 8% of GBC (n = 237) patients were positive for PD-L1 expressed by tumor cells using IHC, respectively.

Treatment of GBC patients with PD-1/PD-L1 inhibitors has emerged as a promising strategy for targeted therapies. In the KEYNOTE-028 (NCT02054806; phase I) and KEYNOTE-158 (NCT02628067; phase II) studies, BTC patients received pembrolizumab treatment, and results suggest that PD-1/PD-L1 inhibitors are safe and effective in treating GBC (125). Durvalumab, a therapeutic human monoclonal antibody that binds to and blocks the activity of the immunosuppressive programmed death-ligand 1 (PD-L1) protein, is approved in the USA, EU, Japan, and other countries in combination with gemcitabine and cisplatin for adults with advanced biliary tract cancer. In the pivotal phase 3 TOPAZ-1 trial, durvalumab plus gemcitabine and cisplatin significantly prolonged overall survival and progression-free survival compared with placebo plus gemcitabine and cisplatin in adults with advanced biliary tract cancer. Benefit from durvalumab was seen irrespective of primary tumor location, disease status at diagnosis (unresectable or recurrent), or initial levels of PD-L1 expression. The tolerability of durvalumab plus gemcitabine and cisplatin was manageable. Overall, the addition of durvalumab to gemcitabine and cisplatin is a valuable new treatment option for adults with advanced biliary tract cancer (126).

DNA damage repair (DDR) can execute full repairing or elimination of damaged cells to protect host organisms against possible carcinogenesis. Four major DDR pathways have been identified in the cells, including base excision repair, nucleotide excision repair (NER), double-strand break repair, and mismatch repair (MMR) (127). To date, clinical trials clarifying the efficacy of DDR inhibitors in GBC patients have not been reported, but poly (ADP-ribose) polymerase (PARP) inhibitors have been developed in monotherapy that is engaged for patients with HR deficiency (i.e., BRCA mutation) and also in BRCA-like tumors.

Personalized medicine, which involves tailoring treatment based on the genetic and molecular profile of the tumor, holds great potential for improving outcomes in GBC. Combining targeted therapies, immunotherapies, and conventional treatments such as chemotherapy may provide synergistic effects and enhance therapeutic efficacy.

11.9 CONCLUSION

Gallbladder cancer remains a formidable challenge due to its aggressive nature, late-stage diagnosis, and limited treatment options. However, significant progress has been made in understanding the molecular mechanisms underlying gallbladder carcinogenesis, and new diagnostic tools and therapeutic strategies are being developed. Continued research into the genetic, epigenetic, and environmental factors that contribute to GBC, coupled with innovations in imaging, molecular profiling, and targeted therapies, holds great promise for improving early detection and treatment outcomes. The future of gallbladder cancer management lies in personalized and precision medicine, where targeted therapies and immunotherapies may offer new hope to patients with this deadly malignancy.

REFERENCES

1. Dutta U, Bush N, Kalsi D, Popli P, Kapoor VK. Epidemiology of gallbladder cancer in India. *Chinese Clinical Oncology.* 2019;8(4):33.
2. Wistuba II, Gazdar AF. Gallbladder cancer: Lessons from a rare tumour. *Nature Reviews Cancer.* 2004;4(9):695–706.
3. Dutta U. Gallbladder cancer: Can newer insights improve the outcome? *Journal of Gastroenterology and Hepatology.* 2012;27(4):642–53.
4. Batra Y, Pal S, Dutta U, Desai P, Garg PK, Makharia G, et al. Gallbladder cancer in India: A dismal picture. *Journal of Gastroenterology and Hepatology.* 2005;20(2):309–14.
5. Tanwar P, Minocha S, Gupta I. A Comprehensive narrative review of transcriptomics and epigenomics of gallbladder cancer. *Journal of Cancer Research and Therapeutics.* 2023;19(Suppl. 2):S499–S507.
6. Rawla P, Sunkara T, Thandra KC, Barsouk A. Epidemiology of gallbladder cancer. *Clinical and Experimental Hepatology.* 2019;5(2):93–102.
7. Koshiol J, Gao Y-T, Dean M, Egner P, Nepal C, Jones K, et al. Association of aflatoxin and gallbladder cancer. *Gastroenterology.* 2017;153(2):488–94. e1.
8. Hundal R, Shaffer EA. Gallbladder cancer: Epidemiology and outcome. *Clinical Epidemiology.* 2014:99–109.
9. Jain K, Sreenivas V, Velpandian T, Kapil U, Garg PK. Risk factors for gallbladder cancer: A case–control study. *International Journal of Cancer.* 2013;132(7):1660–6.

10. Baskaran V, Vij U, Sahni P, Tandon R, Nundy S. Do the progesterone receptors have a role to play in gallbladder cancer? *International Journal of Gastrointestinal Cancer*. 2005;35:61–8.
11. Huang J, Patel HK, Boakye D, Chandrasekar VT, Koulaouzidis A, Lucero-Prisno III DE, et al. Worldwide distribution, associated factors, and trends of gallbladder cancer: A global country-level analysis. *Cancer Letters*. 2021;521:238–51.
12. Shukla V, Prakash A, Tripathi B, Reddy D, Singh S. Biliary heavy metal concentrations in carcinoma of the gall bladder: Case-control study. *BMJ*. 1998;317(7168):1288–9.
13. Shukla Y, Arora A. Enhancing effects of mustard oil on preneoplastic hepatic foci development in Wistar rats. *Human & Experimental Toxicology*. 2003;22(2):51–5.
14. Randi G, Franceschi S, La Vecchia C. Gallbladder cancer worldwide: Geographical distribution and risk factors. *International Journal of Cancer*. 2006;118(7):1591–602.
15. Pandey M, Vishwakarma RA, Khatri AK, Roy SK, Shukla VK. Bile, bacteria, and gallbladder carcinogenesis. *Journal of Surgical Oncology*. 1995;58(4):282–3.
16. Kumar S, Kumar S, Kumar S. Infection as a risk factor for gallbladder cancer. *Journal of Surgical Oncology*. 2006;93(8):633–9.
17. Zatonski W, Przewozniak K, Lowenfels A, Boyle P, Maisonneuve P, Bueno de Mesquita H, et al. Epidemiologic aspects of gallbladder cancer: A case-control study of the SEARCH Program of the International Agency for Research on Cancer. *Journal of the National Cancer Institute*. 1997;89(15):1132–8.
18. Piovani D, Nikolopoulos GK, Aghemo A, Lleo A, Alqahtani SA, Hassan C, et al. Environmental risk factors for gallbladder cancer: Field-wide systematic review and meta-analysis. *Clinical Gastroenterology and Hepatology*. 2025;23(9):1500–1513. doi: 10.1016/j.cgh.2024.07.046. Epub 2024 Oct 5. PMID: 39370088.
19. Puschel K, Sullivan S, Montero J, Thompson B, Díaz A. Cost-effectiveness analysis of a preventive program for gallbladder disease in Chile. *Revista medica de Chile*. 2002;130(4):447–59.
20. Kang CM, Kim KS, Choi JS, Lee WJ, Kim BR. Gallbladder carcinoma associated with anomalous pancreaticobiliary duct junction. *Canadian Journal of Gastroenterology and Hepatology*. 2007;21(6):383–7.
21. Dutta U, Kuchhangi Sureshchandra Poornachandra P. Gallbladder polyps: How to pick up the sinister ones. *Journal of Gastroenterology and Hepatology*. 2009:175–8. Wiley Online Library.
22. Kamaya A, Fung C, Szpakowski J-L, Fetzer DT, Walsh AJ, Alimi Y, et al. Management of incidentally detected gallbladder polyps: Society of radiologists in ultrasound consensus conference recommendations. *Radiology*. 2022;305(2):277–89.
23. DesJardins H, Duy L, Scheirey C, Schnelldorfer T. Porcelain gallbladder: Is observation a safe option in select populations? *Journal of the American College of Surgeons*. 2018;226(6):1064–9.
24. CR V. Carcinoma in choledochal cysts. *Archieves of Surgery*. 1983;118:986–8.
25. Bansal VK, Misra MC, Chaubal G, Datta Gupta S, Das B, Ahuja V, et al. Helicobacter pylori in gallbladder mucosa in patients with gallbladder disease. *Indian Journal of Gastroenterology*. 2012;31:57–60.
26. Nagaraja V, Eslick G. Systematic review with meta-analysis: The relationship between chronic S almonella typhi carrier status and gall-bladder cancer. *Alimentary Pharmacology & Therapeutics*. 2014;39(8):745–50.
27. Aune D, Vatten LJ, Boffetta P. Tobacco smoking and the risk of gallbladder disease. *European Journal of Epidemiology*. 2016;31:643–53.
28. Nogueira L, Foerster C, Groopman J, Egner P, Koshiol J, Ferreccio C, et al. Association of aflatoxin with gallbladder cancer in Chile. *JAMA*. 2015;313(20):2075–7.
29. Yadav A, Gupta P, Gupta P, Patil AN, Das CK, Hooda H, et al. Aflatoxin exposure is associated with an increased risk of gallbladder cancer. *International Journal of Cancer*. 2025;156(2):322–30.
30. Sharma A, Sharma KL, Gupta A, Yadav A, Kumar A. Gallbladder cancer epidemiology, pathogenesis and molecular genetics: Recent update. *World Journal of Gastroenterology*. 2017;23(22):3978.
31. Brown MA. Tumor suppressor genes. *Advances in Genetics*. 1997;36:45.
32. Wistuba II, Gazdar AF, Roa I, Albores-Saavedra J. p53 protein overexpression in gallbladder carcinoma and its precursor lesions: An immunohistochemical study. *Human Pathology*. 1996;27(4):360–5.
33. Yokoyama N, Hitomi J, Watanabe H, Ajioka Y, Pruyas M, Serra I, et al. Mutations of p53 in gallbladder carcinomas in high-incidence areas of Japan and Chile. *Cancer Epidemiology, Biomarkers & Prevention: A Publication of the American Association for Cancer Research, Cosponsored by the American Society of Preventive Oncology*. 1998;7(4):297–301.
34. Feng Z, Chen J, Wei H, Gao P, Shi J, Zhang J, et al. The risk factor of gallbladder cancer: Hyperplasia of mucous epithelium caused by gallstones associates with p16/CyclinD1/CDK4 pathway. *Experimental and Molecular Pathology*. 2011;91(2):569–77.

35. Kuang C, Chen Y. Tumor-derived C-terminal mutations of Smad4 with decreased DNA binding activity and enhanced intramolecular interaction. *Oncogene*. 2004;23(5):1021–9.
36. Wistuba II, Ashfaq R, Maitra A, Alvarez H, Riquelme E, Gazdar AF. Fragile histidine triad gene abnormalities in the pathogenesis of gallbladder carcinoma. *The American Journal of Pathology*. 2002;160(6):2073–9.
37. Saetta A, Lazaris AC, Davaris P. Detection of ras oncogene point mutations and simultaneous proliferative fraction estimation in gallbladder cancer. *Pathology-Research and Practice*. 1996;192(6):532–40.
38. Hanada K, Tsuchida A, Iwao T, Eguchi N, Sasaki T, Morinaka K, et al. Gene mutations of K-ras in gallbladder mucosae and gallbladder carcinoma with an anomalous junction of the pancreaticobiliary duct. *The American Journal of Gastroenterology*. 1999;94(6):1638–42.
39. Kaufman M, Mehrotra B, Limaye S, White S, Fuchs A, Lebowicz Y, et al. EGFR expression in gallbladder carcinoma in North America. *International Journal of Medical Sciences*. 2008;5(5):285.
40. Roa I, Garcia H, Game A, De Toro G, De Aretxabala X, Javle M. Somatic mutations of PI3K in early and advanced gallbladder cancer: Additional options for an orphan cancer. *The Journal of Molecular Diagnostics*. 2016;18(3):388–94.
41. Xie Y, Jiang Y, Yang X-B, Wang A-Q, Zheng Y-C, Wan X-S, et al. Response of BRCA1-mutated gallbladder cancer to olaparib: A case report. *World Journal of Gastroenterology*. 2016;22(46):10254.
42. Goldin RD, Roa JC. Gallbladder cancer: A morphological and molecular update. *Histopathology*. 2009;55(2):218–29.
43. Kanwal R, Gupta S. Epigenetic modifications in cancer. *Clinical Genetics*. 2012;81(4):303–11.
44. House MG, Wistuba II, Argani P, Guo M, Schulick RD, Hruban RH, et al. Progression of gene hypermethylation in gallstone disease leading to gallbladder cancer. *Annals of Surgical Oncology*. 2003;10:882–9.
45. Liu R, Wu J, Guo H, Yao W, Li S, Lu Y, et al. Post-translational modifications of histones: Mechanisms, biological functions, and therapeutic targets. *MedComm*. 2023;4(3):e292.
46. Yang G, Zhang L, Li R, Wang L. The role of microRNAs in gallbladder cancer. *Molecular and Clinical Oncology*. 2016;5(1):7–13.
47. Harjunpää H, Llort Asens M, Guenther C, Fagerholm SC. Cell adhesion molecules and their roles and regulation in the immune and tumor microenvironment. *Frontiers in Immunology*. 2019;10:1078.
48. Kim YS, Gum J, Brockhausen I. Mucin glycoproteins in neoplasia. *Glycoconjugate Journal*. 1996;13:693–707.
49. Vilkin A, Nudelman I, Morgenstern S, Geller A, Dayan YB, Levi Z, et al. Gallbladder inflammation is associated with increase in mucin expression and pigmented stone formation. *Digestive Diseases and Sciences*. 2007;52:1613–20.
50. Verma K, Dixit R, Singh J, Tiwary SK, Khanna AK, Narayan G, et al. Molecular genetics changes in gallbladder carcinoma. *International Journal of Molecular and Immuno Oncology*. 2020;5(2):49–61.
51. Anderson NM, Simon MC. The tumor microenvironment. *Current Biology*. 2020;30(16):R921–R5.
52. Ebata N, Fujita M, Sasagawa S, Maejima K, Okawa Y, Hatanaka Y, et al. Molecular classification and tumor microenvironment characterization of gallbladder cancer by comprehensive genomic and transcriptomic analysis. *Cancers*. 2021;13(4):733.
53. Espinoza JA, Bizama C, García P, Ferreccio C, Javle M, Miquel JF, et al. The inflammatory inception of gallbladder cancer. *Biochimica et Biophysica Acta (BBA)-Reviews on Cancer*. 2016;1865(2):245–54.
54. Letelier P, Garcia P, Leal P, Ili C, Buchegger K, Riquelme I, et al. Immunohistochemical expression of vascular endothelial growth factor A in advanced gallbladder carcinoma. *Applied Immunohistochemistry & Molecular Morphology*. 2014;22(7):530–6.
55. Xie F, Xu X, Xu A, Liu C, Liang F, Xue M, et al. Aberrant activation of Sonic hedgehog signaling in chronic cholecystitis and gallbladder carcinoma. *Human Pathology*. 2014;45(3):513–21.
56. Matsushita S, Onishi H, Nakano K, Nagamatsu I, Imaizumi A, Hattori M, et al. Hedgehog signaling pathway is a potential therapeutic target for gallbladder cancer. *Cancer Science*. 2014;105(3):272–80.
57. Hay N, Sonenberg N. Upstream and downstream of mTOR. *Genes & Development*. 2004;18(16):1926–45.
58. LoPiccolo J, Blumenthal GM, Bernstein WB, Dennis PA. Targeting the PI3K/Akt/mTOR pathway: Effective combinations and clinical considerations. *Drug Resistance Updates*. 2008;11(1–2):32–50.
59. Courtney KD, Corcoran RB, Engelman JA. The PI3K pathway as drug target in human cancer. *Journal of Clinical Oncology*. 2010;28(6):1075–83.
60. Kumari N, Corless CL, Warrick A, Beadling C, Nelson D, Neff T, et al. Mutation profiling in gallbladder cancer in Indian population. *Indian Journal of Pathology and Microbiology*. 2014;57(1):9–12.
61. Deshpande V, Nduaguba A, Zimmerman SM, Kehoe SM, MacConaill LE, Lauwers GY, et al. Mutational profiling reveals PIK3CA mutations in gallbladder carcinoma. *BMC Cancer*. 2011;11:1–7.

62. Liu D-C, Yang Z-L. Overexpression of EZH2 and loss of expression of PTEN is associated with invasion, metastasis, and poor progression of gallbladder adenocarcinoma. *Pathology-Research and Practice*. 2011;207(8):472–8.
63. Wu Q, Kiguchi K, Kawamoto T, Ajiki T, Traag J, Carbajal S, et al. Therapeutic effect of rapamycin on gallbladder cancer in a transgenic mouse model. *Cancer Research*. 2007;67(8):3794–800.
64. El Khatib M, Bozko P, Palagani V, Malek NP, Wilkens L, Plentz RR. Activation of Notch signaling is required for cholangiocarcinoma progression and is enhanced by inactivation of p53 in vivo. *PLoS One*. 2013;8(10):e77433.
65. Kawamoto T, Krishnamurthy S, Tarco E, Trivedi S, Wistuba II, Li D, et al. HER receptor family: Novel candidate for targeted therapy for gallbladder and extrahepatic bile duct cancer. *Gastrointestinal Cancer Research: GCR*. 2007;1(6):221.
66. Peluso I, Yarla NS, Ambra R, Pastore G, Perry G, editors. MAPK signalling pathway in cancers: Olive products as cancer preventive and therapeutic agents. *Seminars in Cancer Biology*. 2019:185–95. Elsevier.
67. You Z, Liu SP, Du J, Wu YH, Zhang SZ. Advancements in MAPK signaling pathways and MAPK-targeted therapies for ameloblastoma: A review. *Journal of Oral Pathology & Medicine*. 2019;48(3):201–5.
68. Kim Y-T, Kim J, Jang YH, Lee WJ, Ryu JK, Park Y-K, et al. Genetic alterations in gallbladder adenoma, dysplasia and carcinoma. *Cancer Letters*. 2001;169(1):59–68.
69. Bergers G, Benjamin LE. Tumorigenesis and the angiogenic switch. *Nature Reviews Cancer*. 2003;3(6):401–10.
70. Sun X-N, Cao W-G, Wang X, Wang Q, Gu B-X, Yang Q-C, et al. Prognostic impact of vascular endothelial growth factor-A expression in resected gallbladder carcinoma. *Tumor Biology*. 2011;32:1183–90.
71. Tian Y, Ding R-Y, Zhi Y-H, Guo R-X, Wu S-D. Analysis of p53 and vascular endothelial growth factor expression in human gallbladder carcinoma for the determination of tumor vascularity. *World Journal of Gastroenterology: WJG*. 2006;12(3):415.
72. Roa JC, García P, Kapoor VK, Maithel SK, Javle M, Koshiol J. Gallbladder cancer. *Nature Reviews Disease Primers*. 2022;8(1):69.
73. Kawakami S, Takano S, Fukasawa M, Shindo H, Takahashi E, Fukasawa Y, et al. Stepwise correlation of TP53 mutations from pancreaticobiliary maljunction to gallbladder carcinoma: A retrospective study. *BMC Cancer*. 2021;21:1–9.
74. Gupta P, Dutta U, Rana P, Singhal M, Gulati A, Kalra N, et al. Gallbladder reporting and data system (GB-RADS) for risk stratification of gallbladder wall thickening on ultrasonography: An international expert consensus. *Abdominal Radiology*. 2022:1–12.
75. Badea R, Zaro R, Opincaru I, Chiorean L. Ultrasound in the examination of the gallbladder–a holistic approach: Grey scale, Doppler, CEUS, elastography, and 3D. *Medical Ultrasonography*. 2014;16(4):345–55.
76. Deqing L-C, Zhang J-W, Yang J. Endoscopic diagnosis and management of gallbladder carcinoma in minimally invasive era: New needs, new models. *World Journal of Gastrointestinal Oncology*. 2024;16(11):4333.
77. Sidhu PS, Cantisani V, Dietrich CF, Gilja OH, Saftoiu A, Bartels E, et al. The EFSUMB guidelines and recommendations for the clinical practice of contrast-enhanced ultrasound (CEUS) in non-hepatic applications: Update 2017 (long version). *Ultraschall in der Medizin-European Journal of Ultrasound*. 2018;39(02):e2–e44.
78. Nishiyama Y, Yamamoto Y, Fukunaga K, Kimura N, Miki A, Sasakawa Y, et al. Dual-time-point 18F-FDG PET for the evaluation of gallbladder carcinoma. *Journal of Nuclear Medicine*. 2006;47(4):633–8.
79. García P, Lamarca A, Díaz J, Carrera E, Roa JC, Consortium E-LAE. Current and new biomarkers for early detection, prognostic stratification, and management of gallbladder cancer patients. *Cancers*. 2020;12(12):3670.
80. Agrawal S, Saxena R. Raised carbohydrate antigen 19–9 levels detect recurrences and impacts overall and disease-free survival in radically resected gallbladder cancer: A simple surveillance marker? *Journal of Cancer Research and Therapeutics*. 2023;19(2):273–7.
81. Rana S, Dutta U, Kochhar R, Rana SV, Gupta R, Pal R, et al. Evaluation of CA 242 as a tumor marker in gallbladder cancer. *Journal of Gastrointestinal Cancer*. 2012;43:267–71.
82. Huang L, Chen W, Liang P, Hu W, Zhang K, Shen S, et al. Serum CYFRA 21–1 in biliary tract cancers: A reliable biomarker for gallbladder carcinoma and intrahepatic cholangiocarcinoma. *Digestive Diseases and Sciences*. 2015;60:1273–83.
83. Kumari S, Tewari S, Husain N, Agarwal A, Pandey A, Singhal A, et al. Quantification of circulating free DNA as a diagnostic marker in gall bladder cancer. *Pathology & Oncology Research*. 2017;23:91–7.

84. Kumari S, Husain N, Agarwal A, Neyaz A, Gupta S, Chaturvedi A, et al. Diagnostic value of circulating free DNA integrity and global methylation status in gall bladder carcinoma. *Pathology & Oncology Research.* 2019;25:925–36.
85. Awasthi NP, Kumari S, Neyaz A, Gupta S, Agarwal A, Singhal A, et al. EpCAM-based flow cytometric detection of circulating tumor cells in gallbladder carcinoma cases. *Asian Pacific Journal of Cancer Prevention: APJCP.* 2017;18(12):3429.
86. Chandra V, Kim JJ, Mittal B, Rai R. MicroRNA aberrations: An emerging field for gallbladder cancer management. *World Journal of Gastroenterology.* 2016;22(5):1787.
87. Mishra S, Srivastava P, Pandey A, Shukla S, Agarwal A, Husain N. Diagnostic utility of next-generation sequencing in circulating free DNA and a comparison with matched tissue in gallbladder carcinoma. *Laboratory Investigation.* 2024;104(2):100301.
88. Baichan P, Naicker P, Augustine TN, Smith M, Candy G, Devar J, et al. Proteomic analysis identifies dysregulated proteins and associated molecular pathways in a cohort of gallbladder cancer patients of African ancestry. *Clinical Proteomics.* 2023;20(1):8.
89. Letelier P, Brebi P, Tapia O, Roa JC. DNA promoter methylation as a diagnostic and therapeutic biomarker in gallbladder cancer. *Clinical Epigenetics.* 2012;4:1–11.
90. Sehrawat A, Gopi VP, Gupta A. A systematic review on role of deep learning in CT scan for detection of gall bladder cancer. *Archives of Computational Methods in Engineering.* 2024:1–9.
91. Gupta P, Rana P, Ganeshan B, Kalage D, Irrinki S, Gupta V, et al. Computed tomography texture-based radiomics analysis in gallbladder cancer: Initial experience. *Clinical and Experimental Hepatology.* 2021;7(4):406–14.
92. Gupta P, Basu S, Rana P, Dutta U, Soundararajan R, Kalage D, et al. Deep-learning enabled ultrasound based detection of gallbladder cancer in northern India: A prospective diagnostic study. *The Lancet Regional Health-Southeast Asia.* 2024;24.
93. Chen T, Tu S, Wang H, Liu X, Li F, Jin W, et al. Computer-aided diagnosis of gallbladder polyps based on high resolution ultrasonography. *Computer methods and Programs in Biomedicine.* 2020;185:105118.
94. Saxena M, Bohara S, Gupta V, Goel VK. Role of morphology and immunohistochemistry in the diagnosis of incidental cancers in gallstone disease. *Acta Medica International.* 2023;10(2):143–7.
95. Kalekou H, Miliaras D. Cytokeratin 7 and 20 expression in gallbladder carcinoma. *Polish Journal of Pathology.* 2011;62(1):25–30.
96. Xu S, Zhan M, Wang J. Epithelial-to-mesenchymal transition in gallbladder cancer: From clinical evidence to cellular regulatory networks. *Cell Death Discovery.* 2017;3(1):1–11.
97. Dongqing X, Jianwen L, Jiang F, Kaishuang C, Guangxue R. The effect and mechanism of vascular endothelial growth factor (VEGF) on tumor angiogenesis in gallbladder carcinoma. *Iranian Journal of Public Health.* 2019;48(4):713.
98. Roa I, de Toro G, Schalper K, de Aretxabala X, Churi C, Javle M. Overexpression of the HER2/neu gene: A new therapeutic possibility for patients with advanced gallbladder cancer. *Gastrointestinal Cancer Research: GCR.* 2014;7(2):42.
99. Goeppert B, Roessler S, Renner M, Loeffler M, Singer S, Rausch M, et al. Low frequency of mismatch repair deficiency in gallbladder cancer. *Diagnostic Pathology.* 2019;14:1–6.
100. Azizi AA, Lamarca A, McNamara MG, Valle JW. Chemotherapy for advanced gallbladder cancer (GBC): A systematic review and meta-analysis. *Critical Reviews in Oncology/Hematology.* 2021;163:103328.
101. Valle J, Wasan H, Palmer DH, Cunningham D, Anthoney A, Maraveyas A, et al. Cisplatin plus gemcitabine versus gemcitabine for biliary tract cancer. *New England Journal of Medicine.* 2010;362(14):1273–81.
102. Primrose JN, Fox RP, Palmer DH, Malik HZ, Prasad R, Mirza D, et al. Capecitabine compared with observation in resected biliary tract cancer (BILCAP): A randomised, controlled, multicentre, phase 3 study. *The Lancet Oncology.* 2019;20(5):663–73.
103. Song X, Hu Y, Li Y, Shao R, Liu F, Liu Y. Overview of current targeted therapy in gallbladder cancer. *Signal Transduction and Targeted Therapy.* 2020;5(1):230.
104. Cai Q, Wang X, Wang S, Jin L, Ding J, Zhou D, et al. Gallbladder cancer progression is reversed by nanomaterial-induced photothermal therapy in combination with chemotherapy and autophagy inhibition. *International Journal of Nanomedicine.* 2020:253–62.
105. Lee YT, Tan YJ, Oon CE. Molecular targeted therapy: Treating cancer with specificity. *European Journal of Pharmacology.* 2018;834:188–96.
106. Oh D-Y, Bang Y-J. HER2-targeted therapies—a role beyond breast cancer. *Nature Reviews Clinical Oncology.* 2020;17(1):33–48.

107. Gasser M, Waaga-Gasser AM. Therapeutic antibodies in cancer therapy. In *Protein Targeting Compounds: Prediction, Selection and Activity of Specific Inhibitors*. 2016:95–120. Advances in Experimental Medicine and Biology. doi: 10.1007/978-3-319-32805-8_6. PMID: 27236554.
108. Seebacher N, Stacy A, Porter G, Merlot A. Clinical development of targeted and immune based anticancer therapies. *Journal of Experimental & Clinical Cancer Research*. 2019;38(1):156.
109. Nam A-R, Kim J-W, Cha Y, Ha H, Park JE, Bang J-H, et al. Therapeutic implication of HER2 in advanced biliary tract cancer. *Oncotarget*. 2016;7(36):58007.
110. Meric-Bernstam F, Beeram M, Hamilton E, Oh DY, Hanna DL, Kang YK, et al. Zanidatamab, a novel bispecific antibody, for the treatment of locally advanced or metastatic HER2-expressing or HER2-amplified cancers: a phase 1, dose-escalation and expansion study. *Lancet Oncology*. 2022;23(12):1558–1570.
111. Inagaki C, Maeda D, Kimura A, Otsuru T, Iwagami Y, Nishida N, et al. Gallbladder cancer harboring ERBB2 mutation on the primary and metastatic site: A case report. *World Journal of Gastrointestinal Oncology*. 2019;11(9):761.
112. Ramanathan RK, Belani CP, Singh DA, Tanaka M, Lenz H-J, Yen Y, et al. A phase II study of lapatinib in patients with advanced biliary tree and hepatocellular cancer. *Cancer Chemotherapy and Pharmacology*. 2009;64:777–83.
113. Prieto M, Gastaca M, Ruiz P, Ventoso A, Palomares I, Perfecto A, et al. Long term recurrence free survival in a stage IV gallbladder cancer treated with chemotherapy plus trastuzumab and salvage liver resection. *Annals of Hepato-biliary-pancreatic Surgery*. 2019;23(4):403–7.
114. Hida Y, Morita T, Fujita M, Miyasaka Y, Horita S, Fujioka Y, et al. Vascular endothelial growth factor expression is an independent negative predictor in extrahepatic biliary tract carcinomas. *Anticancer Research*. 1999;19(3B):2257–60.
115. Zhu AX, Meyerhardt JA, Blaszkowsky LS, Kambadakone AR, Muzikansky A, Zheng H, et al. Efficacy and safety of gemcitabine, oxaliplatin, and bevacizumab in advanced biliary-tract cancers and correlation of changes in 18-fluorodeoxyglucose PET with clinical outcome: A phase 2 study. *The Lancet Oncology*. 2010;11(1):48–54.
116. Tomas A, Futter CE, Eden ER. EGF receptor trafficking: Consequences for signaling and cancer. *Trends in Cell Biology*. 2014;24(1):26–34.
117. Capdevila J, Elez E, Macarulla T, Ramos FJ, Ruiz-Echarri M, Tabernero J. Anti-epidermal growth factor receptor monoclonal antibodies in cancer treatment. *Cancer Treatment Reviews*. 2009;35(4):354–63.
118. Tan C-S, Kumarakulasinghe NB, Huang Y-Q, Ang YLE, Choo JR-E, Goh B-C, et al. Third generation EGFR TKIs: Current data and future directions. *Molecular Cancer*. 2018;17:1–14.
119. Li M, Zhang Z, Li X, Ye J, Wu X, Tan Z, et al. Whole-exome and targeted gene sequencing of gallbladder carcinoma identifies recurrent mutations in the ErbB pathway. *Nature Genetics*. 2014;46(8):872–6.
120. Ikeda M, Ioka T, Fukutomi A, Morizane C, Kasuga A, Takahashi H, et al. Efficacy and safety of trametinib in Japanese patients with advanced biliary tract cancers refractory to gemcitabine. *Cancer Science*. 2018;109(1):215–24.
121. Ghosh M, Sakhuja P, Singh S, Agarwal AK. p53 and beta-catenin expression in gallbladder tissues and correlation with tumor progression in gallbladder cancer. *Saudi Journal of Gastroenterology*. 2013;19(1):34–9.
122. Duan J, Wang Y, Jiao S. Checkpoint blockade-based immunotherapy in the context of tumor microenvironment: Opportunities and challenges. *Cancer Medicine*. 2018;7(9):4517–29.
123. Lin J, Long J, Wan X, Chen J, Bai Y, Wang A, et al. Classification of gallbladder cancer by assessment of CD8+ TIL and PD-L1 expression. *BMC Cancer*. 2018;18:1–10.
124. Weinberg BA, Xiu J, Lindberg MR, Shields AF, Hwang JJ, Poorman K, et al. Molecular profiling of biliary cancers reveals distinct molecular alterations and potential therapeutic targets. *Journal of Gastrointestinal Oncology*. 2019;10(4):652.
125. Piha-Paul SA, Oh DY, Ueno M, Malka D, Chung HC, Nagrial A, et al. Efficacy and safety of pembrolizumab for the treatment of advanced biliary cancer: Results from the KEYNOTE-158 and KEYNOTE-028 studies. *International Journal of Cancer*. 2020;147(8):2190–2198.
126. Oh D-Y, Ruth He A, Qin S, Chen L-T, Okusaka T, Vogel A, et al. Durvalumab plus gemcitabine and cisplatin in advanced biliary tract cancer. *NEJM Evidence*. 2022;1(8):EVIDoa2200015.
127. Reuvers TG, Kanaar R, Nonnekens J. DNA damage-inducing anticancer therapies: From global to precision damage. *Cancers*. 2020;12(8):2098.

12 Leveraging Artificial Intelligence and Multiomics in Gallbladder Cancer Diagnosis and Management

Nouredin Messaoudi,[1] *Aishwarya Boini,*[2] *and Andrew Gumbs*[3,4*]

[1]Department of Hepatopancreatobiliary Surgery, Vrije Universiteit Brussel (VUB), Universitair Ziekenhuis Brussel (UZ Brussel) and Europe Hospitals, Belgium

[2]Davao Medical School Foundation, Philippines

[3]Service de Chirurgie Digestive Minimale Invasive, Hôpital Antoine Béclère, Assistance Publique-Hôpitaux de Paris, France

[4]Department of General, Visceral, Vascular and Transplantation Surgery, University of Magdeburg, Germany

*Corresponding author: aagumbs@gmail.com

12.1 INTRODUCTION

Gallbladder cancer (GBC) is a relatively rare malignancy but has a high mortality rate because of its aggressive nature and late diagnosis (1). It originates from the epithelial cells lining the gallbladder and is often discovered incidentally during cholecystectomy or imaging for other conditions. The annual incidence of GBC varies globally, with higher rates observed in regions such as South America, Eastern Europe, and Southeast Asia (2, 3). Common symptoms, including jaundice, abdominal pain, and weight loss, usually appear when the disease is advanced, making early detection a significant challenge (4).

Routine screening for GBC is not currently recommended because of its low incidence and the limited accuracy of existing diagnostic tools (5). Commonly used tumor markers, such as carcinoembryonic antigen and carbohydrate antigen 19–9 (CA 19–9), lack specificity and sensitivity for early-stage detection (6). Imaging techniques such as ultrasound (Figure 12.1), computed tomography (CT), and magnetic resonance imaging (MRI) (Figure 12.2) are useful but are often insufficient for detecting small or early-stage tumors (5). This necessitates the exploration of more precise and efficient diagnostic methods.

AI has the potential to revolutionize the diagnosis and management of GBC by increasing the accuracy and efficiency of existing methods (7). AI algorithms can analyze vast amounts of data from various sources, including histological, cytological, and radiological images, to identify patterns indicative of GBC. Machine learning models, particularly those involving decision trees and neural networks, can refine diagnostic criteria and improve the detection rates of early-stage tumors (7).

In addition to diagnosis, AI can assist in treatment planning by predicting patient-specific responses to different therapies and identifying optimal treatment regimens. This personalized approach is particularly valuable in oncology, where individual variations in tumor biology significantly impact treatment outcomes (8–10).

DOI: 10.1201/9781003506201-12

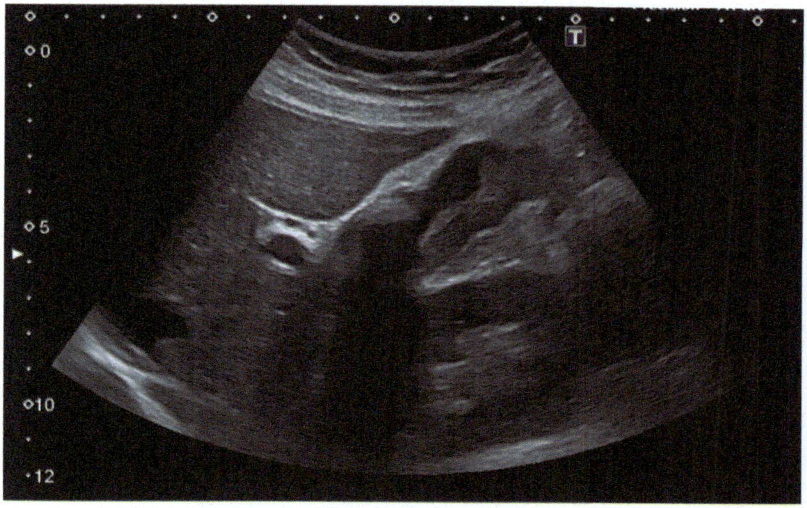

FIGURE 12.1 Transabdominal ultrasound of gallbladder cancer.

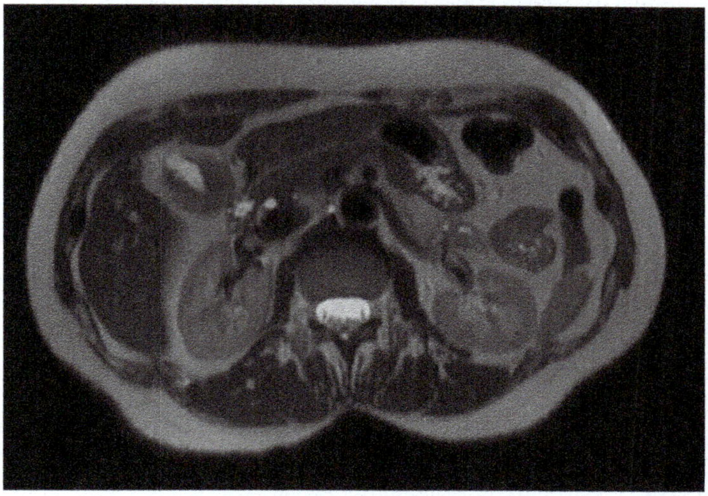

FIGURE 12.2 MR image of gallbladder cancer.

12.2 EARLY DETECTION OF GALLBLADDER CANCER AIDED BY AI

Early detection of GBC is a significant clinical challenge because of its typically asymptomatic early stages and the absence of specific biomarkers (7). Given the importance of early diagnosis for improving survival rates, recent studies have focused on the use of AI and various biomarkers to increase detection accuracy and improve patient outcomes.

Gallbladder polyps, which are common findings in clinical practice, necessitate differentiation between benign and malignant forms to guide appropriate management decisions. Traditional guidelines recommend surgical removal for polyps exceeding 10 mm in diameter, reflecting an increased risk of malignancy. However, recent advancements in AI-driven predictive models have offered more refined risk assessments. Liu et al. (2024) discussed the application of different AI algorithms to predict polyp malignancy, facilitating gallbladder-preserving procedures and optimizing patient outcomes (11).

While CA 19–9 remains a widely used serum biomarker for pancreatic and biliary tract cancers, its limitations in sensitivity and specificity have prompted Shi et al. to investigate plasma cell-free DNA (cfDNA) fragmentomics as an alternative approach (12). Their study revealed that features such as fragment size, end motifs, and nucleosome footprints can be leveraged to construct robust machine learning models for cancer detection. Integrating CA 19–9 into these models significantly enhances their performance, achieving an impressive area under the curve (AUC) of 0.995 for cancer detection.

Dou et al. explored the efficacy of FTIR spectroscopy combined with machine learning algorithms to differentiate GBC from related diseases (13). These findings highlight the effectiveness of specific spectral regions, particularly proteins, in distinguishing between healthy individuals and those with GBC. The radial basis kernel function support vector machine (RBF-SVM) model demonstrated superior accuracy, sensitivity, and specificity in diagnosing GBC.

Additionally, Zhou et al. utilized radiological features to develop a diagnostic model capable of distinguishing xanthogranulomatous cholecystitis (XGC) from GBC (14). The use of machine learning algorithms on CT and MRI data yielded promising results in improving diagnostic accuracy and aiding in clinical decision-making.

To investigate the potential of volatile organic compounds (VOCs) in bile as a detection tool for GBC, Zhang et al. employed gas chromatography-ion mobility spectrometry (GC-IMS) (15). Using this novel approach involving metabolomics, they identified a panel of 12 VOCs that outperformed CA 19–9 in diagnosing GBC when integrated with machine learning algorithms.

Furthermore, comprehensive omics analyses of bile, including lipidomics and metagenomics, have revealed distinct molecular signatures associated with GBC. Sharma et al. reported significant alterations in the bile lipidome and metaproteome profiles, which were validated through machine learning algorithms, indicating high accuracy in detecting early-stage GBC (16).

These studies underscore the potential of AI and diverse biomarkers—cfDNA fragmentomics, FTIR spectroscopy, radiomics, VOCs, lipidomics, and metagenomics—as promising tools for the early detection of GBC (11–17). While offering potential for more accurate and noninvasive screening methods, further research aided by AI is crucial in understanding the complex risk factors for GBC, such as gallstone disease, porcelain gallbladder, sex differences, intrapapillary neoplasia, and choledochal cysts. AI has the potential to analyze vast datasets to identify patterns and correlations within these risk factors, thereby improving early diagnosis, personalized treatment strategies, and preventive measures for those at high risk.

12.3 DIAGNOSIS OF GALLBLADDER CANCER VIA AI

The diagnosis of GBC is particularly challenging because of its asymptomatic nature in the early stages and the lack of specific biomarkers. Additionally, GBC often appears similar to benign gallbladder lesions on imaging, complicating accurate diagnosis. However, advancements in AI, including radiomics, deep learning, and pathomics, have shown promise in improving diagnostic accuracy and patient outcomes.

Radiomics involves extracting many features from medical images, which are then analyzed via machine learning algorithms. A study by Yin et al. demonstrated that CT-based radiomic features, when analyzed with machine learning classifiers such as XG boosting, can effectively distinguish between benign gallbladder diseases and GBC (18). In their study, the XG boosting classifier achieved an AUC of 0.81 and an accuracy rate of 73%. When combined with traditional radiological visual interpretation, the diagnostic performance improved significantly, achieving an AUC of 0.98, a sensitivity of 91%, a specificity of 93%, and an accuracy of 92%. This integrated approach underscores the potential of radiomics in clinical practice (18).

To further enhance the diagnostic process, deep learning methods such as convolutional neural networks (CNNs) have been explored. Another study by Yin et al. investigated the effectiveness of a CNN in differentiating GBC from benign lesions (19). By including adjacent liver parenchyma in the analysis, the CNN's performance improved significantly, achieving an AUC of 0.81. Although combining CNN outputs with radiological interpretation did not yield additional improvements, this study highlights the valuable role of adjacent tissue information in enhancing AI model accuracy for gallbladder lesion characterization (19).

Positron emission tomography/computed tomography (PET/CT) is another critical tool in oncology for distinguishing between benign and malignant lesions. Arslan and Cermik reviewed the application of PET/CT in the liver, biliary tract, and pancreas, noting its high diagnostic value (20). Despite the potential for technical and physiological pitfalls, advancements in AI and new radiopharmaceuticals are expected to further increase the accuracy and efficiency of PET/CT imaging, reduce misinterpretations, and improve diagnostic reliability (20).

AI has also been applied to analyze immunohistochemical markers, revealing significant insights into the prognosis of GBC. Tittarelli et al. investigated the co-expression of the markers MRP2, CXCR4, and PD-L1 in gallbladder tumors (21). These findings indicated that high co-expression of these markers is correlated with prolonged patient survival, with an AUC of 0.85. This study illustrates the potential of AI-driven pathomic analysis to identify crucial prognostic markers, thereby enhancing the clinical management of GBC (21).

The integration of AI into various diagnostic methodologies, including radiomics, deep learning, PET/CT imaging, and pathomics, is revolutionizing the diagnosis of GBC. These technologies offer more accurate and often noninvasive diagnostic methods, significantly improving cancer staging (18–22). Continued research and validation are essential to fully harness these advancements in clinical practice.

12.4 AI-DRIVEN TREATMENT PLANNING FOR GALLBLADDER CANCER

Surgery is currently the only curative option for GBC (Figure 12.3), but the extent and nature of the surgical intervention—whether it is a radical cholecystectomy (Figure 12.4) with partial hepatectomy, central hepatectomy, or the number of lymph nodes to be removed (Figure 12.5)—are not always clear-cut. Recent advancements in AI offer promising solutions to enhance treatment planning and improve patient outcomes.

One of the key factors in determining the extent of surgical resection is predicting serosal involvement in GBC. Zhou et al. conducted a pivotal study in which radiomics-based models derived from

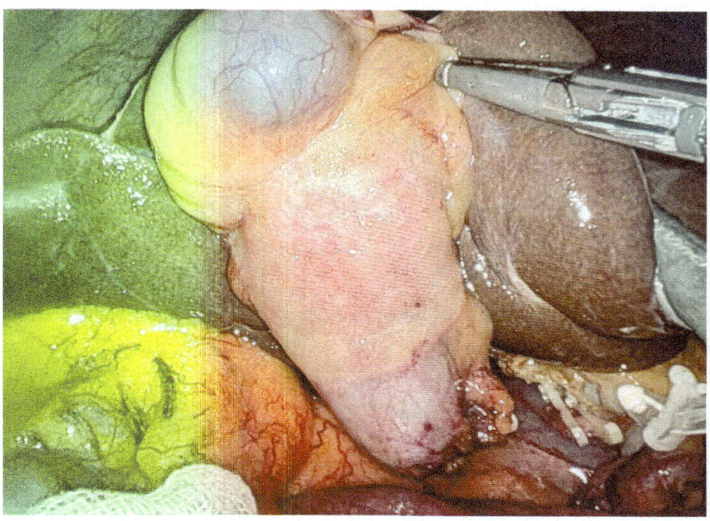

FIGURE 12.3 Robotic surgical resection of gallbladder cancer.

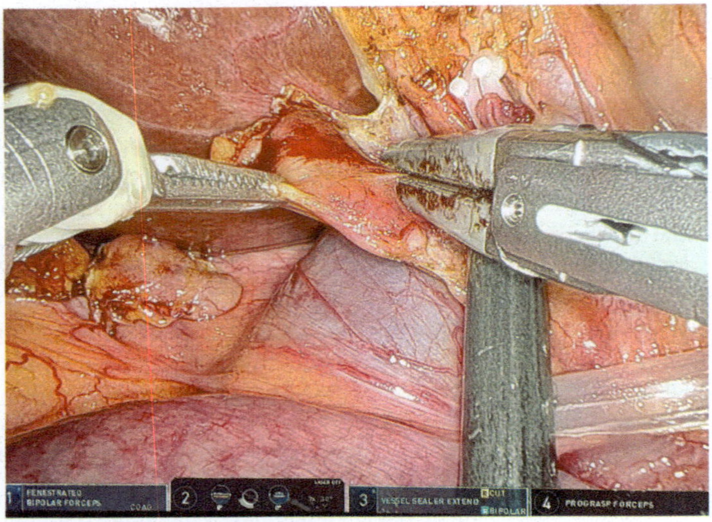

FIGURE 12.4 Lymphadenectomy in a gallbladder cancer patient.

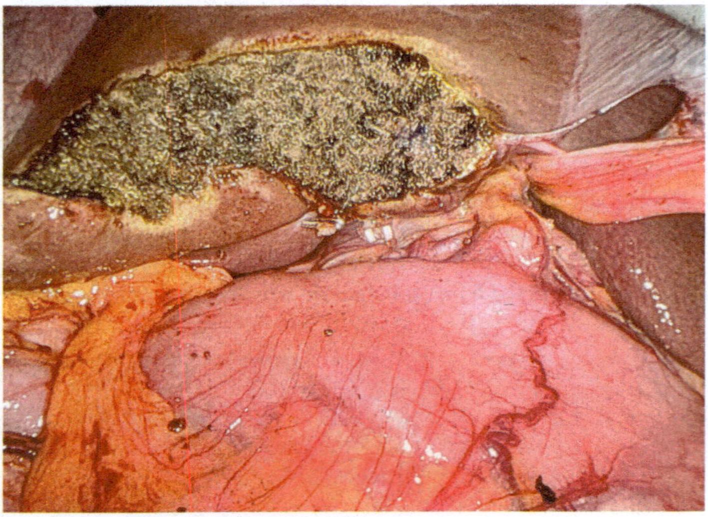

FIGURE 12.5 Radical cholecystectomy in gallbladder cancer.

contrast-enhanced computed tomography were used to predict serosal involvement in GBC patients (23). In this study, CT images from 152 patients were analyzed, and 412 radiomic features were extracted from regions of interest marked by radiologists. Various machine learning algorithms and a deep learning model were used to construct predictive models. Among these models, the logistic regression model achieved the highest performance, with an AUC of 0.944, while the deep learning model also showed a strong predictive ability, with an AUC of 0.916. These results indicate that radiomics can effectively predict serosal involvement, thereby guiding surgeons in determining the necessary extent of resection, which is critical for achieving clear margins and improving patient outcomes (23).

Another crucial aspect of GBC surgical planning involves lymphadenectomy, which is essential for accurate staging and effective treatment. Tsilimigras et al. explored the optimal number of lymph nodes to evaluate during GBC surgery (24). An analysis of data from 6,531 patients revealed that the median number of lymph nodes evaluated was two, with only 21.1% of patients having six or more nodes evaluated. Their findings revealed that evaluating fewer than four lymph nodes was associated with poor prognosis, whereas evaluating four to seven nodes provided the best balance for staging and survival outcomes. This study suggests that while the American Joint Committee on Cancer recommends evaluating six lymph nodes, achieving a practical target of at least four nodes can still ensure optimal patient outcomes, challenging current practice guidelines (24).

One cornerstone study influencing the treatment of biliary tract cancers, including GBC, is the BILCAP study (25). Additionally, the emerging role of neoadjuvant systemic therapy, coupled with AI advancements, is reshaping the approach to managing this challenging disease. The BILCAP study, a phase III randomized controlled trial, investigated the efficacy of adjuvant capecitabine in patients with resected biliary tract cancers, which include GBC (25). Compared with observation alone, this study demonstrated a significant improvement in overall survival with the use of adjuvant capecitabine. Specifically, the median overall survival was 53 months in the capecitabine group versus 36 months in the observation group. These results establish adjuvant capecitabine as a standard of care, underscoring the importance of systemic therapy following surgery to eliminate micrometastatic disease and improve long-term outcomes (25).

The role of neoadjuvant systemic therapy in GBC is an area of active investigation. Neoadjuvant therapy, administered before surgical resection, aims to downstage tumors, potentially making previously unresectable tumors resectable, and to treat micrometastatic disease early. By reducing the tumor burden preoperatively, neoadjuvant therapy can facilitate more complete resections and improve surgical outcomes. AI can play a crucial role in optimizing neoadjuvant treatment strategies by predicting and assessing the tumor response to chemotherapy and identifying the most effective regimens on the basis of tumor biology and patient-specific factors.

Integrating AI into the treatment planning process for GBC extends beyond preoperative decision-making to include perioperative and postoperative phases. During surgery, the use of surgomics—AI applications that analyze intraoperative data—can assist surgeons in making real-time decisions, including the extent of resection and the number of lymph nodes to be removed. Surgomics may offer real-time intraoperative guidance and postoperative prognostic insights, creating a continuum of AI-driven support throughout the treatment process.

These AI-enhanced technologies collectively contribute to more detailed insights into tumor characteristics and treatment responses, facilitating personalized and precise treatment allocation. By improving the accuracy of preoperative assessments, selecting neoadjuvant strategies, guiding intraoperative decisions, and informing postoperative management, AI has the potential to significantly enhance the outcomes of GBC patients, paving the way for more effective and tailored therapeutic approaches.

12.5 REFINING PROGNOSTICATION WITH AI

In recent years, advancements in the prognostication of GBC have been promoted by the integration of AI into clinical research and practice. AI-driven approaches have revolutionized the understanding of GBC prognosis by harnessing the predictive power of genetic, radiomic, and clinical data, offering tailored prognostic models that surpass traditional histological classifications.

A pivotal study by Song et al. introduced the hidden-genome classifier, a supervised machine learning algorithm designed to quantify the genetic heterogeneity of intrahepatic cholangiocarcinoma (IHC) and its implications for prognosis (26). By analyzing the genetic similarities among IHCs, extrahepatic cholangiocarcinomas (EHCs), GBCs, hepatocellular carcinomas (HCCs), and biphenotypic tumors, this study identified distinct genetic profiles associated with varying prognoses. Notably, tumors exhibiting genetic alignment with EHC/GBC, termed the "biliary class," were associated with significantly poorer overall survival than those resembling HCC, underscoring the classifier's superiority over histological subtypes in prognostication (26).

Similarly, Zhou et al. developed multiple clinical index classifiers (MIC1 and MIC2) combined with AI algorithms to predict recurrence and survival in GBC patients (27). By analyzing the correlations between clinical indicators and patient outcomes, their models achieved high sensitivity and specificity, outperforming traditional prognostic methods (27). These advancements enable clinicians to stratify patients more accurately on the basis of individualized risk profiles, guiding personalized treatment strategies and improving survival outcomes.

To further increase predictive accuracy, Meng et al. proposed an integrated radiomics and deep learning model using contrast-enhanced CT imaging for survival prediction after surgical resection of GBC (28). By extracting radiomic features and applying deep learning techniques, their model demonstrated superior performance in prognosticating patient survival, facilitating the identification of high-risk individuals who may benefit from intensified therapeutic interventions (28).

Moreover, Cotter et al. employed machine learning to stratify GBC patients into distinct prognostic groups using preoperative variables such as tumor size, biomarker levels (e.g., CA 19–9), and inflammatory markers (e.g., the neutrophil–lymphocyte ratio) (29). Their classification and regression tree model identified four prognostic groups with significantly different median survival rates, highlighting the potential of AI to refine prognostic stratification and optimize treatment planning on the basis of individual patient characteristics (29).

Finally, Liu et al. leveraged CT-based machine learning to predict survival in GBC patients, emphasizing the role of radiomics in enhancing prognostic accuracy (30). By integrating radiomic features with clinical data, their model accurately differentiated between high- and low-risk patient groups, aiding clinicians in making informed decisions regarding treatment intensity and follow-up care.

As such, AI-driven approaches, including genetic classifiers, multi-index clinical models, radiomics, and machine learning algorithms, are reshaping the prognosis of GBC (26–30). These innovations not only enhance prognostic accuracy beyond traditional methods but also pave the way for personalized treatment strategies tailored to individual patient needs. By integrating AI into clinical practice, clinicians can navigate the complexities of GBC management more effectively, ultimately improving patient prognosis.

12.6 THE AIRGOS PROJECTION: INTEGRATION OF A MULTIOMICS AI MODEL

As advancements in medical technology continue to evolve, the integration of AI has emerged as a transformative force in reshaping the landscape of oncological management in general. The AiRGOS (Artificial Intelligence, Radiomics, Genomics, Oncopathomics, and Surgomics) project, initiated for pancreatic cancer, can represent a transformative approach in the management of GBC, integrating advanced AI technologies with multiomics data to redefine patient care (31). This pioneering project leverages radiomic imaging, genomic sequencing, and comprehensive clinical data to tailor precision therapeutic strategies and improve surgical outcomes.

At its core, AiRGOS utilizes AI algorithms within the oncological framework to process vast datasets, detect subtle patterns, and explore complex relationships across radiomics, genomics, pathomics, and surgomics (31). By overcoming the limitations posed by the sheer volume and complexity of surgical data, AI-driven analyses enable clinicians to make informed decisions with unprecedented accuracy and efficiency. This integrated approach not only enhances surgical decision-making but also optimizes treatment planning and postoperative management through a deeper understanding of tumor biology, patient-specific characteristics, and therapeutic options.

In the realm of oncology, the significance of multiomics approaches cannot be overstated. By combining insights from genomic, epigenomic, transcriptomic, proteomic, and metabolomic data, these methodologies facilitate early cancer detection, accurate diagnosis, precise prognostication, and the identification of biomarkers crucial for guiding therapeutic interventions (31). AI models, particularly those employing deep learning techniques such as neural networks, excel in integrating and interpreting multiomics data, thereby improving prediction accuracy and personalized treatment strategies tailored to each patient's unique molecular profile (31).

In the specific context of gallbladder carcinoma, genomic studies have identified critical mutations such as CTNNB1, ELF3, TP53, and ERBB2, shedding light on potential targets for therapeutic intervention (32). These insights not only inform targeted therapies but also pave the way for immunotherapeutic approaches targeting neoantigens derived from mutated genes. Additionally, genomic profiling reveals alterations in pathways such as KEAP1/NFE2L2 and WNT, providing deeper insights into disease mechanisms and offering opportunities for more effective treatment modalities (32).

Radiomics, another integral component of AiRGOS, enhances diagnostic accuracy and treatment planning through detailed analysis of radiological images (31). By extracting quantitative features from imaging studies such as CT and MRI, radiomics aids in early detection, precise staging, and evaluation of treatment responses. This data-driven approach enables clinicians to stratify patients on the basis of disease severity and predict therapeutic outcomes with greater certainty.

Furthermore, oncopathomics extends the scope of AiRGOS by employing advanced computational techniques to analyze histopathological images (31). By quantifying subtle morphometric and architectural characteristics from digitized tissue samples, oncopathomics enhances traditional histopathological assessments, offering deeper insights into tumor behavior and informing more precise treatment decisions.

The application of surgomics within AiRGOS encompasses the integration of genomic, radiomic, pathomic, and real-time surgical data to optimize patient outcomes. Surgomics analyzes genetic variants influencing surgical risk, uses radiomics to predict disease progression and plan interventions, and employs pathomics to identify predictive markers from histopathological data (31, 33). This comprehensive approach empowers surgeons with real-time insights during procedures, facilitating personalized surgical strategies and improving patient outcomes.

The continued advancement of AI technologies holds immense promise for further revolutionizing GBC management. Collaborations between surgeons, radiologists, oncologists, geneticists, and AI experts are essential to harnessing these innovations effectively. By leveraging AiRGOS, clinicians can anticipate earlier tumor detection, more precise surgical interventions, and tailored treatment regimens that optimize therapeutic efficacy and patient quality of life.

As such, AiRGOS represents a paradigm shift in the management of GBC, where AI-driven multiomics approaches pave the way for personalized, data-driven medicine. Through the integration of advanced AI technologies with radiomics, genomics, and pathomics, AiRGOS promises to enhance diagnostic accuracy, refine treatment strategies, and ultimately improve outcomes for patients battling this challenging disease.

12.7 CONCLUSION

The integration of AI into the management of GBC holds great promise for improving early detection, diagnosis, treatment planning, and prognosis. While challenges remain, ongoing advancements in AI technology and computing power, such as the AiRGOS initiative, are likely to overcome these obstacles, paving the way for more effective and personalized GBC care.

REFERENCES

1. Dulskas A, Cerkauskaite D, Patasius A, Smailyte G. Incidence, mortality, and survival trends in cancer of the gallbladder and extrahepatic bile ducts in Lithuania. *Medicina (Kaunas)*. 2023;59(4).
2. Wi Y, Woo H, Won YJ, Jang JY, Shin A. Trends in gallbladder cancer incidence and survival in Korea. *Cancer Res Treat*. 2018;50(4):1444–51.
3. Malhotra RK, Manoharan N, Shukla NK, Rath GK. Gallbladder cancer incidence in Delhi urban: A 25-year trend analysis. *Indian J Cancer*. 2017;54(4):673–7.

4. Ito K, Koshita S, Kanno Y, Ogawa T. Symptoms, physical findings, and laboratory data of gallbladder cancer. *Nihon Rinsho*. 2015;73(Suppl. 3):553–6.
5. Sandrasegaran K, Menias CO. Imaging and screening of cancer of the gallbladder and bile ducts. *Radiol Clin North Am*. 2017;55(6):1211–22.
6. Strom BL, Maislin G, West SL, Atkinson B, Herlyn M, Saul S, et al. Serum CEA and CA 19–9: Potential future diagnostic or screening tests for gallbladder cancer? *Int J Cancer*. 1990;45(5):821–4.
7. Gupta P, Basu S, Rana P, Dutta U, Soundararajan R, Kalage D, et al. Deep-learning enabled ultrasound based detection of gallbladder cancer in northern India: A prospective diagnostic study. *Lancet Reg Health Southeast Asia*. 2024;24:100279.
8. Peng J, Wang W, Jin H, Qin X, Hou J, Yang Z, et al. Develop and validate a radiomics space-time model to predict the pathological complete response in patients undergoing neoadjuvant treatment of rectal cancer: An artificial intelligence model study based on machine learning. *BMC Cancer*. 2023;23(1):365.
9. Ke J, Jin C, Tang J, Cao H, He S, Ding P, et al. A Longitudinal MRI-based artificial intelligence system to predict pathological complete response after neoadjuvant therapy in rectal cancer: A multicenter validation study. *Dis Colon Rectum*. 2023;66(12):e1195–e206.
10. Irajizad E, Wu R, Vykoukal J, Murage E, Spencer R, Dennison JB, et al. Application of artificial intelligence to plasma metabolomics profiles to predict response to neoadjuvant chemotherapy in triple-negative breast cancer. *Front Artif Intell*. 2022;5:876100.
11. Liu H, Lu Y, Shen K, Zhou M, Mao X, Li R. Advances in the management of gallbladder polyps: Establishment of predictive models and the rise of gallbladder-preserving polypectomy procedures. *BMC Gastroenterol*. 2024;24(1):7.
12. Shi X, Guo S, Duan Q, Zhang W, Gao S, Jing W, et al. Detection and characterization of pancreatic and biliary tract cancers using cell-free DNA fragmentomics. *J Exp Clin Cancer Res*. 2024;43(1):145.
13. Dou J, Dawuti W, Li J, Zhao H, Zhou R, Zhou J, et al. Rapid detection of serological biomarkers in gallbladder carcinoma using Fourier transform infrared spectroscopy combined with machine learning. *Talanta*. 2023;259:124457.
14. Zhou QM, Liu CX, Zhou JP, Yu JN, Wang Y, Wang XJ, et al. Machine learning-based radiological features and diagnostic predictive model of xanthogranulomatous cholecystitis. *Front Oncol*. 2022;12:792077.
15. Zhang X, Gui X, Zhang Y, Liu Q, Zhao L, Gao J, et al. A panel of bile volatile organic compounds servers as a potential diagnostic biomarker for gallbladder cancer. *Front Oncol*. 2022;12:858639.
16. Sharma N, Yadav M, Tripathi G, Mathew B, Bindal V, Falari S, et al. Bile multi-omics analysis classifies lipid species and microbial peptides predictive of carcinoma of gallbladder. *Hepatology*. 2022;76(4):920–35.
17. Dawuti W, Dou J, Li J, Zhang R, Zhou J, Maimaitiaili M, et al. Label-free surface-enhanced Raman spectroscopy of serum with machine-learning algorithms for gallbladder cancer diagnosis. *Photodiagnosis Photodyn Ther*. 2023;42:103544.
18. Yin Y, Yakar D, Slangen JJG, Hoogwater FJH, Kwee TC, de Haas RJ. Optimal radiological gallbladder lesion characterization by combining visual assessment with CT-based radiomics. *Eur Radiol*. 2023;33(4):2725–34.
19. Yin Y, Yakar D, Slangen JJG, Hoogwater FJH, Kwee TC, de Haas RJ. The value of deep learning in gallbladder lesion characterization. *Diagnostics (Basel)*. 2023;13(4).
20. Arslan E, Cermik TF. PET/CT variants and pitfalls in liver, biliary tract, gallbladder and pancreas. *Semin Nucl Med*. 2021;51(5):502–18.
21. Tittarelli A, Barria O, Sanders E, Bergqvist A, Brange DU, Vidal M, et al. Co-expression of immunohistochemical markers MRP2, CXCR4, and PD-L1 in gallbladder tumors is associated with prolonged patient survival. *Cancers (Basel)*. 2023;15(13).
22. Gupta P, Basu S, Arora C. Applications of artificial intelligence in biliary tract cancers. *Indian J Gastroenterol*. 2024;43(4):717–28. https://doi.org/10.1007/s12664-024-01518-0
23. Zhou S, Han S, Chen W, Bai X, Pan W, Han X, et al. Radiomics-based machine learning and deep learning to predict serosal involvement in gallbladder cancer. *Abdom Radiol (NY)*. 2024;49(1):3–10.
24. Tsilimigras DI, Hyer JM, Paredes AZ, Moris D, Beal EW, Merath K, et al. The optimal number of lymph nodes to evaluate among patients undergoing surgery for gallbladder cancer: Correlating the number of nodes removed with survival in 6531 patients. *J Surg Oncol*. 2019;119(8):1099–107.
25. Primrose JN, Fox RP, Palmer DH, Malik HZ, Prasad R, Mirza D, et al. Capecitabine compared with observation in resected biliary tract cancer (BILCAP): A randomized, controlled, multicenter, phase 3 study. *Lancet Oncol*. 2019;20(5):663–73.

26. Song Y, Boerner T, Drill E, Shin P, Kumar S, Sigel C, et al. A novel approach to quantify heterogeneity of intrahepatic cholangiocarcinoma: The hidden-genome classifier. *Clin Cancer Res.* 2024;15;30(16):3499–511. https://doi.org/10.1158/1078-0432.CCR-24-0657
27. Zhou Y, Chen S, Wu Y, Li L, Lou Q, Chen Y, et al. Multiclinical index classifier combined with AI algorithm model to predict the prognosis of gallbladder cancer. *Front Oncol.* 2023;13:1171837.
28. Meng FX, Zhang JX, Guo YR, Wang LJ, Zhang HZ, Shao WH, et al. Contrast-enhanced CT-based deep learning radiomics nomogram for the survival prediction in gallbladder cancer postoperative. *Acad Radiol.* 2023;31(6):2356–2366. doi:10.1016/j.acra.2023.11.027.
29. Cotter G, Beal EW, Poultsides GA, Idrees K, Fields RC, Weber SM, et al. Using machine learning to preoperatively stratify prognosis among patients with gallbladder cancer: A multi-institutional analysis. *HPB (Oxford).* 2022;24(11):1980–8.
30. Liu Z, Zhu G, Jiang X, Zhao Y, Zeng H, Jing J, et al. Survival prediction in gallbladder cancer using CT based machine learning. *Front Oncol.* 2020;10:604288.
31. Gumbs AA, Croner R, Abu-Hilal M, Bannone E, Ishizawa T, Spolverato G, et al. Surgomics and the artificial intelligence, radiomics, genomics, oncopathomics and surgomics (AiRGOS) project. *Artificial Intelligence Surgery.* 2023;3(3):180–5.
32. D'Afonseca V, Arencibia AD, Echeverria-Vega A, Cerpa L, Cayun JP, Varela NM, et al. Identification of altered genes in gallbladder cancer as potential driver mutations for diagnostic and prognostic purposes: A computational approach. *Cancer Inform.* 2020;19:1176935120922154.
33. Wagner M, Brandenburg JM, Bodenstedt S, Schulze A, Jenke AC, Stern A, et al. Surgomics: Personalized prediction of morbidity, mortality and long-term outcome in surgery using machine learning on multimodal data. *Surg Endosc.* 2022;36(11):8568–91.

13 Proteomic Insights in Gallbladder Cancer

Techniques for Biomarker Identification

Doutrina Das,[1]* *Ruhi Dixit,*[2] *and Manoj Pandey*[1,2]
[1]Department of Surgical Oncology, Institute of Medical Sciences, Banaras Hindu University, Uttar Pradesh, India
[2]Health Technology Assessment in India (HTAIn) Nodal Centre, Institute of Medical Sciences, Banaras Hindu University, Uttar Pradesh, India
*Corresponding author: doutrinadas2014@gmail.com

13.1 INTRODUCTION

Gallbladder cancer (GBC), also referred to as the "New King of Cancer," is the most common bile tract malignancy (1). Global GBC incidence projections for 2022 from the GLOBOCAN database for 185 countries are compared to reported data from the most current volume of *Cancer Incidence in Five Continents*. The prevalence of primary gallbladder cancer varies globally, with the highest rates found in Poland, Northern India, Native, and some South American populations. Every year, around 17,000–18,000 cases of GBC are recognized in India, with a noticeable yearly fatality rate (2).

Identification and confirmatory studies of proteins linked to the transition of inflammatory gallbladder lesions to invasive cancer are still not clear, and therefore there is a need to find more markers associated with the disease for its better diagnosis (3). Several genes linked to GBC have been recognized through the human genome project. Many of their functions, interconnection, and the activities of their protein products in different molecular pathways have been elucidated. Although investigations on differential expression of mRNA are important, they do not always necessarily correspond with protein concentrations, as proteins frequently undergo proteolytic cleavage and post-translational modifications. In addition to genetic and epigenetic modifications, translational and post-translational modifications and inappropriate positioning within the cell contribute to cancer initiation and progression (2).

13.1.1 Sample Types Used in Proteomic Approaches for GBC

i. Tissue samples collected from the resected gallbladder of patients during surgery, placed in 1X PBS vial, and immediately stored at -80 °C (3)
ii. Blood samples of GBC patients (4).
iii. Proteins in the preoperative serum are also used for the analysis (5).

13.2 TECHNIQUES USED FOR PROTEOMICS STUDY

13.2.1 Immunohistochemistry

IHC is one of the most prevalent immunostaining methods, combining concepts from histology, immunology, and biochemistry. It uses monoclonal and polyclonal antibodies to identify certain

DOI: 10.1201/9781003506201-13

proteins or antigens inside tissue samples (6). The bound antigen-antibody complex is then seen using enzymatic labels, like alkaline phosphatase and peroxidase, for the detection of the antigens of interest. IHC can be performed on formalin-fixed paraffin-embedded (FFPE) or processed fresh tissues. It may detect many antigens at once, or may be successively stained to identify multiple antigens. These techniques are widely used in pathology laboratories for diagnostic purposes. Many biomarkers detected by this technique are also used for patient staging, therapy allocation, and prognosis (7). Studies have used IHC on GBC tissue samples and cell lines to get the expression of SOAT1 (sterol O-acyltransferase 1) protein, and the results show that it has a significant increase in both GBC cell lines and tissue samples (8). The expression of Her-2/neu was found in around 235 carcinoma gall bladder tissues by IHC. The results have shown an association between Her-2/neu positivity and poor survival, which was statistically significant, suggesting that anti-Her-2/neu therapy can serve as a potential treatment option for this cancer (9).

13.2.2 SDS-PAGE

Sodium dodecyl sulfate polyacrylamide gel electrophoresis (SDS-PAGE) is a technique for separating proteins based on their ability to move within an electrical current. SDS detergent is added to the gel to remove secondary and tertiary protein structures and to maintain the proteins as polypeptide chains. The most widely used gel system for separating a broad range of proteins by SDS-PAGE is the Laemmli system, which uses tris-glycine gels comprised of a stacking gel component and the resolving gel where varying acrylamide gel percentages are used to separate the proteins based on their mass weight. All proteins retain only their primary structure, and all have a large negative charge, which means they will all migrate toward the positive pole when placed in an electric field. SDS binds strongly to proteins, with approximately one detergent molecule binding to two amino acids when SDS is present at 0.1% (10).

13.2.3 Western Blot

It is a powerful and important procedure for the immunodetection of proteins post-electrophoresis, particularly proteins that are of low abundance. Since the inception of the protocol for protein transfer from an electrophoresed gel to a membrane in 1979, protein blotting has evolved greatly (11). This is an important and routine method for protein analysis that depends on the specificity of antibody-antigen interaction and is useful for the qualitative or semiquantitative identification of specific proteins and their molecular weight from a complex mixture (12). WB is a powerful technique to monitor changes in expression or post-translational modifications (13). Sodium dodecyl sulfate-polyacrylamide gel electrophoresis (SDS-PAGE)-separated proteins are electrophoretically transferred to a polyvinylidene fluoride (PVDF) membrane, which is then incubated with specific antibodies, then developed to show the protein of interest (14). Protein expression analysis by WB for ERK1/2, p-ERK1/2, EGFR, ERBB2, and ERBB3 was performed in GBC cell lines, where higher ERK1/2 and p-ERK1/2 expression was found in GBC tissues compared to chronic cholecystitis (CC) tissues (15). WB analysis was also used to determine the alteration among the levels of several key proteins of the PI3K-AKT pathway that facilitates GBC progression. Thus, western blot helps in GBC biomarker detection.

13.2.4 2D Polyacrylamide Gel Electrophoresis

Two-dimensional gel electrophoresis can separate around 10,000 proteins at the same time in a single gel based on two parameters, that is, charge (isoelectric point) and mass of protein. Charge is the first dimension, and mass of protein is the second (16). This is applicable to nearly any type of protein-containing specimen, like eukaryotic extracts derived from eukaryotic tissues, cells and organelles, body fluids like plasma or serum, cerebrospinal fluid (CSF), urine, saliva, tears, milk, nasal mucus, amniotic fluid, semen; prokaryotic organisms such as bacteria, fungi, and parasites; seeds, vegetal matter, and plants. Recent advances in sample preparation methodology for 2-DE have focused on improvements in sample prefractionation, results visualization, and sample buffer constituents to achieve better representation of poorly soluble proteins (17). 2D-PAGE can show hundreds of proteins at once on a single gel because it has a high-resolution power and can hold a lot of samples. This gives a clear picture of the whole proteome of a sample at a certain point in time. 2D gel electrophoresis has also been employed to detect post-translational modifications such as glycosylation and phosphorylation. Each protein spot on the gel can be excised, processed, and analyzed further using LC/MS (16). It is important to remember that a fully automatic 2D-PAGE machine is still in the works and that the one we have now has some issues that need to be thought about when designing any proteomics-based analysis that will use it. Some proteins might not be well separated by standard 2D-PAGE for a number of reasons, such as not being able to dissolve well (membrane proteins, clumps), having an extremely high or low pI, or being too big or too small (18). Serum samples from GBC patients and healthy volunteers were subjected to 2-DE. After isoelectric focusing, the equilibrated IPG strips were loaded onto pre-cast 12.5% homogeneous polyacrylamide gels for electrophoresis. The silver-stained gels were scanned, and resulting images were analyzed using specific software for spot detection, quantification, as well as comparative analyses. The gel image from the GBC group was compared with that from the healthy volunteer group. Sixty-four differentially expressed protein spots that changed >2-fold were found from these two separate comparisons (19).

13.2.5 2D Difference in Gel Electrophoresis (2D-DIGE)

As 2-DE became more reproducible, there were still issues with gel-to-gel repeatability that needed technical replicates to get rid of the effects of experimental variation. The ability to combine multiple fluorescently tagged proteins on the same gel marked a significant advancement in 2D gel electrophoresis. 2D difference gel electrophoresis (2D-DIGE) is the name given to this method. Specifically modified cyanine dyes that match in molecular weight and charge and offer a usable dynamic range of up to four orders of magnitude are the fluorescent dyes that are employed. One dye is employed to mark a combined internal standard in order to get beyond the issues caused by gel-to-gel fluctuation. With the use of multiplexing and a virtual reduction of gel-to-gel variance, biological replicates may now be performed, drastically lowering the number of gels needed in comparison to traditional detection methods. It is feasible to identify potential proteins of interest by running samples with multiple labels on a single gel and examining the pictures that appear (20). The 2D-DIGE technology can work on FFPE (formalin-fixed, paraffin-embedded) to detect the proteins that are differentially present and are identified later using MS (21).

13.2.6 Mass Spectrometry

Mass Spectrometry (MS) has become a powerful tool in protein analysis and the key technology in the emerging field of proteomics (22). The use of mass spectrometry has been completely transformed by the development of two ionization techniques: electrospray (ESI) and matrix-assisted laser desorption/ionization (MALDI), applicable to almost any biological molecule (23). Mass spectrometers are used either to measure simply the molecular mass of a polypeptide or to determine additional structural features, including the amino acid sequence or the site of attachment and type of post-translational modifications (24). It separates ionized particles such as atoms, molecules, and clusters by using differences in the ratios of their charges to their respective masses and can be used to determine the molecular weight of the particles. The components of an MS are as follows: an ion source that separates the molecules in the sample into ions; a mass analyzer that uses electromagnetic fields to sort the ions based on their masses; a detector that determines the amount of an indicator quantity and provides information for determining the quantity of each ion present; and a computer that controls the mass analyzer and handles the data obtained from the detector (25). According to some studies, proteins in complex biological samples are first converted into peptides by proteolytic digestion. The resulting peptide mixture can then be separated by LC (liquid chromatography) and ionized by electrospray before entering the MS. A full MS spectrum is acquired for the peptides that are eluting from the LC column at any given time; peptides are then isolated and fragmented to obtain the MS pattern of its fragments (26). A study (27) found the concentrations of bile acids and bile salts in aqueous-phase samples that were analyzed using liquid chromatography–mass spectrometry (LC-MS) in GBC patients. Results indicate that concentrations of bile salts in bile decrease with the development of GBC, and this decrease in concentration is greater for primary bile salts. The differentially expressed proteins obtained from GBC tissues using 2-DE were identified using a MALDI-TOF mass spectrometer (MS) and database searches, helping to find out the proteins that are overexpressed or decreased in their expression in GBC. Huang et al. (28) in their study prepared and isolated proteins from FFPE tissue blocks of GBC patients. The processed samples were subjected to high-pH liquid chromatography and thereby fractionated and injected afterward into the MS, identifying that CEACAM6 is one of the significantly most upregulated proteins in GBC. MS analysis, along with other techniques, was performed to confirm that TRIM-47 protein is an essential biomarker in GBC and is one of the responsible factors for reduced overall survival of GBC patients (29). MS was performed on GBC tissue samples of patients to analyze their proteome, and GPRCA5 protein was found to be highly up-regulated and strongly associated with tumor progression and poor survival (30).

13.2.6.1 i-TRAQ

The iTRAQ (isobaric tags for relative and absolute quantification) technique, a novel MS-based approach, is widely employed in proteomic workflows requiring relative quantification (31). High-throughput proteome analysis is made possible by shotgun proteomic techniques that use isobaric tagging of peptides. Using tandem mass spectrometry, iTRAQ reagents enable concurrent identification and measurement of proteins in four distinct samples (32). The process relies on the derivatization of primary amino groups in intact proteins using iTRAQ (33). Proteomic analysis of gallbladder cancer cell lines was carried out using iTRAQ. The expression of macrophage migration inhibitory factor was analyzed in gallbladder adenocarcinoma tissues and was found to be overexpressed, thus acting as a potential therapeutic target in GBC (34). An iTRAQ-based quantitative proteomic analysis of tumor tissues from early-stage GBC cases was carried out, and 357 differentially expressed proteins were identified based on unique peptides. This analytical technique can thus be used for drug-induced protein expression, discovery and elucidation of disease markers, and protein–protein interactions in multi-protein complexes (35).

13.2.6.2 "Tandem Mass Tags"

An innovative MS/MS-based analytical method that makes use of isotopomer tags, or "tandem mass tags" (TMTs), has been developed. The purpose of the new tags is to guarantee that identical peptides marked with distinct TMTs precisely migrate together in every separation. Tandem mass spectrometry is needed to analyze the tags using new quantification techniques. Compared to previous techniques, the new tags and analytic techniques make it easier and more accurate to identify peptides from various samples based on their level of abundance (36). Tandem mass tag (TMT) reagents have been employed in a wide range of different clinically oriented serum and plasma proteomics studies. In contrast to earlier works, which struggled to quantify more than a few hundred proteins, increasingly these studies have provided deeper insight into the plasma proteome, extending the numbers of quantified proteins to over a thousand (37). This quantitative proteomic analysis was performed to determine the differentially expressed proteins (DEPs) in cerebrospinal fluid (CSF) samples collected from patients with MS (multiple sclerosis) and non-inflammatory neurological controls (38). A study using TMT found that DKK3, a protein belonging to a family of Wnt antagonists, was overexpressed in three GBC cell lines. The results report for the first time that DKK3 overexpression leads to alteration in the phosphorylation of around 14 kinases (39). To identify aberrantly activated signaling events in GBC, TMT-quantitative phosphoproteomic analysis of five GBC cell lines was carried out, where proline-rich Akt substrate 40 kDa (PRAS40) was found to be one of the proteins that was hyperphosphorylated in all the invasive GBC cell lines, elucidating its role as a potential biomarker for GBC (40).

13.2.7 Flow Cytometry

Flow cytometry allows for rapid multiparametric examination of single cells in solution. It uses lasers as light sources to generate dispersed and fluorescent light signals, which are detected by photodiodes. These visual signals are then translated into electronic signals, which are then analyzed by a computer and saved in a standardized format data file. Different fluorescent reagents are used herein, like nucleic acid-binding dyes, fluorescently conjugated antibodies, ion indicator dyes, viability dyes, and fluorescent expression proteins (41). Flow cytometry analyses have shown there was an upregulation in MPZL1 protein, which had an antiapoptotic effect on GBC cell lines (42). Survival of tumor cells due to the influence of SPOCK-1 protein—bringing about an antitumor effect by inactivating the PI3K/Akt pathway—was determined using flow cytometry in a study (43). Novel methods of flow cytometry are becoming important for diagnostic purposes, e.g., identifying multiple clinical biomarkers for a wide range of diseases, as well as developing novel therapies like elucidating drug mechanisms. In addition, it is used for multiplex analysis and has the potential to significantly enhance proteomics research, particularly the analysis of post-translational modifications of proteins, on a large scale (44)

Sl.no	Sample type	Proteomics technique used	Proteins up/down regulated in GBC
1.	Tissue and cell-line	IHC	Myelin protein 0-like 1 (MPZL1) overexpression (42)
			High EGFR expression in poorly differentiated carcinomas and HER-2/NEU expression is high in well-differentiated carcinomas (45)
			Shh (Sonic Hedgehog) Ptch1 (Patched receptor) and Gli1 (glioma-associated oncogene (Gli) transcription factors (46).
			CD97 and CD55 up-regulated (47)
			Upregulation of fatty acid binding protein 1 (FABP1) (48).
			Positive expression of CK7, CK19, E-cadherin, MMP-2, CD44, SOX2, and TP53 in GBC cell line (49).
			Tripartite motif containing 47 (TRIM47) up-regulated (50).
			High expression of Astrocyte elevated gene-1 (AEG-1) (51).
			Stathmin 1 (STMN1) highly expressed (52)
			High expression of S100A10 and haptoglobin proteins (19)
2.	Tissue samples and Cell lines.	Western Blot	PLEK2 (pleckstrin-2) had higher expression (53)
			High ERK1/2 and p-ERK1/2 expression (15)
			Wnt inhibitory factor 1 (WIF-1) low expression (54)
			High level of expression of Tumor necrosis factor-α (TNF-α) (55)
			SPOCK1, a novel proteoglycan increases in expression (43).
			KLF5 and PDGFA expression up-regulated (56)
			Tripartite motif containing 47 (TRIM47) up-regulated (Xu et al., 2025) (29)
			(FBXO33) expression is significantly elevated (57)
			FOXA2 was under-expressed (58)
			CEP_{55} found to be overexpressed (59)
			AP endonuclease 1 (APE1) was up-regulated (60)

(Continued)

Sl.no	Sample type	Proteomics technique used	Proteins up/down regulated in GBC
3.	Tissue	ELISA	High expression of Survivin (61) TNF-α mediated upregulation of VEGF-D (62) IL-1β protein increased significantly (63) cIAP2 expression was elevated (64)
4.	Tissue and Cell lines	2D gel electrophoresis	Overexpression of annexinA3 (65) CLIC1 up-regulated; Programmed cell death protein 8 and Phosphoglycerate kinase 1 was down-regulated (66) Vimentin overexpressed (67) ANXA1 protein up-regulated (68)
5.	Serum samples	2D-DIGE	Increased LRG1 (leucine-rich α-2-glycoprotein), CA19–9 and IL-6 levels in serum from patients with BTC (69).
6.	Tissue and cell-line	i-TRAQ	MIF is over-expressed (34) KRT7 and SRI over-expressed (70)
7.	Tissue samples and cell-lines	TMT	Overexpressed DKK3 (39)
8.	Tissue samples	MS	TRPM2-AS over-expressed (71) CEACAM6 up-regulated (28) S100A8 and S100A9 were down-regulated (72) S100A8 was overexpressed (73) (Tandem MS was used specifically) ANK3 and FHL1 were downregulated in GBC (74)
9.	Tissue samples	Flow cytometry	PSMC2 up-regulated (75)

13.3 CONCLUSION

The most common biliary system cancer, gallbladder carcinoma (GBC), has a poor prognosis and substantial morbidity. Despite recent improvements in the detection and treatment of gastrointestinal malignancies, GBC is still difficult to treat and has a dismal prognosis. There are currently no trustworthy molecular markers for GBC. Proteomic analysis can overcome some of the drawbacks of the methods used to clarify the molecular processes of GBC and has been applied to a variety of tumor types. This method may be applied to find novel biomarkers for gallbladder cancer. In order to discover biomarkers and perform early illness diagnosis, several proteomic techniques are being used, such as Western blot, 2D-DIGE, IHC, 2D gel electrophoresis, mass spectroscopy investigations, and iTRAQ-based quantitative proteomics research.

Gallbladder cancer's molecular causes are not well known, which suggests that more research and analysis in the field may be necessary. Protein dysregulation and other molecular alterations are important in the development and course of GBC. It is necessary to look at these changes across the various groups, since the research now available indicates that the molecular alterations linked to GBC may differ among various geographic and ethnic groups. Since proteins are comparatively stable identifiers, measuring them can be useful in determining these molecular alterations during the state of illness and in locating potential biomarkers. Bloodstream-circulating biomarkers are ideal because they offer a less intrusive way to find GBC biomarkers. Crucially, proteins present in tumors and plasma may offer a strong indication that the markers contribute to the development of cancers. Additionally, measured protein alterations may be examined in the context of enhanced biological pathways to gain a better understanding of the mechanism behind the development and progression of GBC. Preoperative evaluation of plasma proteins may reliably distinguish cholecystitis from cancer, according to recent studies, which support the creation of a non-invasive test to aid in preoperative decision-making when GBC is suspected.

Even while studies on mRNA differential expression are crucial, protein concentrations and mRNA differential expression do not always match, since proteins are often subject to post-translational changes and proteolytic cleavage. Along with genetic and epigenetic changes, translational and post-translational changes, as well as improper cell placement, all have a role in the development and spread of cancer. To sum up, GBC's proteome characterization can be used as a resource for further biomarker research, aiding in the disease's early diagnosis.

REFERENCES

1. Zhao J, Shi L, Yang Y, Zhu J, Zhou Z, Dong P, et al. *Wu-Mei-Wan Promotes Ferroptosis in Gallbladder Cancer Through Stat3 Negative Regulation: An Integrated Hplc, Proteomics, Network Pharmacology, and Experimental Validation Study* [Internet]. Rochester, NY: Social Science Research Network, 2025 [cited 2025 Feb 22]. Available from: https://papers.ssrn.com/abstract=5120520
2. Huang HL, Yao HS, Wang Y, Wang WJ, Hu ZQ, Jin KZ. Proteomic identification of tumor biomarkers associated with primary gallbladder cancer. *World J Gastroenterol WJG*. 2014 May 14;20(18):5511–8.
3. Rawal N, Hariprasad G, Bandyopadhyay S, Dash NR, Kumar S, Das P, et al. Molecular biomarkers involved in the progression of gallbladder inflammatory lesions to invasive cancer: A proteomic approach. *Biomol Biomed*. 2025 Jan 1;25(1):115–43.
4. Rao D, Li J, Zhang M, Huang S, Meng L, Song G, et al. Multi-model analysis of gallbladder cancer reveals the role of OxLDL-absorbing neutrophils in promoting liver invasion. *Exp Hematol Oncol*. 2024 May 31;13(1):58.
5. Nouairia G, Cornillet M, Jansson H, Bergquist A, Sparrelid E. 319P Serum based test to assist preoperative decision making in suspected gallbladder cancer. *Ann Oncol*. 2024 Jun;35:S134.
6. Mebratie DY, Dagnaw GG. Review of immunohistochemistry techniques: Applications, current status, and future perspectives. *Semin Diagn Pathol*. 2024 May 1;41(3):154–60.
7. Hussaini HM, Seo B, Rich AM. Immunohistochemistry and immunofluorescence. *Methods Mol Biol Clifton NJ*. 2023;2588:439–50.
8. Hong Y, Abudukeremu X, She F, Chen Y. SOAT1 in gallbladder cancer: Clinicopathological significance and avasimibe therapeutics. *J Biochem Mol Toxicol*. 2024;38(6):e23733.
9. Dash S, Anirvan P, Samantaray S, Swain PK, Parida PK, Rout N, et al. Human epidermal growth factor receptor-2/neu expression in gallbladder cancer is significantly associated with clinicopathological parameters and survival. *Indian J Gastroenterol* [Internet]. 2025 Feb 3 [cited 2025 Feb 28]. Available from: https://doi.org/10.1007/s12664-024-01723-x
10. Al-Tubuly AA. SDS-PAGE and Western Blotting. In: George AJT, Urch CE, editors. *Diagnostic and Therapeutic Antibodies* [Internet]. Totowa, NJ: Humana Press, 2000 [cited 2025 Mar 4]. p. 391–405. Available from: https://doi.org/10.1385/1-59259-076-4:391
11. Kurien BT, Scofield RH. Western blotting. *Methods*. 2006 Apr 1;38(4):283–93.
12. Hnasko TS, Hnasko RM. The Western Blot. In: Hnasko R, editor. *ELISA: Methods and Protocols* [Internet]. New York, NY: Springer, 2015 [cited 2025 Mar 1]. p. 87–96. Available from: https://doi.org/10.1007/978-1-4939-2742-5_9
13. Gwozdz T, Dorey K. Chapter 6—Western blot. In: Jalali M, Saldanha FYL, Jalali M, editors. *Basic Science Methods for Clinical Researchers* [Internet]. Boston, MA: Academic Press, 2017 [cited 2025 Mar 1]. p. 99–117. Available from: www.sciencedirect.com/science/article/pii/B9780128030776000060
14. Kim B. Western Blot Techniques. In: Espina V, editor. *Molecular Profiling: Methods and Protocols* [Internet]. New York, NY: Springer, 2017 [cited 2025 Mar 1]. p. 133–9. Available from: https://doi.org/10.1007/978-1-4939-6990-6_9
15. Buchegger K, Silva R, López J, Ili C, Araya JC, Leal P, et al. The ERK/MAPK pathway is overexpressed and activated in gallbladder cancer. *Pathol—Res Pract*. 2017 May 1;213(5):476–82.
16. Chandrasekharan G, Unnikrishnan M. High throughput methods to study protein-protein interactions during host-pathogen interactions. *Eur J Cell Biol*. 2024 Jun 1;103(2):151393.
17. Shaw MM, Riederer BM. Sample preparation for two-dimensional gel electrophoresis. *Proteomics*. 2003;3(8):1408–17.

18. Carrette O, Burkhard PR, Sanchez JC, Hochstrasser DF. State-of-the-art two-dimensional gel electrophoresis: A key tool of proteomics research. *Nat Protoc.* 2006 Aug;1(2):812–23.
19. Tan Y, Ma SY, Wang FQ, Meng HP, Mei C, Liu A, et al. Proteomic-based analysis for identification of potential serum biomarkers in gallbladder cancer. *Oncol Rep.* 2011 Oct 1;26(4):853–9.
20. Beckett P. The Basics of 2D DIGE. In: Cramer R, Westermeier R, editors. *Difference Gel Electrophoresis (DIGE): Methods and Protocols* [Internet]. Totowa, NJ: Humana Press, 2012 [cited 2025 Mar 1]. p. 9–19. Available from: https://doi.org/10.1007/978-1-61779-573-2_2
21. Viswanathan S, Ünlü M, Minden JS. Two-dimensional difference gel electrophoresis. *Nat Protoc.* 2006 Aug;1(3):1351–8.
22. Mann M, Hendrickson RC, Pandey A. Analysis of proteins and proteomes by mass spectrometry. *Annu Rev Biochem.* 2001 Jul 1;70(Volume 70, 2001):437–73.
23. Glish GL, Vachet RW. The basics of mass spectrometry in the twenty-first century. *Nat Rev Drug Discov.* 2003 Feb;2(2):140–50.
24. Domon B, Aebersold R. Mass spectrometry and protein analysis. *Science.* 2006 Apr 14;312(5771):212–7.
25. Murayama C, Kimura Y, Setou M. Imaging mass spectrometry: Principle and application. *Biophys Rev.* 2009 Sep 1;1(3):131–9.
26. Xie F, Liu T, Qian WJ, Petyuk VA, Smith RD. Liquid chromatography-mass spectrometry-based quantitative proteomics*. *J Biol Chem.* 2011 Jul 22;286(29):25443–9.
27. Jang E, Jung S, Sohng W, Choi D, Hwang GS, Chung H. Screening of gall bladder cancer through infrared analysis of bile and examination of varied bile constituent composition by the disease. *Spectrochim Acta A Mol Biomol Spectrosc.* 2023 Feb 5;286:122030.
28. Sugiyanto RN, Metzger C, Inal A, Truckenmueller F, Gür K, Eiteneuer E, et al. Proteomic profiling reveals CEACAM6 function in driving gallbladder cancer aggressiveness through integrin receptor, PRKCD and AKT/ERK signaling. *Cell Death Dis.* 2024 Oct 28;15(10):1–15.
29. Xu M, Deng C, Man Z, Zhu H. TRIM47 is a prognostic biomarker for gallbladder cancer and promotes tumor progression through regulating K63-linked ubiquitination of PARP1. *Transl Oncol.* 2025 Jan 1;51:102164.
30. Yang J, Li X, Chen S, Li G, Pu P, Yang Y, et al. GPRC5A promotes gallbladder cancer metastasis by upregulating TNS4 via the JAK2–STAT3 pathway. *Cancer Lett.* 2024 Aug 28;598:217067.
31. Evans C, Noirel J, Ow SY, Salim M, Pereira-Medrano AG, Couto N, et al. An insight into iTRAQ: Where do we stand now? *Anal Bioanal Chem.* 2012 Sep 1;404(4):1011–27.
32. Aggarwal K, Choe LH, Lee KH. Shotgun proteomics using the iTRAQ isobaric tags. *Brief Funct Genomics.* 2006 Jun 1;5(2):112–20.
33. Wiese S, Reidegeld KA, Meyer HE, Warscheid B. Protein labeling by iTRAQ: A new tool for quantitative mass spectrometry in proteome research. *Proteomics.* 2007;7(3):340–50.
34. Subbannayya T, Leal-Rojas P, Barbhuiya MA, Raja R, Renuse S, Sathe G, et al. Macrophage migration inhibitory factor—a therapeutic target in gallbladder cancer. *BMC Cancer.* 2015 Nov 4;15(1):843.
35. Zieske LR. A perspective on the use of iTRAQ™ reagent technology for protein complex and profiling studies. *J Exp Bot.* 2006 Apr 1;57(7):1501–8.
36. Thompson A, Schäfer J, Kuhn K, Kienle S, Schwarz J, Schmidt G, et al. Tandem mass tags: a novel quantification strategy for comparative analysis of complex protein mixtures by MS/MS. *Anal Chem.* 2003 Apr 1;75(8):1895–904.
37. Moulder R, Bhosale SD, Goodlett DR, Lahesmaa R. Analysis of the plasma proteome using iTRAQ and TMT-based Isobaric labeling. *Mass Spectrom Rev.* 2018;37(5):583–606.
38. Shi Y, Ding Y, Li G, Wang L, Osman RA, Sun J, et al. Discovery of novel biomarkers for diagnosing and predicting the progression of multiple sclerosis using TMT-based quantitative proteomics. *Front Immunol* [Internet]. 2021 Aug 20 [cited 2025 Mar 3];12. Available from: www.frontiersin.org/journals/immunology/articles/10.3389/fimmu.2021.700031/full
39. Gondkar K, Sathe G, Joshi N, Nair B, Pandey A, Kumar P. Integrated proteomic and phosphoproteomics analysis of DKK3 signaling reveals activated kinase in the most aggressive gallbladder cancer. *Cells.* 2021 Mar;10(3):511.
40. Subbannayya T, Leal-Rojas P, Zhavoronkov A, Ozerov IV, Korzinkin M, Babu N, et al. PIM1 kinase promotes gallbladder cancer cell proliferation via inhibition of proline-rich Akt substrate of 40 kDa (PRAS40). *J Cell Commun Signal.* 2019 Jun 1;13(2):163–77.
41. McKinnon KM. Flow cytometry: An overview. *Curr Protoc Immunol.* 2018;120(1):5.1.1–5.1.11.
42. Liu X, Huang J, Liu L, Liu R. MPZL1 is highly expressed in advanced gallbladder carcinoma and promotes the aggressive behavior of human gallbladder carcinoma GBC-SD cells. *Mol Med Rep.* 2019 Sep;20(3):2725–33.

43. Shu YJ, Weng H, Ye YY, Hu YP, Bao RF, Cao Y, et al. SPOCK1 as a potential cancer prognostic marker promotes the proliferation and metastasis of gallbladder cancer cells by activating the PI3K/AKT pathway. *Mol Cancer.* 2015 Jan 27;14(1):12.
44. Krishhan VV, Khan IH, Luciw PA. Multiplexed microbead immunoassays by flow cytometry for molecular profiling: Basic concepts and proteomics applications. *Crit Rev Biotechnol.* 2009 Mar;29(1):29–43.
45. Das C, Mukhopadhyay M, Subba S, Saha AK, Mukhopadhyay B. Role of EGFR and HER-2/NEU expression in gall bladder carcinoma (GBC). *J Lab Physicians.* 2021 May 19;13:29–35.
46. Skoda AM, Simovic D, Karin V, Kardum V, Vranic S, Serman L. The role of the Hedgehog signaling pathway in cancer: A comprehensive review. *Bosn J Basic Med Sci.* 2018 Feb;18(1):8–20.
47. Wu J, Lei L, Wang S, Gu D, Zhang J. Immunohistochemical expression and prognostic value of CD97 and its ligand CD55 in primary gallbladder carcinoma. *BioMed Res Int.* 2012 [cited 2025 Mar 20]. Available from: https://onlinelibrary.wiley.com/doi/10.1155/2012/587672
48. Qiu S, Liu Z, Hu J, Wang Z, Yue Z, Jia Z, et al. Fatty acid binding protein 1 is an independent prognostic biomarker for gallbladder cancer with direct hepatic invasion. *Int J Med Sci.* 2024 Mar 25;21(5):862–73.
49. Chai C, Tang H, Miao X, Chen T, Su Y, Li L, et al. Establishment and characterization of a novel human gallbladder cancer cell line, GBC-X1. *Sci Rep.* 2024 Sep 13;14(1):21439.
50. Xu M, Deng C, Man Z, Zhu H. TRIM47 is a prognostic biomarker for gallbladder cancer and promotes tumor progression through regulating K63-linked ubiquitination of PARP1. *Transl Oncol.* 2025 Jan 1;51:102164.
51. Sun W, Fan YZ, Xi H, Lu XS, Ye C, Zhang JT. Astrocyte elevated gene-1 overexpression in human primary gallbladder carcinomas: An unfavorable and independent prognostic factor. *Oncol Rep.* 2011 Nov;26(5):1133–42.
52. Wang J, Yao Y, Ming Y, Shen S, Wu N, Liu J, et al. Downregulation of stathmin 1 in human gallbladder carcinoma inhibits tumor growth in vitro and in vivo. *Sci Rep.* 2016 Jun 28;6(1):28833.
53. Shen H, He M, Lin R, Zhan M, Xu S, Huang X, et al. PLEK2 promotes gallbladder cancer invasion and metastasis through EGFR/CCL2 pathway. *J Exp Clin Cancer Res CR.* 2019 Jun 10;38:247.
54. Huang Y, Du Q, Wu W, She F, Chen Y. Rescued expression of WIF-1 in gallbladder cancer inhibits tumor growth and induces tumor cell apoptosis with altered expression of proteins. *Mol Med Rep.* 2016 Sep 1;14(3):2573–81.
55. Zhu G, Du Q, Wang X, Tang N, She F, Chen Y. TNF-α promotes gallbladder cancer cell growth and invasion through autocrine mechanisms. *Int J Mol Med.* 2014 Jun;33(6):1431–40.
56. Lu X, Hu K, Zhang D, Yin X, Nie J, Zhao K. KLF5 silencing restrains proliferation, invasion, migration and angiogenesis of gallbladder carcinoma cells by transcriptional regulation of PDGFA. *J Cancer Res Clin Oncol.* 2024 Dec 20;151(1):11.
57. Wu Z, Peng Y, Chen W, Xia F, Song T, Ke Q. Lactylation-driven transcriptional activation of FBXO33 promotes gallbladder cancer metastasis by regulating p53 polyubiquitination. *Cell Death Dis.* 2025 Feb 28;16(1):1–15.
58. Hong L, Chen M, Huang M, Chen W, Abudukeremu X, She F, et al. FOXA2 suppresses gallbladder carcinoma cell migration, invasion, and epithelial-mesenchymal transition by targeting SERPINB5. *Environ Toxicol.* 2024;39(2):708–22.
59. Huang M, Zhong F, Chen M, Hong L, Chen W, Abudukeremu X, et al. CEP55 as a promising biomarker and therapeutic target on gallbladder cancer. *Front Oncol.* 2023;13:1156177.
60. Kumar M, Shukla VK, Misra PK, Raman MJ. Dysregulated expression and subcellular localization of base excision repair (BER) pathway enzymes in gallbladder cancer. *Int J Mol Cell Med.* 2018 Aug 18;7(2):119.
61. Nigam J, Chandra A, Kazmi HR, Parmar D, Singh D, Gupta V. Prognostic significance of survivin in resected gallbladder cancer. *J Surg Res.* 2015 Mar;194(1):57–62.
62. Hong H, Jiang L, Lin Y, He C, Zhu G, Du Q, et al. TNF-alpha promotes lymphangiogenesis and lymphatic metastasis of gallbladder cancer through the ERK1/2/AP-1/VEGF-D pathway. *BMC Cancer.* 2016 Mar 19;16(1):240.
63. Guo R, Qin Y, Shi P, Xie J, Chou M, Chen Y. IL-1β promotes proliferation and migration of gallbladder cancer cells via Twist activation. *Oncol Lett.* 2016 Dec 1;12(6):4749–55.
64. Jiang X, Li C, Lin B, Hong H, Jiang L, Zhu S, et al. cIAP2 promotes gallbladder cancer invasion and lymphangiogenesis by activating the NF-κB pathway. *Cancer Sci.* 2017 Jun;108(6):1144–56.
65. Tan Y, Meng HP, Wu Q, Wang FQ, Wu HR. Proteomic study of gallbladder cancer, with special reference on the expression and significance of annexin A3. *Zhonghua Bing Li Xue Za Zhi.* 2010 Jun;39(6):382–6.

66. Wang JW, Peng SY, Li JT, Wang Y, Zhang ZP, Cheng Y, et al. Identification of metastasis-associated proteins involved in gallbladder carcinoma metastasis by proteomic analysis and functional exploration of chloride intracellular channel 1. *Cancer Lett.* 2009 Aug 18;281(1):71–81.
67. Dong P, He X wei, Gu J, Wu W guang, Li M lan, Yang J hua, et al. Vimentin significantly promoted gallbladder carcinoma metastasis. *Chin Med J (Engl).* 2011 Dec;124(24):4236.
68. Akhtar J, Priya R, Jain V, Sakhuja P, Agarwal AK, Goyal S, et al. Immunoproteomics approach revealed elevated autoantibody levels against ANXA1 in early stage gallbladder carcinoma. *BMC Cancer.* 2020 Dec 1;20(1):1175.
69. Sandanayake NS, Sinclair J, Andreola F, Chapman MH, Xue A, Webster GJ, et al. A combination of serum leucine-rich α-2-glycoprotein 1, CA19–9 and interleukin-6 differentiate biliary tract cancer from benign biliary strictures. *Br J Cancer.* 2011 Oct;105(9):1370–8.
70. Jain V, Akhtar J, Priya R, Sakhuja P, Goyal S, Agarwal AK, et al. Tissue proteome analysis for profiling proteins associated with lymph node metastasis in gallbladder cancer. *BMC Cancer.* 2023 May 4;23(1):402.
71. He Z, Zhong Y, Regmi P, Lv T, Ma W, Wang J, et al. Exosomal long non-coding RNA TRPM2-AS promotes angiogenesis in gallbladder cancer through interacting with PABPC1 to activate NOTCH1 signaling pathway. *Mol Cancer.* 2024 Mar 27;23(1):65.
72. Baichan P, Naicker P, Augustine TN, Smith M, Candy G, Devar J, et al. Proteomic analysis identifies dysregulated proteins and associated molecular pathways in a cohort of gallbladder cancer patients of African ancestry. *Clin Proteomics.* 2023 Mar 1;20(1):8.
73. Wang W, Ai KX, Yuan Z, Huang XY, Zhang HZ. Different expression of S100A8 in malignant and benign gallbladder diseases. *Dig Dis Sci.* 2013 Jan;58(1):150–62.
74. Truckenmueller F, Goeppert B, Pusch S, Heinze I, Kirkpatrick J, Schirmacher P, et al. Mass-spectrometric analysis of deregulated proteins in gallbladder carcinoma. *J Hepatol* [Internet]. 2019 [cited 2025 Apr 1]. Available from: www.postersessiononline.eu/173580348_eu/congresos/ILC2019/aula/-THU_498_ILC2019.pdf
75. Zhu D, Gu X, Lin Z, Yu D, Wang J. High expression of PSMC2 promotes gallbladder cancer through regulation of GNG4 and predicts poor prognosis. *Oncogenesis.* 2021 May 20;10(5):1–12.

14 Molecular Profile of Gallbladder Cancer

Tanuj Chawla[1]* *and Lovenish Bains*[2]
[1]Consultant, Department of Medical Oncology, Advanced Centre for Treatment Research and Education in Cancer, Tata Memorial Centre, Navi Mumbai, India
[2]Professor, Department of Surgery, Maulana Azad Medical College, Delhi, India
*Corresponding author: tanujucms@yahoo.com

14.1 INTRODUCTION

Gallbladder cancer (GBC) is a rare but aggressive malignancy characterized by a complex molecular landscape. Recent studies have provided insights into the genetic and molecular alterations associated with GBC, offering potential avenues for targeted therapies. GBC is one of the most common types of biliary tract cancer (BTC) and is particularly more prevalent in women and in certain areas of the world.

Incidence of GBC varies geographically; for instance, it is high in Chile and Northern India, intermediate in Shanghai, China, and rare in Western countries. Although 5-year survival rates across all tumor stages range from 5% to 20%, radical cholecystectomy can result in 5-year survival up to 80% in T2 tumors.[(1)]

As per the GLOBOCAN 2022 data,[(2)] incidence of GBC in India, Chile, and some Asian countries is high (IARC). It is the 10th, 12th, 14th, and 15th most common cause of cancer mortality in Chile, South Korea, China (females), and India, respectively. China is leading in incidence with 31,132 new cases, closely followed by India at 21,780, and then Japan with incidence of 10,814, as derived from GLOBOCAN 2022.[(2)] The prevalence of GBC is much less in regions where there are high proportions of persons with European ancestry, that is, Canada, the United States, New Zealand, Australia, and the United Kingdom. This indicates an influence of geography in the etiology of this cancer.

Development of GBC is mainly associated with chronic inflammation led by gallstones, hyalinizing cholecystitis, chronic infection with gram-negative organisms (e.g., *Salmonella typhi* and *Helicobacter bilis*), obesity, and congenital developmental pancreatobiliary abnormalities.[(3)]

Large-scale studies assessing the molecular mechanisms underlying the pathogenesis of GBC are still limited due to the low incidence of this tumor. Two main molecular pathways have been described: one governed by TP53 alterations and enriched in patients with a history of cholelithiasis, and the other governed by KRAS alterations mostly in patients with pancreatobiliary duct maljunction. A poorly understood combination of predisposing factors makes gallbladder cancer a unique malignancy. Some of the known factors causing GBC but without significant scientific evidence are congenital developmental abnormalities, obesity, a personal or family history of gallstones, chronic infection and inflammation in the gallbladder, and ill-defined genetic variants.

GBC, if detected at an early stage, should be treated with surgery. Depending on the stage of the disease, radiotherapy and chemotherapy can be added to improve survival rates. Understanding the genomic and mutational landscape of this highly aggressive disease has become easier with the advent of next-generation sequencing (NGS). It has helped in getting a deeper look at the molecular

DOI: 10.1201/9781003506201-14

TABLE 14.1
Common Genetic Mutations in Gallbladder Cancer

Gene	Mutation frequency (%)	Role in cancer	References
TP53	40–60	Tumor suppressor	4, 5
KRAS	11%	Oncogene (promotes growth)	15
c-erb-B2	10–46%	Growth Factor receptor (Oncogenic signaling)	20
SMAD4	–	TGF-β pathway (oncogenic signaling)	23, 24, 25
PIK3CA	33%	Oncogenic signaling (PI3K pathway)	27
ARID1A	~20%	Disruption in transcription process	29

structure and possible cause of this elusive cancer. This chapter highlights the most relevant molecular signatures and genetic alterations (Table 14.1) which play a substantial role in the development and prognostication of GBCs.

14.2 KEY GENETIC ALTERATIONS

14.2.1 TP53

TP53 is one of the most frequently mutated genes in gallbladder cancers. Studies from Asia and Europe found prevalence of altered TP53 in up to 60% samples.[4, 5] Genomic integrity and stability, apoptosis, cell cycle arrest, autophagy, DNA dysregulation and repair, inhibition of angiogenesis and metabolism, and so on, are regulated by this protein which primarily functions as a transcription factor.[6] The afflictions by TP53 mutations provide a nidus for the cancer development of various types.

Nepal et al. identified a significant burden (29.3%) of TP53 mutations. The mutations were distributed across the gene, where 93% (25/27) of them are predicted to be nocuous. In addition, 96% (26/27) of TP53 mutations overlapped with previously reported hotspot loci within the gene.[7]

14.2.2 CDKN2A

CDKN2A is also among the most frequently damaged cancer genes, with loss of function (LoF) reported in at least 35 different tumor types across 12 organ systems.[8] Similar to TP53, it promotes cellular senescence, induces cell cycle arrest, as well as prevents angiogenesis and metastasis. This gene encoding p16 protein negatively regulates cell cycle progression. p16 protein binds to CDK4/6 inhibiting their ligation to cyclin D. Methylation of CDKN2A has been described in multiple studies with varied frequencies (14.5–80%), but there is no reported association with prognosis or clinicopathologic variables.[9–11]

Yang et al. reported similar frequency of altered CDKN2A gene, that is, 25% among patients from China and US. Despite the high prevalence of CDKN2A alterations, the target remains elusive and GBC survival remains abysmal.[12]

14.2.3 KRAS

Kirsten rat sarcoma viral oncogene homologue (KRAS) is the best-known oncogene with the highest mutation rate among all cancers. There is a significant association of its mutation with non-small cell lung cancer, pancreatic ductal adenocarcinoma, and colorectal cancers. The KRAS gene encodes two highly related protein isoforms, KRAS-4B and KRAS-4A, which consist of 188 and 189 amino acids, respectively, due to different clipping of the fourth exon.[13] This gene is involved in the RAS/MAPK pathways and PIK3CA pathway. Despite its GTPase activity, the K-Ras protein is essential for controlling signals via several regulatory pathways.[14] Mutations in KRAS are observed in about 11% of tumors, while ERBB2 (HER2) overexpression is reported in 15% of cases, indicating its role in GBC carcinogenesis.[15]

14.2.4 c-erb-B2

The activity of ErbB2 (HER2) receptor tyrosine kinases has been established in the development of various cancers. The HER2 receptor blockers have been shown to improve survival in cancers of breast,[16] gastric,[17] colorectal, and GBCs,[18, 19] and so on. Kiguchi et al.,[20] reviewed their cases and found that immunohistochemical expression of c-erb-B2 was positive in 10%–46% of GBCs. Moreover, the expression of HER2/neu in about 28% of GBCs was directly correlated with advanced stage of cancer.

In a study from India,[21] c-erbB2 was frequently expressed in well-differentiated and stage II to stage IV in about 9.4% of GBC cases. Another prospective case-control study from India showed a total of 10 (20%) cases having positive (3+), 19 (38%) equivocal (2+) expression, while 21 (42%) were negative on HER2/ERBB2 immunocytochemistry.[22]

Response to anti-HER2 therapies in biliary tract cancers, including GBCs, has been perfectly summarized by Haaft et al.[18] Various other studies, including a phase 2 multicenter trial by Ostwal et al.,[19] have shown how trastuzumab (anti-HER2) has improved progression-free survival in GBCs.

14.2.5 SMAD4

The SMAD4 gene plays an important role in the transforming growth factor-β pathway (TGF-β). It regulates cellular proliferation, differentiation, cell death, and migration. Its mutation in colorectal and pancreatic cancer indicates advanced disease with poor prognosis.[23, 24] It is another highly mutated gene found in GBC. The data above highlight its potential prognostic value.

A study from India published in 2024,[25] mentioned SMAD4 as the second most frequently mutated gene. There were 5 patients with SMAD4 mutations who also had cholelithiasis along with gallbladder cancer. p.D537Y, p.D493N, p.E330K, p.W509R, p.R361H, and p.G352V mutations were discovered to be related with SMAD4 in the GBC cohort.

14.2.6 PIK3CA

The PIK3CA gene encodes the catalytic subunit of class I phosphatidylinositol-3-kinase (PI3K). Utilizing the PI3K/AKT/mTOR signaling pathway, its responsibility is to regulate cell growth, proliferation, migration, and survival. Mutated *PIK3CA* gene leads to the continuously activated PI3K/AKT signaling pathway, which may predispose the cell to an unchecked cell division.[26] Mishra et al.,[27] in their study demonstrated that GBCs commonly exhibit PIK3CA mutations in exons 9, 20, and 21. They observed PIK3CA mutations in 33.33% of cases, with pathogenic potential in 6 out of 11 cases. The mutation in exon 21 was most prevalent, followed by exon 10.

14.2.7 ARID1A

The ARID1A (AT-rich interactive domain-containing protein 1A) regulates the transcription process and provides a DNA damage response. It encodes a subunit of the BAF (barrier to autointegration factor) chromatin-remodeling complex. Mutations in this gene lead to inactivation or loss of expression in at least 5 cancer types.[28] Data from public repositories of Japan, the United States,

Chile, and China showed that ARID1A was the third most mutated gene in GBC samples. It showed nonsense mutations, frameshift insertion, and deletion, indicating transformation of around 20.7%, that is, 31 mutations. All mutations of ARID1A were annotated as oncogenic alterations.[29]

14.3 CLINICAL IMPLICATIONS

The identification of actionable mutations has important implications for treatment. Approximately 35% of patients exhibit molecular alterations that are clinically actionable, which could guide targeted therapies or enrollment in clinical trials.[30] Various molecules have been developed but only a few showed durable response and improved progression-free survival and overall survival, namely HER2 inhibitors,[19] and PD1/PDL-1 inhibitors.[31] Understanding genetic alterations is crucial for developing targeted therapies and improving prognostic assessments and survival for patients with GBC.

Genetic alterations in the RTK-RAS pathway can significantly influence treatment options for GBC by identifying potential targets for therapy and informing the use of existing treatments. Given the complexity of signaling pathways involved in GBC, combination therapies that target multiple pathways (e.g., RTK-RAS and PI3K/AKT) may enhance treatment efficacy and overcome resistance mechanisms.

14.4 CONCLUSION

The molecular profile of gallbladder cancer reveals a heterogeneous landscape marked by significant genetic alterations. Understanding these changes not only aids in the diagnosis and prognosis but also opens avenues for targeted therapies that could improve patient outcomes. Continued research into the molecular characteristics of GBC is essential for developing effective treatment strategies. Identifying actionable mutations not only informs therapeutic decisions but also enhances patient outcomes through targeted interventions and participation in clinical trials. As research progresses, further elucidation of these pathways will likely lead to more effective treatments for GBC.

REFERENCES

1. Goetze TO, Paolucci V. Adequate extent in radical re-resection of incidental gallbladder carcinoma: Analysis of the German Registry. *Surg Endosc.* 2010 Sep;24(9):2156–2164. doi: 10.1007/s00464-010-0914-4. Epub 2010 Feb 23. PMID: 20177938.
2. https://gco.iarc.who.int/en (Last accessed on 27 September 2025).
3. Schmidt MA, Marcano-Bonilla L, Roberts LR. Gallbladder cancer: Epidemiology and genetic risk associations. *Chin Clin Oncol.* 2019 Aug;8(4):31. doi: 10.21037/cco.2019.08.13. PMID: 31484487.
4. Oda T, Tsutsumi K, Obata T, Ueta E, Kikuchi T, Ako S, Fujii Y, Yamazaki T, Uchida D, Matsumoto K, Horiguchi S, Kato H, Okada H, Chijimatsu R, Otsuka M. MicroRNA-34a-5p: A pivotal therapeutic target in gallbladder cancer. *Mol Ther Oncol.* 2024 Feb 8;32(1):200765. doi: 10.1016/j.omton.2024.200765. PMID: 38596294; PMCID: PMC10963938.
5. de Bitter TJJ, de Reuver PR, de Savornin Lohman EAJ, Kroeze LI, Vink-Börger ME, van Vliet S, Simmer F, von Rhein D, Jansen EAM, Verheij J, van Herpen CML, Nagtegaal ID, Ligtenberg MJL, van der Post RS. Comprehensive clinicopathological and genomic profiling of gallbladder cancer reveals actionable targets in half of patients. *NPJ Precis Oncol.* 2022 Nov 5;6(1):83. doi: 10.1038/s41698-022-00327-y. PMID: 36335173; PMCID: PMC9637208.
6. Wang H, Guo M, Wei H, Chen Y. Targeting p53 pathways: Mechanisms, structures, and advances in therapy. *Signal Transduct Target Ther.* 2023 Mar 1;8(1):92. doi: 10.1038/s41392-023-01347-1. PMID: 36859359; PMCID: PMC9977964.
7. Nepal C, Zhu B, O'Rourke CJ, Bhatt DK, Lee D, Song L, Wang D, Van Dyke AL, Choo-Wosoba H, Liu Z, Hildesheim A, Goldstein AM, Dean M, LaFuente-Barquero J, Lawrence S, Mutreja K, Olanich ME, Lorenzo Bermejo J; CGR Exome Studies Group; Ferreccio C, Roa JC, Rashid A, Hsing AW, Gao YT, Chanock SJ, Araya JC, Andersen JB, Koshiol J. Integrative molecular characterisation of gallbladder cancer reveals micro-environment-associated subtypes. *J Hepatol.* 2021 May;74(5):1132–1144. doi: 10.1016/j.jhep.2020.11.033. Epub 2020 Dec 1. PMID: 33276026; PMCID: PMC8058239.

8. Ganguli P, Basanta CC, Acha-Sagredo A, Misetic H, Armero M, Mendez A, Zahra A, Devonshire G, Kelly G, Freeman A, Green M, Nye E, Bichisecchi A, Bonfanti P; Oesophageal Cancer Clinical and Molecular Stratification (OCCAMS) Consortium; Rodriguez-Justo M, Spencer J, Fitzgerald RC, Ciccarelli FD. Context-dependent effects of CDKN2A and other 9p21 gene losses during the evolution of esophageal cancer. *Nat Cancer.* 2025 Jan;6(1):158–174. doi: 10.1038/s43018-024-00876-0. Epub 2025 Jan 3. PMID: 39753721; PMCID: PMC11779637.
9. Klump B, Hsieh CJ, Dette S, et al. Promoter methylation of INK4a/ARF as detected in bile-significance for the differential diagnosis in biliary disease. *Clin Cancer Res.* 2003;9(5):1773–1778. PubMed: 12738733.
10. Tozawa T, Tamura G, Honda T, et al. Promoter hypermethylation of DAP-kinase is associated with poor survival in primary biliary tract carcinoma patients. *Cancer Sci.* 2004;95(9):736–740. PubMed: 15471559.
11. Ueki T, Hsing AW, Gao YT, et al. Alterations of p16 and prognosis in biliary tract cancers from a population–based study in China. *Clin Cancer Res.* 2004;10(5):1717–1725. PubMed: 15014024.
12. Yang P, Javle M, Pang F, Zhao W, Abdel-Wahab R, Chen X, Meric-Bernstam F, Chen H, Borad MJ, Liu Y, Zou C, Mu S, Xing Y, Wang K, Peng C, Che X. Somatic genetic aberrations in gallbladder cancer: Comparison between Chinese and US patients. *Hepatobiliary Surg Nutr.* 2019 Dec;8(6):604–614.
13. Huang L, Guo Z, Wang F, Fu L. KRAS mutation: From undruggable to druggable in cancer. *Signal Transduct Target Ther.* 2021 Nov 15;6(1):386. doi: 10.1038/s41392-021-00780-4. PMID: 34776511; PMCID: PMC8591115.
14. Kumar M, Kumar A, Srivastav A, Ghosh A, Kumar D. Genomic and molecular landscape of gallbladder cancer elucidating pathogenic mechanisms novel therapeutic targets and clinical implications. *Mutat Res.* 2024 Dec 26;830:111896. doi: 10.1016/j.mrfmmm.2024.111896. Epub ahead of print. PMID: 39754821.
15. Giraldo NA, Drill E, Satravada BA, Dika IE, Brannon AR, Dermawan J, Mohanty A, Ozcan K, Chakravarty D, Benayed R, Vakiani E, Abou-Alfa GK, Kundra R, Schultz N, Li BT, Berger MF, Harding JJ, Ladanyi M, O'Reilly EM, Jarnagin W, Vanderbilt C, Basturk O, Arcila ME. Comprehensive molecular characterization of gallbladder carcinoma and potential targets for intervention. *Clin Cancer Res.* 2022 Dec 15;28(24):5359–5367. doi: 10.1158/1078-0432.CCR-22-1954. PMID: 36228155; PMCID: PMC9772093.
16. Lluch-Gómez J, Núñez-Álvarez V, de la Torre-Hita C, Bernal-Gómez M, Campini-Bermejo A, Perdomo-Zaldívar E, Rodríguez-Pérez L, Calvete-Candenas J, Martínez-Bautista MJ, Benítez-Rodríguez E, Baena-Cañada JM. Real world evidence of adjuvant trastuzumab in HER2 positive early breast cancer. *Sci Rep.* 2023 May 3;13(1):7168. doi: 10.1038/s41598-023-34429-9. PMID: 37137963; PMCID: PMC10156809.
17. Gunturu KS, Woo Y, Beaubier N, Remotti HE, Saif MW. Gastric cancer and trastuzumab: First biologic therapy in gastric cancer. *Ther Adv Med Oncol.* 2013 Mar;5(2):143–151. doi: 10.1177/1758834012469429. PMID: 23450234; PMCID: PMC3556876.
18. Ten Haaft BH, Pedregal M, Prato J, Klümpen HJ, Moreno V, Lamarca A. Revolutionizing anti-HER2 therapies for extrahepatic cholangiocarcinoma and gallbladder cancer: Current advancements and future perspectives. *Eur J Cancer.* 2024 Mar;199:113564. doi: 10.1016/j.ejca.2024.113564. Epub 2024 Jan 20. PMID: 38266541.
19. Vikas O, Mandavkar S, Bhargava P, et al. Trastuzumab plus gemcitabine-cisplatin for treatment-naïve human epidermal growth factor receptor 2–positive biliary tract adenocarcinoma: A multicenter, open-label, phase II study (TAB). *JCO.* 2024;42:800–807.
20. Kiguchi K, Carbajal S, Chan K, Beltrán L, Ruffino L, Shen J, Matsumoto T, Yoshimi N, DiGiovanni J. Constitutive expression of ErbB-2 in gallbladder epithelium results in development of adenocarcinoma. *Cancer Res.* 2001;61:6971–6976.
21. Kumari N, Kapoor VK, Krishnani N, Kumar K, Baitha DK. Role of C-erbB2 expression in gallbladder cancer. *Indian J Pathol Microbiol.* 2012;55:75–79.
22. Verma P, Gupta P, Gupta N, Srinivasan R, Gupta P, Dutta U, Sharma S, Uppal R, Nada R, Lal A. HER2/ERBB2 overexpression in advanced gallbladder carcinoma: Comprehensive evaluation by immunocytochemistry and fluorescence in situ hybridisation on fine-needle aspiration cytology samples. *J Clin Pathol.* 2024 Aug 16;77(9):614–621. doi: 10.1136/jcp-2023-208940. PMID: 37221046)
23. Sarshekeh AM, Advani S, Overman MJ, et al. Association of SMAD4 mutation with patient demographics, tumor characteristics, and clinical outcomes in colorectal cancer. *PLoS ONE.* 2017;12:e0178275. doi: 10.1371/journal.pone.0173345.

24. Herman JM, Fan KY, Wild AT, et al. Correlation of Smad4 status with outcomes in patients receiving erlotinib combined with adjuvant chemoradiation and chemotherapy after resection for pancreatic adenocarcinoma. *Int J Radiat Oncol Biol Phys.* 2013;87:458–459. doi: 10.1016/j.ijrobp.2013.06.2039.
25. Awasthi S, Kumar R, Pradhan D, Rawal N, Goel H, Sahu P, Sisodiya S, Rana R, Kumar S, Dash NR, Das P, Agrawal U, Rath GK, Kaur T, Dhaliwal RS, Hussain S, Saluja SS, Tanwar P. Genomic landscape of gallbladder cancer: Insights from whole exome sequencing. *Int J Surg.* 2024 Nov 1;110(11):6883–6897. doi: 10.1097/JS9.0000000000002031. PMID: 39166960; PMCID: PMC11573093.
26. Samuels Y, Waldman T. Oncogenic mutations of PIK3CA in human cancers. *Curr Top Microbiol Immunol.* 2010;347:21–41. doi: 10.1007/82_2010_68. PMID: 20535651; PMCID: PMC3164550.
27. Mishra S, Kumari S, Srivastava P, Pandey A, Shukla S, Husain N. Genomic profiling of gallbladder carcinoma: Targetable mutations and pathways involved. *Pathol. Res Pract.* 2022;232:153806. doi: 10.1016/j. prp.2022.153806.
28. Chang MT, Asthana S, Gao SP, Lee BH, Chapman JS, Kandoth C, Gao J, Socci ND, Solit DB, Olshen AB, Schultz N, Taylor BS. Identifying recurrent mutations in cancer reveals widespread lineage diversity and mutational specificity. *Nat Biotechnol.* 2016 Feb;34(2):155–163. doi: 10.1038/nbt.3391. Epub 2015 Nov 30. PMID: 26619011; PMCID: PMC4744099.
29. D'Afonseca V, Arencibia AD, Echeverría-Vega A, Cerpa L, Cayún JP, Varela NM, Salazar M, Quiñones LA. Identification of altered genes in gallbladder cancer as potential driver mutations for diagnostic and prognostic purposes: A computational approach. *Cancer Inform.* 2020 May 25;19:1176935120922154. doi: 10.1177/1176935120922154. PMID: 32546937; PMCID: PMC7249562.
30. Giraldo NA, Drill E, Satravada BA, Dika IE, Brannon AR, Dermawan J, Mohanty A, Ozcan K, Chakravarty D, Benayed R, Vakiani E, Abou-Alfa GK, Kundra R, Schultz N, Li BT, Berger MF, Harding JJ, Ladanyi M, O'Reilly EM, Jarnagin W, Vanderbilt C, Basturk O, Arcila ME. Comprehensive molecular characterization of gallbladder carcinoma and potential targets for intervention. *Clin Cancer Res.* 2022 Dec 15;28(24):5359–5367. doi: 10.1158/1078-0432.CCR-22-1954. PMID: 36228155; PMCID: PMC9772093.
31. Fung S, Syed YY. Durvalumab: A review in advanced biliary tract cancer. *Target Oncol.* 2023 Nov;18(6):965–972. doi: 10.1007/s11523-023-01007-y. Epub 2023 Nov 9. Erratum in: Target Oncol. 2024 Jan;19(1):129. doi: 10.1007/s11523-023-01021-0. PMID: 37943483; PMCID: PMC10667376.

Index

For Product Safety Concerns and Information please contact our EU representative GPSR@taylorandfrancis.com Taylor & Francis Verlag GmbH, Kaufingerstraße 24, 80331 München, Germany

Printed and bound by CPI Group (UK) Ltd, Croydon, CR0 4YY
14/09/2026
02216265-0001